Grundlehren der mathematischen Wissenschaften 299

A Series of Comprehensive Studies in Mathematics

Editors

M. Artin S. S. Chern J. Coates J. M. Fröhlich
H. Hironaka F. Hirzebruch L. Hörmander
C. C. Moore J. K. Moser M. Nagata W. Schmidt
D. S. Scott Ya. G. Sinai J. Tits M. Waldschmidt
S. Watanabe

Managing Editors

M. Berger B. Eckmann S. R. S. Varadhan

Ch. Pommerenke

Boundary Behaviour of Conformal Maps

With 76 Figures

Springer-Verlag

Berlin Heidelberg New York
London Paris Tokyo
Hong Kong Barcelona
Budapest

Christian Pommerenke
Fachbereich Mathematik
Technische Universität
1000 Berlin 12, FRG

Mathematics Subject Classification (1991):
30Cxx, 30D40, 30D45, 30D50, 28A78

ISBN 978-3-642-08129-3

Library of Congress Cataloging-in-Publication Data
Pommerenke, Christian.
Boundary behaviour of conformal maps / Ch. Pommerenke.
p. cm. - (Grundlehren der mathematischen Wissenschaften; 299)
Includes bibliographical references and indexes.

1. Conformal mapping. 2. Boundary value problems. I. Title II. Series.
QA360.P66 1992 515'.9-dc20 92-10365 CIP

This work is subject to copyright. All rights are reserved, whether the whole or part of the material is concerned, specifically the rights of translation, reprinting, reuse of illustrations, recitation, broadcasting, reproduction on microfilm or in any other way, and storage in data banks. Duplication of this publication or parts thereof is permitted only under the provisions of the German Copyright Law of September 9, 1965, in its current version, and permission for use must always be obtained from Springer-Verlag. Violations are liable for prosecution under the German Copyright Law.

© Springer-Verlag Berlin Heidelberg 2010
Printed in the United States of America

41/3140-543210 Printed on acid-free paper

Preface

We study the boundary behaviour of a conformal map of the unit disk onto an arbitrary simply connected plane domain. A principal aim of the theory is to obtain a one-to-one correspondence between analytic properties of the function and geometric properties of the domain.

In the classical applications of conformal mapping, the domain is bounded by a piecewise smooth curve. In many recent applications however, the domain has a very bad boundary. It may have nowhere a tangent as is the case for Julia sets. Then the conformal map has many unexpected properties, for instance almost all the boundary is mapped onto almost nothing and vice versa.

The book is meant for two groups of users.

(1) Graduate students and others who, at various levels, want to learn about conformal mapping. Most sections contain exercises to test the understanding. They tend to be fairly simple and only a few contain new material. Prerequisites are general real and complex analyis including the basic facts about conformal mapping (e.g. Ahl66a).

(2) Non-experts who want to get an idea of a particular aspect of conformal mapping in order to find something useful for their work. Most chapters therefore begin with an overview that states some key results avoiding technicalities.

The book is not meant as an exhaustive survey of conformal mapping. Several important aspects had to be omitted, e.g. numerical methods (see e.g. Gai64, Gai83, Hen86, Tre86, SchLa91), probabilistic methods and multiply connected domains, also the new approach of Thurston, Rodin and Sullivan via circle packings (see e.g. RoSu87, He91).

Several principles guided the selection of material, first of all the importance for the theory and next its applicability (e.g. Chapter 3). If there is a good modern treatment of a topic then the book tries to select aspects that are of particular importance for conformal mapping or are as yet not covered in books; see e.g. Chapters 5, 7 and 9. On the other hand, many of the results in Chapters 8, 10 and 11 can be found only in research papers. It is clear that the author's bias played a major part in the choice of subjects.

There are references to further (mainly more recent) papers that contain additional material. But this is only a selection of the extensive literature.

Many more results and references can be found e.g. in the books of Gattegno-Ostrowski, Lelong-Ferrand, Carathéodory and Golusin.

The figures are sketches to illustrate the concepts and proofs. They do not represent exact conformal maps.

I want to thank many mathematicians for their generous help. Above all I am grateful to Jochen Becker and Steffen Rohde who read the manuscript, eliminated a lot of errors and suggested many improvements. Jim Langley kindly advised me on language problems.

I want to thank H. Schiemanowski for typing the manuscript and its many changes and U. Graeber for drawing the figures.

My gratitude also goes to the Technical University Berlin and the Centre de Recerca Matemàtica in Barcelona for making it possible for me to write this book.

Berlin, December 1991   *Christian Pommerenke*

Contents

Contents

Chapter 1. Some Basic Facts

1.1 Sets and Curves

1. Throughout the book we shall use the following notation:

$$D(a,r) = \{z \in \mathbb{C} : |z - a| < r\} \qquad (a \in \mathbb{C},\ r > 0),$$
$$\mathbb{D} = \{z \in \mathbb{C} : |z| < 1\} = D(0,1) \qquad \text{(unit disk)},$$
$$\mathbb{T} = \{z \in \mathbb{C} : |z| = 1\} = \partial\mathbb{D} \qquad \text{(unit circle)},$$
$$\widehat{\mathbb{C}} = \mathbb{C} \cup \{\infty\} \qquad \text{(Riemann sphere)},$$
$$\mathbb{D}^* = \{z \in \mathbb{C} : |z| > 1\} \cup \{\infty\} \qquad \text{(exterior of unit circle)}.$$

The *spherical metric* (or chordal metric) in $\widehat{\mathbb{C}}$ is defined by

$$(1) \qquad d^{\#}(z,w) = \frac{|z - w|}{\sqrt{(1 + |z|^2)(1 + |w|^2)}} \qquad \text{for} \quad z, w \in \mathbb{C}$$

and $d^{\#}(z, \infty) = 1/\sqrt{1 + |z|^2}$, $d^{\#}(\infty, \infty) = 0$. It is invariant under *rotations of the sphere*, the Möbius transformations

$$(2) \qquad \tau(z) = (az - b)/(\bar{b}z + \bar{a}), \quad a, b \in \mathbb{C}, \quad |a|^2 + |b|^2 > 0.$$

We shall use this spherical metric when we deal with sets in $\widehat{\mathbb{C}}$. Thus $\operatorname{diam}^{\#} A$ will denote the spherical diameter of A. When we deal with sets in $\mathbb{C}$ we will try to avoid the awkward spherical metric and use the euclidean metric $|z - w|$. Thus $\operatorname{diam} A$ will denote the euclidean diameter. The spherical and euclidean metrics are equivalent on bounded sets in $\mathbb{C}$.

2. The set $A \subset \widehat{\mathbb{C}}$ is called *connected* if there is no partition

$$A = A_1 \cup A_2 \quad \text{with} \quad A_1 \neq \emptyset, \quad A_2 \neq \emptyset, \quad \overline{A}_1 \cap A_2 = A_1 \cap \overline{A}_2 = \emptyset.$$

If a connected set A intersects an arbitrary set E and also its complement $\widehat{\mathbb{C}} \backslash E$ then it intersects its boundary ∂E. The continuous image of a connected set is connected.

A closed set is connected if and only if it cannot be written as the union of two disjoint non-empty closed sets. A compact connected set with more than one point is called a *continuum*. It has uncountably many points.

A connected open set G is called a *domain*. If G is open the following three conditions are equivalent:

(i) G is a domain;

(ii) G cannot be written as the union of two disjoint non-empty open sets;

(iii) any two points in G can be connected by a polygonal arc in G.

Every open set G can be decomposed as

(3)
$$G = \bigcup_k G_k, \quad G_k \text{ disjoint domains}, \ \partial G_k \subset \partial G.$$

The countably many domains G_k are called the *components* of G.

3. A *curve* C is given by a parametric representation

(4)
$$C : \varphi(t), \quad \alpha \leq t \leq \beta \quad \text{with } \varphi \text{ continuous on } [\alpha, \beta].$$

We write $C \subset E$ if $\varphi(t) \in E$ for all t. It is called *closed* if $\varphi(\alpha) = \varphi(\beta)$. A closed curve can also be parametrized as

(5)
$$C : \psi(\zeta), \quad \zeta \in \mathbb{T} \quad \text{with} \quad \psi \text{ continuous on } \mathbb{T}.$$

We call C a *Jordan arc* if (4) holds for an injective (= one-to-one) function φ, and a *Jordan curve* if (5) holds with an injective ψ. Thus a Jordan curve is closed whereas a Jordan arc has distinct endpoints. An *open Jordan arc* has a parametrization

(6)
$$C : \varphi(t), \quad \alpha < t < \beta \quad \text{with } \varphi \text{ continuous and injective on } (\alpha, \beta).$$

Jordan Curve Theorem. *If J is a Jordan curve in $\widehat{\mathbb{C}}$ then $\widehat{\mathbb{C}} \setminus J$ has exactly two components G_0 and G_1, and these satisfy $\partial G_0 = \partial G_1 = J$.*

A domain G bounded by a Jordan curve J is called a *Jordan domain*. If $J \subset \mathbb{C}$ the bounded component of $\mathbb{C} \setminus J$ will be called the *inner domain* of J.

We say that two points in $\widehat{\mathbb{C}}$ are *separated* by the closed set A if they lie in different components of $\widehat{\mathbb{C}} \setminus A$. Thus they are not separated by A if it is possible to connect them by a curve without meeting A. The following result is very useful for giving rigorous proofs of "obvious" facts in plane topology (New64, p. 110; Pom75, p. 31).

Janiszewski's Theorem. *Let A and B be closed sets in $\widehat{\mathbb{C}}$ such that $A \cap B$ is connected. If two points are neither separated by A nor by B, then they are not separated by $A \cup B$.*

4. Let $C : \varphi(t)$, $\alpha \leq t \leq \beta$ be a curve in $\mathbb{C}$. Its *length* $\Lambda(C)$ is defined by

(7)
$$\Lambda(C) = \sup \sum_{\nu=1}^{n} |\varphi(t_\nu) - \varphi(t_{\nu-1})|$$

where the supremum is taken over all partitions $\alpha = t_0 < t_1 < \cdots < t_n = \beta$ and all $n \in \mathbb{N}$. We call C *rectifiable* if $\Lambda(C) < \infty$. It is clear from (7) that

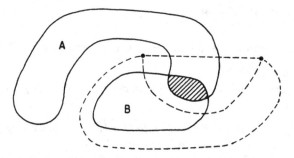

Fig. 1.1. Janiszewski's theorem

$$(8) \qquad \qquad \operatorname{diam} C \le \Lambda(C).$$

If φ is piecewise continuously differentiable then

$$(9) \qquad \qquad \Lambda(C) = \int_\alpha^\beta |\varphi'(t)|\, dt.$$

In the case of a closed curve $C : \psi(\zeta)$, $\zeta \in \mathbb{T}$ with piecewise continuously differentiable ψ we can write

$$(10) \qquad \qquad \Lambda(C) = \int_\mathbb{T} |\psi'(\zeta)|\, |d\zeta|.$$

If $C_1, C_2, \ldots$ are disjoint curves we define $\Lambda(\bigcup_k C_k) = \sum_k \Lambda(C_k)$, in particular $\Lambda(\emptyset) = 0$. The concept of length and its generalization, linear measure, will be of great importance.

Now let G be an open set in $\mathbb{C}$ and let C be a *halfopen curve* in G given by

$$(11) \qquad C : \varphi(t), \quad \alpha \le t < \beta \quad \text{with } \varphi \text{ continuous in } [\alpha, \beta).$$

Its length is

$$(12) \qquad \Lambda(C) = \lim_{\tau \to \beta} \Lambda(C_\tau), \quad C_\tau : \varphi(t), \quad \alpha \le t \le \tau.$$

We say that C *ends* at b if

$$(13) \qquad \qquad b = \lim_{t \to \beta} \varphi(t) \in \overline{G} \subset \widehat{\mathbb{C}}$$

exists. Then $\overline{C} = C \cup \{b\}$ is a curve in $G \cup \{b\}$. For simplicity we shall refer to C as a curve in G ending at b.

Proposition 1.1. *If C is a halfopen curve in G and if $\Lambda(C) < \infty$ then C ends at a point $b \in \overline{G}$.*

Proof. Let $\alpha < \tau < \tau' < \beta$. Then

$$|\varphi(\tau') - \varphi(\tau)| \le \Lambda(C_{\tau'}) - \Lambda(C_\tau) \to 0 \quad \text{as} \quad \tau, \tau' \to \beta$$

by (7) and (12). It follows that the limit (13) exists. $\qquad \qquad \square$

1.2 Conformal Maps

1. We say that f maps the domain $H \subset \widehat{\mathbb{C}}$ *conformally onto* $G \subset \widehat{\mathbb{C}}$ if

(i) f is meromorphic in H;
(ii) f is *injective* (one-to-one), that is $z, z' \in H$, $z \neq z' \Rightarrow f(z) \neq f(z')$;
(iii) $f(H) = G$.

We say that f maps H *conformally into* G if (iii) is replaced by $f(H) \subset G$. A *univalent function* is the same as a conformal map.

Note that we define conformal maps only for (connected) open sets. Here are some elementary properties; see e.g. Ahl66a.

(a) The inverse is also a conformal map.
(b) A conformal map is a *homeomorphism*, i.e. a continuous injective map with continuous inverse.
(c) Conformal maps are *locally univalent*, i.e. the derivative does not vanish and there are only simple poles.
(d) Angles between curves including their orientation are preserved by conformal maps.
(e) The conformal image of a (measurable) subset A has the *area*

$$\text{area } f(A) = \iint_A |f'(z)|^2 \, dx \, dy \, .$$

In particular, if $f(z) = a_0 + a_1 z + a_2 z^2 + \ldots$ is a conformal map of $\mathbb{D}$ into $\mathbb{C}$ then

$$\text{area } f(\mathbb{D}) = \pi \sum_{n=1}^{\infty} n |a_n|^2 \, .$$

Reflection on *circles* in $\widehat{\mathbb{C}}$, i.e. ordinary circles or straight lines, is a powerful tool for analytic continuation (Ahl66a, p. 171).

Reflection Principle. *Let H, G be domains and H^*, G^* their reflections on the circles C, B in $\widehat{\mathbb{C}}$ and suppose that $H \cap C = \emptyset$, $G \cap B = \emptyset$. If f is a conformal map of H onto G such that*

$$f(z) \rightarrow B \quad \text{as} \quad z \rightarrow C, \quad z \in H$$

then f can be extended to a conformal map of the set of interior points of $H \cup H^* \cup (C \cap \partial H)$.

2. The domain G in $\widehat{\mathbb{C}}$ is called *simply connected* if the closed set $\widehat{\mathbb{C}} \setminus G$ is connected.

Riemann Mapping Theorem. *Let $G \subsetneq \mathbb{C}$ be a simply connected domain and let $w_0 \in G$, $0 \leq \alpha < 2\pi$. Then there is a unique conformal map of $\mathbb{D}$ onto G such that $f(0) = w_0$ and $\arg f'(0) = \alpha$.*

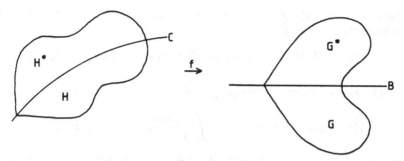

Fig. 1.2. The reflection principle

The Riemann mapping theorem is one of the most important results of complex analysis. It implies e.g. that any simply connected domain G in $\mathbb{C}$ is homeomorphic to $\mathbb{C}$ and thus that any closed curve in G is homotopic to a point.

If G is a simply connected domain with $\infty \in G$ that has more than one boundary point, then the transformation $\widetilde{w} = 1/(w - a)$, $a \notin G$, leads to a domain $\widetilde{G}$ with $0 \in \widetilde{G} \subsetneqq \mathbb{C}$. Hence there is a conformal map $\widetilde{f}$ of $\mathbb{D}$ onto $\widetilde{G}$ with $\widetilde{f}(0) = 0$. Thus $f = a + 1/\widetilde{f}$ maps $\mathbb{D}$ onto G such that $f(0) = \infty$. If $\infty \in G$ it is often more convenient to consider the exterior $\mathbb{D}^*$ of the unit circle instead of $\mathbb{D}$. Then we obtain a conformal map

$$(1) \qquad g(\zeta) = f(\zeta^{-1}) = b\zeta + b_0 + b_1\zeta^{-1} + \dots \quad (|\zeta| > 1)$$

of $\mathbb{D}^*$ onto G with $g(\infty) = \infty$. It is uniquely determined by $\arg b$.

The Riemann mapping theorem has no analogue in higher dimensions: A theorem of Liouville (see e.g. Nev60, Geh62) states that every conformal map of a domain in $\mathbb{R}^n$ with $n > 2$ is a Möbius transformation.

3. The *Möbius transformations*

$$\tau(z) = \frac{az + b}{cz + d}, \quad a, b, c, d \in \mathbb{C}, \quad ad - bc \neq 0$$

are the only conformal maps of $\widehat{\mathbb{C}}$. They form a group with respect to composition that we denote by Möb. Möbius transformations map circles in $\widehat{\mathbb{C}}$ onto circles in $\widehat{\mathbb{C}}$ where a line is considered as a circle through ∞.

The conformal maps of $\mathbb{D}$ onto $\mathbb{D}$ are of the form

$$(2) \qquad \tau(z) = \frac{cz + z_0}{1 + \overline{z}_0 cz}, \quad |z_0| < 1, \quad |c| = 1.$$

They form a subgroup Möb($\mathbb{D}$) so that Möb($\mathbb{D}$) = $\{\tau \in$ Möb $: \tau(\mathbb{D}) = \mathbb{D}\}$. It is easy to see that

$$(3) \qquad (1 - |z|^2)|\tau'(z)| = 1 - |\tau(z)|^2 \quad \text{for} \quad z \in \mathbb{D}, \quad \tau \in \text{Möb}(\mathbb{D}).$$

We shall often consider $h = f \circ \tau$ where f is analytic in $\mathbb{D}$ and τ is given by (2). Then $h(0) = f(z_0)$ and

(4) $\qquad (1 - |z|^2)|h'(z)| = (1 - |\tau(z)|^2)|f'(\tau(z))| \qquad \text{for} \quad z \in \mathbb{D}.$

4. The *non-euclidean metric* (hyperbolic metric) is defined by

(5) $\qquad \lambda(z_1, z_2) = \lambda_{\mathbb{D}}(z_1, z_2) = \min_{C} \int_{C} \frac{|dz|}{1 - |z|^2} \qquad \text{for} \quad z_1, z_2 \in \mathbb{D}$

where the minimum is taken over all curves C in $\mathbb{D}$ from z_1 to z_2; it is attained for the *non-euclidean segment* S from z_1 to z_2, that is the arc of the circle through z_1 and z_2 orthogonal to $\mathbb{T}$. A consequence of (5) is the triangle inequality

(6) $\qquad \lambda(z_1, z_3) \leq \lambda(z_1, z_2) + \lambda(z_2, z_3) \qquad \text{for} \quad z_1, z_2, z_3 \in \mathbb{D}$

and, together with (3), the invariance property

(7) $\qquad \lambda(\tau(z_1), \tau(z_2)) = \lambda(z_1, z_2) \qquad \text{for} \quad \tau \in \text{Möb}(\mathbb{D}).$

Transforming one point to 0 we easily see that

(8) $\qquad \lambda(z_1, z_2) = \frac{1}{2} \log \frac{1 + |z_1 - z_2|/|1 - \bar{z}_1 z_2|}{1 - |z_1 - z_2|/|1 - \bar{z}_1 z_2|} \qquad \text{for} \quad z_1, z_2 \in \mathbb{D}.$

It follows that

(9) $\qquad \left| \frac{z_1 - z_2}{1 - \bar{z}_1 z_2} \right| = \frac{e^{2\lambda} - 1}{e^{2\lambda} + 1} = \tanh \lambda.$

Often the non-euclidean metric is defined with a factor 2 in (5).

5. A *Stolz angle* at $\zeta \in \mathbb{T}$ is of the form

(10) $\quad \Delta = \left\{ z \in \mathbb{D} : \left| \arg\left(1 - \bar{\zeta} z\right) \right| < \alpha, |z - \zeta| < \rho \right\} \left(0 < \alpha < \frac{\pi}{2}, \rho < 2\cos\alpha \right).$

The exact shape is not relevant. The important fact is that all points of Δ have bounded non-euclidean distance from the radius $[0, \zeta]$.

Now let f be a function from $\mathbb{D}$ to $\widehat{\mathbb{C}}$. There are two important kinds of limits. We say that f has the *angular limit* $a \in \widehat{\mathbb{C}}$ at $\zeta \in \mathbb{T}$ if

(11) $\qquad f(z) \to a \qquad \text{as} \quad z \to \zeta, \quad z \in \Delta$

for each Stolz angle Δ at ζ; the opening angle 2α of Δ can be any number $< \pi$. We say that f has the *unrestricted limit* $a \in \widehat{\mathbb{C}}$ at ζ if

(12) $\qquad f(z) \to a \qquad \text{as} \quad z \to \zeta, \quad z \in \mathbb{D};$

if we now define $f(\zeta) = a$ then f becomes *continuous* at ζ as a function in $\mathbb{D} \cup \{\zeta\}$.

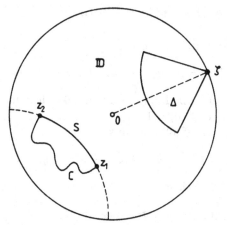

Fig. 1.3. The non-euclidean segment S between $z_1, z_2 \in \mathbb{D}$ and a Stolz angle Δ at $\zeta \in \mathbb{T}$

We shall use the symbol $f(\zeta)$ to denote the angular limit whenever it exists. Its existence does not imply that f has an unrestricted limit. For example,

$$(13) \qquad f_0(z) = \exp[-(1+z)/(1-z)] \quad (z \in \mathbb{D})$$

has the angular limit 0 at 1. But if z tends to 1 on a circle tangential to $\mathbb{T}$ then $\mathrm{Re}[(1+z)/(1-z)]$ and thus $|f_0(z)|$ are constant.

As we shall see, most of the functions we are going to encounter have angular limits at many points but unrestricted limits only in more special circumstances.

6. There are two useful differential invariants for meromorphic functions. The *spherical derivative* of f is defined by

$$(14) \qquad f^{\#}(z) = |f'(z)|/(1 + |f(z)|^2) \,.$$

It is invariant under rotations of the sphere, see (1.1.2).

If f is locally univalent, then the *Schwarzian derivative*

$$(15) \qquad S_f(z) = \frac{d}{dz}\frac{f''(z)}{f'(z)} - \frac{1}{2}\left(\frac{f''(z)}{f'(z)}\right)^2$$

is analytic. It satisfies

$$(16) \qquad S_{\sigma \circ f \circ \tau}(z) = S_f(\tau(z))\tau'(z)^2 \quad \text{for} \quad \sigma, \tau \in \mathrm{M\ddot{o}b}$$

and is thus Möbius-invariant, that is $S_{\sigma \circ f} = S_f$ for $\sigma \in \mathrm{M\ddot{o}b}$. Furthermore (3) and (16) imply that, if $\tau \in \mathrm{M\ddot{o}b}(\mathbb{D})$,

$$(17) \qquad (1 - |z|^2)^2|S_{f \circ \tau}(z)| = \left(1 - |\tau(z)|^2\right)^2|S_f(\tau(z))| \quad (z \in \mathbb{D}).$$

1.3 The Koebe Distortion Theorem

The class S consists of all functions

$$(1) \qquad f(z) = z + a_2 z^2 + a_3 z^3 + \dots \quad (|z| < 1)$$

analytic and univalent in $\mathbb{D}$. Many formulas take their nicest form with this normalization $f(0) = 0$, $f'(0) = 1$.

The class Σ consists of all functions

$$(2) \qquad g(\zeta) = \zeta + b_0 + b_1 \zeta^{-1} + \dots \quad (|\zeta| > 1)$$

univalent in $\mathbb{D}^*$. If $f \in S$ then

$$(3) \qquad g(\zeta) = 1/f(\zeta^{-1}) = \zeta - a_2 + (a_2^2 - a_3)\zeta^{-1} + \dots \quad (|\zeta| > 1)$$

belongs to Σ and omits 0. Conversely if $g \in \Sigma$ and $g(\zeta) \neq 0$ for $\zeta \in \mathbb{D}^*$ then

$$(4) \qquad f(z) = 1/g(z^{-1}) = z - b_0 z^2 + (b_0^2 - b_1)z^3 + \dots \quad (|z| < 1)$$

belongs to S.

It is not difficult to show (see e.g. Dur83, p. 29) that

$$(5) \qquad \text{area}\,(\mathbb{C} \setminus g(\mathbb{D}^*)) = \pi \left(1 - \sum_{n=1}^{\infty} n|b_n|^2 \right) \qquad \text{for} \quad g \in \Sigma.$$

This implies the *area theorem*

$$(6) \qquad |b_1|^2 \leq \sum_{n=1}^{\infty} n|b_n|^2 \leq 1 \qquad \text{for} \quad g \in \Sigma;$$

equality holds if and only if $g(\zeta) \equiv \zeta + b_0 + b_1 \zeta^{-1}$ with $|b_1| = 1$.

If $g \in \Sigma$ and $g(\zeta) \neq w$ for $\zeta \in \mathbb{D}^*$ then

$$(7) \qquad h(\zeta) = \sqrt{g(\zeta^2) - w} = \zeta + \frac{1}{2}(b_0 - w)\zeta^{-1} + \dots$$

is an odd function in Σ. Hence it follows from (6) that

$$(8) \qquad |w - b_0| \leq 2 \qquad \text{for} \quad g \in \Sigma, \quad w \notin g(\mathbb{D}^*)$$

and thus from (3) that (Bie16)

$$(9) \qquad |a_2| \leq 2 \qquad \text{for} \quad f \in S.$$

Equality holds for the *Koebe function*

$$(10) \qquad f_0(z) = \frac{z}{(1-z)^2} = \sum_{n=1}^{\infty} n z^n.$$

The Bieberbach conjecture that

(11) $$|a_n| \leq n \quad \text{for} \quad f \in S, \quad n = 2, 3, \ldots$$

was proved by de Branges (deB85) after many partial results, see e.g. Löw 23, GaSch55a, Hay58a, Ped68, PeSch72. The study of this conjecture led to the development of a great number of different and deep methods that have solved many other problems; see e.g. the books Gol57, Hay58b, Jen65, Pom75, Dur83 and also Section 8.4.

If f is analytic and univalent in $\mathbb{D}$ and if $z_0 \in \mathbb{D}$ then the *Koebe transform*

$$(12) \quad h(z) = \frac{f\left(\frac{z + z_0}{1 + \bar{z}_0 z}\right) - f(z_0)}{(1 - |z_0|^2)\, f'(z_0)} = z + \left(\frac{1}{2}\left(1 - |z_0|^2\right) \frac{f''(z_0)}{f'(z_0)} - \bar{z}_0\right) z^2 + \ldots$$

belongs to S. This fact makes it possible to transfer information at 0 to information at any point of $\mathbb{D}$. Thus we obtain from (9):

Proposition 1.2. *If f maps $\mathbb{D}$ conformally into $\mathbb{C}$ then*

$$(13) \quad \left|(1 - |z|^2)\frac{f''(z)}{f'(z)} - 2\bar{z}\right| \leq 4 \quad \text{for} \quad z \in \mathbb{D}.$$

Equality holds for the Koebe function. Integrating this inequality twice (see e.g. Dur83, p. 32) we obtain the famous *Koebe distortion theorem*.

Theorem 1.3. *If f maps $\mathbb{D}$ conformally into $\mathbb{C}$ and if $z \in \mathbb{D}$ then*

$$(14) \quad |f'(0)|\frac{|z|}{(1 + |z|)^2} \leq |f(z) - f(0)| \leq |f'(0)|\frac{|z|}{(1 - |z|)^2},$$

$$(15) \quad |f'(0)|\frac{1 - |z|}{(1 + |z|)^3} \leq |f'(z)| \leq |f'(0)|\frac{1 + |z|}{(1 - |z|)^3}.$$

The distance to the boundary is an important geometric quantity. We define

$$(16) \quad d_f(z) = \text{dist}(f(z), \partial f(\mathbb{D})) \quad \text{for} \quad z \in \mathbb{D}.$$

Corollary 1.4. *If f maps $\mathbb{D}$ conformally into $\mathbb{C}$ then*

$$(17) \quad \frac{1}{4}(1 - |z|^2)|f'(z)| \leq d_f(z) \leq (1 - |z|^2)|f'(z)| \quad \text{for} \quad z \in \mathbb{D}.$$

Proof. Let h again be the Koebe transform. Since $h \in S$ we obtain from (14) and the minimum principle that

$$\frac{1}{4} \leq \liminf_{|z| \to 1} \left|\frac{h(z)}{z}\right| \leq |h'(0)| = 1$$

and (17) (with z_0 instead of z) follows because

$$d_f(z_0) = \text{dist}(\partial G, f(z_0)) = \liminf_{|\zeta| \to 1} |f(\zeta) - f(z_0)| . \qquad \square$$

We now turn to an invariant consequence of the Koebe distortion theorem in terms of the non-euclidean distance λ given by (1.2.8).

Corollary 1.5. *If f is a conformal map of $\mathbb{D}$ into $\mathbb{C}$ and if $z_0, z_1 \in \mathbb{D}$ then*

$$(18) \qquad \frac{\tanh \lambda(z_0, z_1)}{4} \leq \frac{|f(z_1) - f(z_0)|}{(1 - |z_0|^2)|f'(z_0)|} \leq \exp\left[4\lambda(z_0, z_1)\right] ,$$

$$(19) \qquad |f'(z_0)| \exp[-6\lambda(z_0, z_1)] \leq |f'(z_1)| \leq |f'(z_0)| \exp[6\lambda(z_0, z_1)] .$$

Proof. We consider the Koebe transform (12) with $z = (z_1 - z_0)/(1 - \bar{z}_0 z_1)$ and thus $z_1 = (z + z_0)/(1 + \bar{z}_0 z)$. We obtain from (14) that

$$\frac{|z|}{4} \leq \frac{|f(z_1) - f(z_0)|}{(1 - |z_0|^2)|f'(z_0)|} = |h(z)| \leq \frac{|z|}{(1 - |z|)^2} \leq \left(\frac{1 + |z|}{1 - |z|}\right)^2$$

and (18) follows from (1.2.8). Furthermore we see from (15) that

$$\left|\frac{f'(z_1)}{f'(z_0)}\right| = |1 + \bar{z}_0 z|^2 |h'(z)| \leq \left(\frac{1 + |z|}{1 - |z|}\right)^3$$

which implies the upper estimate (19); the lower one follows from symmetry.
$$\square$$

The following form is more geometric and perhaps easier to apply.

Corollary 1.6. *Let a, b, c be positive constants and let $0 < |z_0| = 1 - \delta < 1$. If*

$$(20) \qquad 0 \leq 1 - b\delta \leq |z| \leq 1 - a\delta , \quad |\arg z - \arg z_0| \leq c\delta$$

and if f maps $\mathbb{D}$ conformally into $\mathbb{C}$ then

$$(21) \qquad |f(z) - f(z_0)| \leq M_1 \delta |f'(z_0)| ,$$

$$(22) \qquad M_2^{-1} |f'(z_0)| \leq |f'(z)| \leq M_2 |f'(z_0)| ,$$

where the constants M_1 and M_2 depend only on a, b, c.

Proof (see Fig. 1.4). The circular arc from z_0 to z_2 has length $< c\delta$. Hence we obtain from (1.2.5) that

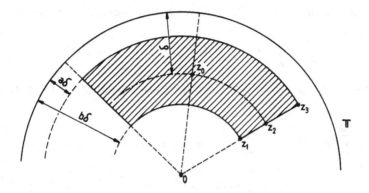

Fig. 1.4. The set of values described by (20) is shaded. The angle at 0 is $2c\delta$

$$\lambda(z_0, z_2) < \frac{c\delta}{1 - (1 - \delta)^2} = \frac{c}{2 - \delta} < c.$$

Furthermore

$$\left| \frac{z_3 - z_1}{1 - \bar{z}_1 z_3} \right| = \frac{(1 - a\delta) - (1 - b\delta)}{1 - (1 - a\delta)(1 - b\delta)} = \frac{b - a}{b + a - ab\delta} \leq \frac{b - a}{b}.$$

Hence we obtain from (1.2.6) and (1.2.8) that

$$\lambda(z_0, z) \leq \lambda(z_0, z_2) + \lambda(z_1, z_3) \leq c + \frac{1}{2} \log \frac{2b - a}{a}.$$

Thus (21) and (22) follow from Corollary 1.5.   □

We show now that the average growth of f is much lower than $(1 - r)^{-2}$; see e.g. Hay58b or Dur83 for more precise results.

Theorem 1.7. *Let f map $\mathbb{D}$ conformally into $\mathbb{C}$. Then*

(23)
$$f(\zeta) = \lim_{r \to 1} f(r\zeta) \neq \infty$$

exists for almost all $\zeta \in \mathbb{T}$ and

(24)
$$\frac{1}{2\pi} \int_0^{2\pi} |f(re^{it}) - f(0)|^{2/5} dt \leq 5|f'(0)|^{2/5} \quad \text{for} \quad 0 \leq r < 1.$$

It follows from this result (Pra27) that f belongs to the Hardy space $H^{2/5}$; in fact $f \in H^p$ holds for every $p < 1/2$ (see Theorem 8.2) but not always for $p = 1/2$ as the Koebe function (10) shows.

Proof. We may assume that $f \in S$. The function

(25)
$$g(z) = f(z^5)^{1/5} = z \left(\frac{f(z^5)}{z^5} \right)^{1/5} = \sum_{n=1}^{\infty} b_n z^n$$

is analytic and univalent in $\mathbb{D}$. It follows from Parseval's formula that

$$\frac{1-r}{2\pi} \int_0^{2\pi} |g'(re^{it})|^2 dt = (1-r) \sum_{n=1}^\infty n^2 |b_n|^2 r^{2n-2}$$

for $0 \le r < 1$. Since $n(1-r)r^{n-1} \le 1$ and since g is univalent, this is bounded by

$$(26) \qquad \sum_1^\infty n|b_n|^2 r^{n-1} = \frac{1}{\pi r} \text{ area } \{g(z) : |z| \le \sqrt{r}\}$$

$$\le \frac{1}{r} \max_{|z| \le \sqrt{r}} |g(z)|^2 \le (1 - r^{5/2})^{-4/5} \le (1-r)^{-4/5}$$

because of (25) and (14). Hence we conclude that

$$(27) \qquad \frac{1}{2\pi} \int_0^1 \int_0^{2\pi} (1-r)^{9/10} |g'(re^{it})|^2 \, dt \, dr \le \int_0^1 (1-r)^{-9/10} \, dr = 10.$$

It follows by Schwarz's inequality that

$$\left(\int_0^1 |g'(re^{it})| \, dr \right)^2 \le \int_0^1 (1-r)^{-9/10} \, dr \int_0^1 (1-r)^{9/10} |g'|^2 \, dr.$$

If we integrate we obtain from (27) that

$$(28) \qquad \frac{1}{2\pi} \int_0^{2\pi} \left(\int_0^1 |g'(re^{it})| \, dr \right)^2 dt \le 100.$$

Hence the inner integral is finite for almost all t and $\lim_{r \to 1} g(re^{it}) \ne \infty$ exists for these t by Proposition 1.1. Hence the limit (23) exists for almost all t by (25).

Finally we see from (25) and Parseval's formula that

$$\frac{1}{2\pi} \int_0^{2\pi} |f(r^5 e^{it})|^{2/5} \, dt = \frac{1}{2\pi} \int_0^{2\pi} |g(re^{it})|^2 \, dt \le \sum_{n=1}^\infty |b_n|^2 \le 5;$$

the last inequality follows from (26) by integrating over $[0,1]$. $\qquad \square$

Exercises 1.3

1. Let f map $\mathbb{D}$ conformally onto a bounded domain. Show that $(1 - |z|)|f'(z)| \to 0$ as $|z| \to 1$.

2. Let f be a conformal map of $\mathbb{D}$ into $\mathbb{C} \setminus \{c\}$ with $f(0) = 0$. Show that

$$|f(z)| \le \frac{4|cz|}{(1-|z|)^2} \qquad \text{for} \quad z \in \mathbb{D}.$$

3. Use the Koebe transform to deduce from (14) that

$$\frac{1-|z|}{1+|z|} \leq \left| \frac{zf'(z)}{f(z)-f(0)} \right| \leq \frac{1+|z|}{1-|z|} \quad \text{for} \quad z \in \mathbb{D}.$$

4. If f maps $\mathbb{D}$ conformally into $\mathbb{C}$ apply the preceding exercise to show that

$$|\log[(f(z)-f(0))/(zf'(0))]| \leq 4\log[(1+|z|)/(1-|z|)]$$

for $z \in \mathbb{D}$ and deduce that e.g.

$$\left| \arg \frac{f(z)-f(z_0)}{z-z_0} - \arg f'(z_0) \right| \leq 8\lambda(z_0, z) + \frac{\pi}{2}.$$

5. If $f \in S$ show that $\log[z^{-1}f(z)]$ belongs to the Hardy space H^2. (Compare the proof of Proposition 3.8.)

6. Show that $|a_2^2 - a_3| \leq 1$ if $f \in S$ and deduce that the Schwarzian derivative satisfies

$$|S_f(z)| \leq 6(1-|z|^2)^{-2}$$

if f maps $\mathbb{D}$ conformally into $\widehat{\mathbb{C}}$.

1.4 Sequences of Conformal Maps

The usual notion of convergence of domain sequences is not adequate for our purposes where the limit has to be a domain. Carathéodory therefore introduced a different concept.

Let $w_0 \in \mathbb{C}$ be given and let G_n be domains with $w_0 \in G_n \subset \mathbb{C}$. We say that

(1) $\qquad\qquad G_n \to G \quad \text{as} \quad n \to \infty \quad$ with respect to w_0

in the sense of *kernel convergence* if

(i) either $G = \{w_0\}$, or G is a domain $\neq \mathbb{C}$ with $w_0 \in G$ such that some neighbourhood of every $w \in G$ lies in G_n for large n;

(ii) for $w \in \partial G$ there exist $w_n \in \partial G_n$ such that $w_n \to w$ as $n \to \infty$.

It is clear that every subsequence also converges to G.

We show now that the limit is uniquely determined. Let also $G_n \to G^*$ as $n \to \infty$ and suppose that $G \not\subset G^*$. Since $w_0 \in G \cap G^*$ and since G is a domain by (i), there would exist $w^* \in G \cap \partial G^*$ and thus, by (ii), points $w_n \in \partial G_n$ with $w_n \to w^*$. This contradicts (i) because $w^* \in G$. Thus $G \subset G^*$ and similarly $G^* \subset G$.

Kernel convergence has surprising properties and depends on the choice of the given point w_0. As an example, if

$$G_n = \mathbb{C} \setminus (-\infty, -1/n] \setminus [1/n, +\infty)$$

then, as $n \to \infty$,

$$G_n \to \begin{cases} \{\operatorname{Im} w > 0\} & \text{if } w_0 = i, \\ \{0\} & \text{if } w_0 = 0, \\ \{\operatorname{Im} w < 0\} & \text{if } w_0 = -i. \end{cases}$$

The usual definition of kernel convergence is formally different; see e.g. Gol57, p. 46 or Pom75, p. 28. Some special cases are given in the exercises.

Theorem 1.8. *Let f_n map $\mathbb{D}$ conformally onto G_n with $f_n(0) = w_0$ and $f_n'(0) > 0$. If $G = \{w_0\}$ let $f(z) \equiv w_0$, otherwise let f map $\mathbb{D}$ conformally onto G with $f(0) = w_0$ and $f'(0) > 0$. Then, as $n \to \infty$,*

$$f_n \to f \quad \text{locally uniformly in } \mathbb{D} \quad \Leftrightarrow \quad G_n \to G \text{ with respect to } w_0.$$

This is the *Carathéodory kernel theorem* (Car12). It is a powerful tool for constructing conformal maps with given boundary behaviour. In Corollary 2.4 we shall give a geometric characterization of uniform convergence.

Proof. (a) Suppose first that $f_n \to f$ locally uniformly in $\mathbb{D}$. If f is constant then $G = \{w_0\}$. Otherwise G is a domain. Let $w \in G$ and let $\overline{D}(w, 2\delta) \subset G$. Then $A = f^{-1}(\overline{D}(w, 2\delta)) \subset \mathbb{D}$ is compact so that (f_n) converges uniformly on A. Hence we can find n_0 such that $|f_n(z) - f(z)| < \delta$ for $z \in A$ and $n \geq n_0$, and it follows from Rouché's theorem that $D(w, \delta) \subset f_n(A) \subset G_n$ for $n \geq n_0$. Thus (i) holds.

Now let $w \in \partial G$. Then there exist $z_k \in \mathbb{D}$ such that $|f(z_k) - w| < 1/k$. We can find an increasing sequence (n_k) such that

$$(2) \qquad |f_n(z_k) - f(z_k)| < 1/k, \quad |f_n'(z_k) - f'(z_k)| < 1/k \quad \text{for} \quad n \geq n_k.$$

We define w_n as the point of ∂G_n nearest to $f_n(z_k)$ if $n_k \leq n < n_{k+1}$. Then

$$|w_n - f_n(z_k)| \leq \left(1 - |z_k|^2\right)|f_n'(z_k)| < (1 - |z_k|^2)|f'(z_k)| + \frac{1}{k} \leq 4d_f(z_k) + \frac{1}{k}$$

by Corollary 1.4 and (2), and since $d_f(z_k) \leq |f(z_k) - w|$ it follows that

$$|w_n - w| \leq 4|f(z_k) - w| + \frac{1}{k} + |f_n(z_k) - f(z_k)| + |f(z_k) - w| < \frac{7}{k}$$

again by (2). It follows that $w_n \to w$ as $n \to \infty$. Hence (ii) holds.

(b) Conversely let $G_n \to G$ as $n \to \infty$. Since $G \neq \mathbb{C}$ it follows by (ii) that $\operatorname{dist}(w_0, \partial G_n) \leq M$ for some constant M and thus, by (1.3.14) and (1.3.17),

$$|f_n(z) - w_0| \leq \frac{4M|z|}{(1 - |z|)^2} \quad \text{for} \quad z \in \mathbb{D}.$$

Hence (f_n) is normal. Suppose now that $f_n \to f (n \to \infty)$ does not hold. Then there is a subsequence with $f_{n_\nu} \to f^* \neq f$ as $\nu \to \infty$ locally uniformly in $\mathbb{D}$. It follows from part (a) that $G_{n_\nu} \to G^* = f^*(\mathbb{D})$ as $\nu \to \infty$. Since also

$G_{n_\nu} \to G$ we conclude from the uniqueness of the limit that $G^* = G$ and thus that $f^* = f$ because of our normalization. This is a contradiction.   □

Exercises 1.4

1. Show that an increasing sequence of domains converges to its union unless this is $= \mathbb{C}$.

2. Let (G_n) be a decreasing sequence of domains and let H be the open kernel of its intersection. If $w_0 \in H$ show that G_n converges (with respect to w_0) to the component of H containing w_0.

3. Let f_n map $\mathbb{D}$ conformally onto the bounded domain G_n with $f_n(0) = w_0$. If $G_n \to G = f(\mathbb{D})$, show that

$$\iint_{|z|<r} |f_n'(z)|^2 \, dx \, dy \to \iint_{|z|<r} |f'(z)|^2 \, dx \, dy \quad \text{as} \quad n \to \infty$$

holds for $r < 1$ but not always for $r = 1$.

1.5 Some Univalence Criteria

We come now to some sufficient conditions for a given analytic function to be injective. The first criterion is of a topological nature and is related to the degree of a map; see e.g. Mil65, p. 28.

Theorem 1.9. *Let f be nonconstant and analytic in the domain $H \subset \widehat{\mathbb{C}}$ and let G be the inner domain of the Jordan curve J. If*

$$(1) \qquad\qquad f(z) \to J \quad \text{as} \quad z \to \partial H$$

then $f(H) = G$. If furthermore f assumes in H some value of G only once (with multiplicity 1) then f is injective and H is simply connected.

We mean by (1) that if (z_n) is any sequence in H whose limit points lie on ∂H then all limit points of $(f(z_n))$ lie on $\partial G = J$.

Proof. (a) We show first that $\partial f(H) \subset J$. If $\omega \in \partial f(H)$ (possibly $\omega = \infty$) then there exist $z_n \in H$ such that $f(z_n) \to \omega$ and $z_n \to \zeta$ for some $\zeta \in \overline{H}$. If $\zeta \in H$ then f would map a neighbourhood of ζ onto a neighbourhood of $\omega = f(\zeta)$ contrary to $\omega \in \partial f(H)$. Hence $\zeta \in \partial H$ and it follows from (1) that $\omega \in J$.

By the Jordan curve theorem, $\widehat{\mathbb{C}} \setminus J$ has a component G^* with $\infty \in G^*$. If $f(H) \not\subset G$ then G^* would meet $f(H)$ and also $\widehat{\mathbb{C}} \setminus f(H)$ because $\infty \notin f(H)$. Hence G^* would meet $\partial f(H) \subset J = \partial G^*$ which is impossible. Thus $f(H) \subset G$ and it follows from $\partial f(H) \subset J$ that $f(H) = G$.

(b) Our further assumption implies that the set

$$H_1 = \{z \in H : f'(z) \neq 0, f(z') \neq f(z) \quad \text{for} \quad z' \in H, z' \neq z\} \tag{2}$$

is non-empty. We claim that $H_1 = H$. Suppose this is false. Since H is a domain there exist $z_0 \in H \cap \partial H_1$ and therefore $z_n \in H \setminus H_1$ with $z_n \to z_0 (n \to \infty)$. Since $z_n \notin H_1$ we can find $z_n' \neq z_n$ such that $f(z_n') = f(z_n)$; we may assume that $z_n' \to z_0' \in \overline{H}$. Since $f(z_n') \to f(z_0) \in G$ it follows from (1) that $z_0' \in H$. Hence f assumes all values near $f(z_0') = f(z_0)$ in a neighbourhood V of z_0 and also in a neighbourhood V' of z_0'; if $z_0 = z_0'$ then $f'(z_0) = 0$ and these values are assumed at least twice in V. This contradicts our assumption that $z_0 \in \partial H_1$. Thus $H_1 = H$ and f is injective in H, by (2).

Hence f is a conformal map of H onto G and in particular a homeomorphism. Since G is simply connected, it follows (New64, p. 156) that H is also simply connected. This completes the proof.   □

The following *boundary principle* is a consequence of Corollary 2.10 below: If f is analytic in $\mathbb{D}$ and continuous and injective on $\mathbb{T}$ then f is injective also in $\mathbb{D}$.

We turn now to analytic univalence criteria.

Proposition 1.10. *Let f be analytic in the convex domain H. If*

$$\operatorname{Re} f'(z) > 0 \quad \text{for} \quad z \in H \tag{3}$$

then f is univalent in H. If $H = \mathbb{D}$ and f is continuous in $\overline{\mathbb{D}}$ then $f(\mathbb{D})$ is a Jordan domain.

Proof. Let z_1 and z_2 be distinct points in H. Since H is convex the segment $[z_1, z_2]$ lies in H. Hence it follows from (3) that

$$\operatorname{Re} \frac{f(z_2) - f(z_1)}{z_2 - z_1} = \int_0^1 \operatorname{Re}[f'(z_1 + (z_2 - z_1)t)] \, dt > 0 \tag{4}$$

and thus that $f(z_1) \neq f(z_2)$. If f is continuous in $\overline{\mathbb{D}}$ and if z_1 and z_2 are distinct points on $\mathbb{T}$ then $(z_1, z_2) \subset \mathbb{D}$ so that $f(z_1) \neq f(z_2)$ by (4).   □

The *Becker univalence criterion* (DuShSh66, Bec72, BePo84) is much deeper.

Theorem 1.11. *Let f be analytic and locally univalent in $\mathbb{D}$. If*

$$(1 - |z|^2) \left| z \frac{f''(z)}{f'(z)} \right| \leq 1 \quad \text{for} \quad z \in \mathbb{D}$$

then f is univalent in $\mathbb{D}$. The constant 1 is best possible.

The proof uses Löwner chains (see e.g. Pom75, p. 172); the univalence ultimately comes from the uniqueness theorem for ordinary differential equations.

An alternative proof (Ahl74; Scho75, p. 169) uses quasiconformal maps; the univalence follows then from topological considerations.

The *Nehari univalence criterion* (Neh49) uses the Schwarzian derivative defined in (1.2.15).

Theorem 1.12. *Let f be meromorphic and locally univalent in $\mathbb{D}$. If*

$$\left(1 - |z|^2\right)^2 |S_f(z)| \leq 2 \quad for \quad z \in \mathbb{D}$$

then f is univalent in $\mathbb{D}$. The bound 2 is best possible. If $f(0) \in \mathbb{C}$ and $f''(0) = 0$ then f has no pole.

The proof uses either the theory of second order linear differential equations (see e.g. Hil62 or Dur83, p. 261) or again quasiconformal maps. See GePo84 or Leh86, p. 91 for the final statement.

The *Epstein univalence criterion* (Eps87) is a common generalization of the last two criteria: Let f be meromorphic and g analytic in $\mathbb{D}$. If both functions are locally univalent and if

$$\left| \frac{1}{2}(1 - |z|^2)^2 [S_f(z) - S_g(z)] + (1 - |z|^2)\bar{z}\frac{g''(z)}{g'(z)} \right| \leq 1 \quad for \quad z \in \mathbb{D},$$

then f is univalent in $\mathbb{D}$. The original proof uses differential geometry and works in a more general context, see Pom86 for a proof using Löwner chains.

Exercises 1.5

1. Let $g(z) = z + b_0 + \dots$ be analytic in $\{1 < |z| < \infty\}$ and continuous in $\{1 \leq |z| < \infty\}$ and let J be a Jordan curve in $\mathbb{C}$. If $g(\mathbb{T}) \subset J$ and if g omits some point $\notin J$, show that g maps $\mathbb{D}^*$ conformally onto the outer domain of J.

2. Let $\sum_2^\infty n|a_n| \leq 1$. Show that $f(z) = z + \sum_2^\infty a_n z^n$ maps $\mathbb{D}$ conformally onto a Jordan domain.

Chapter 2. Continuity and Prime Ends

2.1 An Overview

Let f map the unit disk $\mathbb{D}$ conformally onto $G \subset \widehat{\mathbb{C}} = \mathbb{C} \cup \{\infty\}$. In this chapter, we study the problem whether it is possible to extend f to some or all points $\zeta \in \mathbb{T} = \partial \mathbb{D}$ by defining

$$(1) \qquad\qquad f(\zeta) = \lim_{z \to \zeta} f(z) \in \widehat{\mathbb{C}}.$$

If this limit exists for all $\zeta \in \mathbb{T}$ then f becomes continuous in $\overline{\mathbb{D}}$; if $\infty \in \partial G$ then continuity has to be understood in the spherical metric.

The global extension problem has a completely topological answer. A set is called *locally connected* if nearby points can be joined by a connected subset of small diameter.

Continuity Theorem. *The function f has a continuous extension to $\mathbb{D} \cup \mathbb{T}$ if and only if ∂G is locally connected.*

The problem whether this extension is injective (= one-to-one) on $\overline{\mathbb{D}}$ has again a topological answer; see Section 2.3. The following theorem is often used and has many important consequences.

Carathéodory Theorem. *The function f has a continuous and injective extension to $\mathbb{D} \cup \mathbb{T}$ if and only if ∂G is a Jordan curve.*

If ∂G is locally connected but not a Jordan domain then parts of ∂G are run through several times. Some of the possibilities are indicated in Fig. 2.1 where points with the same letters correspond to each other. Note that e.g. the arcs a_2d_1 and a_1d_2 are both mapped onto the segment ad.

The situation becomes much more complicated if ∂G is not locally connected. Carathéodory introduced the notion of a prime end to describe this situation; see Section 2.4 and 2.5.

Prime End Theorem. *The points ζ on $\mathbb{T}$ correspond one-to-one to the prime ends of G, and the limit $f(\zeta)$ exists if and only if the prime end has only one point.*

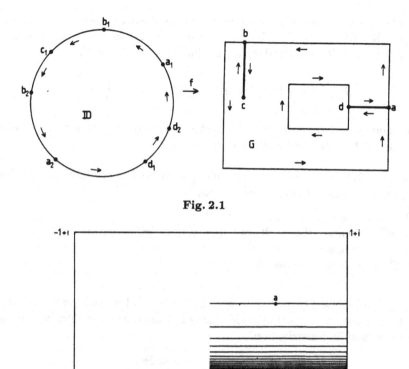

Fig. 2.1

Fig. 2.2. The point a belongs to two prime ends whereas the entire segment $[0,1]$ belongs to a single prime end

In Section 2.6 we shall see that, rather surprisingly, certain situations can occur only a countable number of times. For instance, there are at most countably many points on ∂G that have more than two preimages on $\mathbb{T}$.

2.2 Local Connection

We shall consider bounded domains in $\mathbb{C}$; everything carries over to the general case of domains in $\widehat{\mathbb{C}}$ if the spherical metric is used instead of the euclidean metric.

The closed set $A \subset \mathbb{C}$ is called *locally connected* if for every $\varepsilon > 0$ there is $\delta > 0$ such that, for any two points $a, b \in A$ with $|a - b| < \delta$, we can find a continuum B with

(1) $$ a, b \in B \subset A, \quad \text{diam } B < \varepsilon . $$

The continuous image of a compact locally connected set is again compact and locally connected (New64, p. 89). Hence every curve is locally connected as the continuous image of a (locally connected) segment.

The union of finitely many compact locally connected sets is locally connected (New64, p. 88). This is not generally true for infinitely many sets. A counterexample is the boundary of the domain (see Figure 2.2)

$$(2) \qquad G = \{u + iv : |u| < 1, 0 < v < 1\} \setminus \bigcup_{n=2}^{\infty} \left[\frac{i}{n}, \frac{i}{n} + 1 \right].$$

The points 0 and i/n cannot be connected in ∂G by any connected subset of diameter < 1 so that ∂G is not locally connected.

Theorem 2.1. *Let f map $\mathbb{D}$ conformally onto the bounded domain G. Then the following four conditions are equivalent:*

(i) *f has a continuous extension to $\overline{\mathbb{D}}$;*
(ii) *∂G is a curve, that is $\partial G = \{\varphi(\zeta) : \zeta \in \mathbb{T}\}$ with continuous φ;*
(iii) *∂G is locally connected;*
(iv) *$\mathbb{C} \setminus G$ is locally connected.*

We can rephrase (ii) $\Rightarrow$ (i) as follows: If it is at all possible to parametrize ∂G as a closed curve (perhaps with many self-intersections) then we can use the conformal parametrization

$$(3) \qquad \partial G : f(e^{it}), \quad 0 \le t \le 2\pi.$$

The implication (iii) $\Rightarrow$ (ii) is a special case of the **Hahn-Mazurkiewicz** theorem (New64, p. 89) that every locally connected continuum is a curve; we will not use this theorem.

The proof will be based on the following estimate (Wolff's lemma) about length distortion under conformal mapping; see Chapter 9 and e.g. Ahl73 for the systematic use of this method.

Proposition 2.2. *Let h map the open set $H \subset \mathbb{C}$ conformally into $D(0, R)$. If $c \in \mathbb{C}$ and $C(r) = H \cap \{|z - c| = r\}$ then*

$$(4) \qquad \inf_{\rho < r < \sqrt{\rho}} \Lambda(h(C(r))) \le \frac{2\pi R}{\sqrt{\log 1/\rho}} \quad for \quad 0 < \rho < 1$$

so that there is a decreasing sequence $r_n \to 0$ with

$$(5) \qquad \Lambda(h(C(r_n))) \to 0 \quad as \quad n \to \infty.$$

Proof. We write $l(r) = \Lambda(h(C(r)))$ and obtain from the Schwarz inequality that

$$l(r)^2 = \left(\int_{C(r)} |h'(z)| \, |dz| \right)^2 \le \int_{C(r)} |dz| \int_{C(r)} |h'(z)|^2 |dz|$$

$$\le 2\pi r \int_{c + re^{it} \in H} |h'(c + re^{it})|^2 r \, dt.$$

It follows that

$$(6) \qquad \int_0^\infty l(r)^2 \frac{dr}{r} \le 2\pi \int\!\!\int_H |h'(z)|^2 \, dx \, dy = 2\pi \, \text{area} \, h(H)$$

and since $h(H) \subset D(0, R)$ we conclude that

$$\frac{1}{2} \log \frac{1}{\rho} \inf_{\rho < r < \sqrt{\rho}} l(r)^2 \le \int_\rho^{\sqrt{\rho}} l(r)^2 \frac{dr}{r} \le 2\pi^2 R^2 \,.$$

This implies (4), and (5) is an immediate consequence.   □

Proof of Theorem 2.1. (i) $\Rightarrow$ (ii). If f is continuous in $\overline{\mathbb{D}}$ we can write ∂G as the curve (3).

(ii) $\Rightarrow$ (iii). This is clear because every curve is locally connected.

(iii) $\Rightarrow$ (iv). Let ∂G be locally connected. For $\varepsilon > 0$ we choose δ with $0 < \delta < \varepsilon$ as in the definition of local connection; see (1) with $A = \partial G$. Let now $a, b \in \mathbb{C} \setminus G$ and $|a - b| < \delta$. The case that $[a, b] \cap \partial G = \emptyset$ is trivial. In the other case let a' and b' be the first and last points where $[a, b]$ intersects ∂G. We can find a continuum $B \subset \partial G$ from a' to b' with diam $B < \varepsilon$, and $[a, a'] \cup B \cup [b', b]$ is a continuum of diameter $< 3\varepsilon$ connecting a and b in $\mathbb{C} \setminus G$.

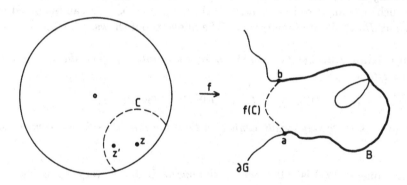

Fig. 2.3

(iv) $\Rightarrow$ (i). We may assume that $f(0) = 0$. Then

$$(7) \qquad\qquad D(0, R_0) \subset G \subset D(0, R)$$

with suitable $R_0 < R$. Let $0 < \varepsilon < R_0$ and choose δ with $0 < \delta < \varepsilon$ as in (1) with $A = \mathbb{C} \setminus G$. Next we choose ρ with $0 < \rho < 1/4$ such that $2\pi R(\log 1/\rho)^{-1/2} < \delta$. Let now $1/2 < |z| < 1$, $1/2 < |z'| < 1$ and $|z - z'| < \rho$. Applying Proposition 2.2 with $H = \mathbb{D}$ and $c = z$, we therefore find r with $\rho < r < 1/2$ such that

$$(8) \qquad\qquad \Lambda(f(C)) < \delta < \varepsilon, \quad C = \mathbb{D} \cap \partial D(z, r) \,.$$

We shall show below that

(9) $$|f(z) - f(z')| < 2\varepsilon .$$

It follows that f is uniformly continuous in $\mathbb{D}$ and therefore that f has a continuous extension to $\overline{\mathbb{D}}$. □

Suppose now that (9) is false and furthermore that $\overline{C} \cap \mathbb{T} \neq \emptyset$; the other case is simpler. Then $\overline{f(C)}$ is a curve with endpoints $a, b \in \partial G \subset \mathbb{C} \setminus G$ by Proposition 1.1. Since $|a - b| < \delta$ by (8), there is a continuum $B \subset \mathbb{C} \setminus G$ with $a, b \in B$ and $\operatorname{diam} B < \varepsilon$; see Fig. 2.3. Hence we see from (7) and (8) that

$$B \cup f(C) \subset D(a, \varepsilon), \quad 0 \notin D(a, \varepsilon).$$

Consequently $f(z')$ (or $f(z)$) is not separated from 0 by $B \cup f(C)$. Since $f(z')$ and 0 are neither separated by $\mathbb{C} \setminus G$ and since the intersection

$$(B \cup f(C)) \cap (\mathbb{C} \setminus G) = B$$

is connected, we conclude from Janiszewski's theorem (Section 1.1) that $f(z')$ and $f(0) = 0$ are not separated by the union $f(C) \cup (\mathbb{C} \setminus G)$. Hence z' and 0 are not separated by $C \cup \mathbb{T}$ which is false because $|z - z'| < \rho < r$.

We turn now to sequences of domains. We say that the closed sets A_n are *uniformly locally connected* if, for every $\varepsilon > 0$, there exists $\delta > 0$ independent of n such that any two points $a_n, b_n \in A_n$ with $|a_n - b_n| < \delta$ can be joined by continua $B_n \subset A_n$ of diameter $< \varepsilon$. The above proof shows:

Proposition 2.3. *Let f_n map $\mathbb{D}$ conformally onto G_n such that $f_n(0) = 0$. Suppose that*

(10) $$D(0, R_0) \subset G_n \subset D(0, R) \quad \text{for all} \quad n.$$

If $\mathbb{C} \setminus G_n$ is uniformly locally connected then the functions f_n are equicontinuous in $\overline{\mathbb{D}}$.

A sequence (f_n) is called *equicontinuous* in $\overline{\mathbb{D}}$ if, for every $\varepsilon > 0$, there is $\rho > 0$ independent of n such that

$$|f_n(z) - f_n(z')| < \varepsilon \quad \text{for} \quad z, z' \in \overline{\mathbb{D}}, \quad |z - z'| < \rho, \quad n \in \mathbb{N}.$$

The Arzelà-Ascoli theorem states that every uniformly bounded equicontinuous sequence contains a uniformly convergent subsequence.

The Carathéodory kernel theorem (Theorem 1.8) gives a necessary and sufficient condition for locally uniform convergence. We can supplement it by the following result about uniform convergence.

Corollary 2.4. *Let f_n map $\mathbb{D}$ conformally onto G_n with $f_n(0) = 0$. Suppose that (10) holds and that $\mathbb{C} \setminus G_n$ is uniformly locally connected. If*

(11) $$f_n(z) \to f(z) \quad \text{as} \quad n \to \infty \quad \text{for each} \quad z \in \mathbb{D}$$

then the convergence is uniform in $\overline{\mathbb{D}}$.

This is an immediate consequence of Proposition 2.3. The converse also holds (Pom75, p. 283): If pointwise convergence implies uniform convergence then $\mathbb{C} \setminus G_n$ is uniformly locally connected.

Exercises 2.2

1. Let f map $\mathbb{D}$ conformally onto a bounded domain. Use the method of Proposition 2.2 to show that almost all radii $[0, \zeta)$ are mapped onto curves of finite length.

2. Let G be a bounded simply connected domain with locally connected boundary. Use the Poisson integral in $\mathbb{D}$ to solve the Dirichlet problem: If φ is real-valued and continuous on ∂G then there is a unique harmonic function u in G that is continuous in $\overline{G}$ and satisfies $u = \varphi$ on ∂G.

3. Let $f(z) = a_0 + a_1 z + \ldots$ map $\mathbb{D}$ conformally onto a bounded domain G and write

$$s_n(z) = \sum_{\nu=0}^{n} a_\nu z^\nu , \quad \varepsilon_n^2 = \sum_{\nu=n}^{\infty} \nu |a_\nu|^2 .$$

Show that $s_n(z) = f(r_n z) + O(\varepsilon_n^{1/2})$ $(n \to \infty)$ uniformly in $\overline{\mathbb{D}}$ where $r_n = 1 - \varepsilon_n/n$; even $O(\varepsilon_n \log 1/\varepsilon_n)$ is true.

4. If furthermore ∂G is locally connected, show that the power series of f converges uniformly on $\mathbb{T}$. This is Fejér's Tauberian theorem (LaGa86, p. 65); the convergence need not be absolute even for Jordan domains (Pir62).

2.3 Cut Points and Jordan Domains

Let E be a locally connected continuum. We say that $a \in E$ is a *cut point* of E if $E \setminus \{a\}$ is no longer connected (see e.g. Why42). For example, a (closed) Jordan curve has no cut points whereas for a Jordan arc every point except the two endpoints is a cut point.

Proposition 2.5. *Let f map $\mathbb{D}$ conformally onto the bounded domain G with locally connected boundary. Let $a \in \partial G$ and*

$$A = f^{-1}(\{a\}) = \{\zeta \in \mathbb{T} : f(\zeta) = a\}, \quad m = \operatorname{card} A \le \infty .$$

Then a is a cut point if and only if $m > 1$, and the components of $\partial G \setminus \{a\}$ are $f(I_k)$ where I_k $(k = 1, \ldots, m)$ are the arcs that make up $\mathbb{T} \setminus A$.

We shall see (Corollary 2.19) that the case $3 \le m \le \infty$ can occur only for countably many points a. The set A may be uncountable but has always zero measure and even zero capacity (Beu40; see Theorem 9.19).

Proof. Since f is continuous in $\overline{\mathbb{D}}$ by Theorem 2.1, we can write

(1)
$$\partial G \setminus \{a\} = f(\mathbb{T} \setminus A) = \bigcup_{k=1}^{m} f(I_k) .$$

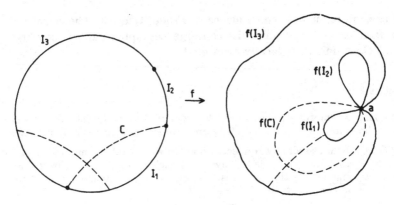

Fig. 2.4

The sets $f(I_k)$ are connected as continuous images of connected sets. It remains to show that $f(I_j)$ and $f(I_k)$ for $j \neq k$ cannot be connected within $\partial G \setminus \{a\}$.

To show this (see Fig. 2.4) consider a circular arc C in $\mathbb{D}$ that connects the endpoints of I_j. Then $J = f(\overline{C})$ is a Jordan curve in $G \cup \{a\}$. By the Jordan curve theorem, $\mathbb{C} \setminus J$ has exactly two components, each bounded by J. If C' is a circular arc in $\mathbb{D}$ from I_j to I_k then C and C' cross exactly once. Hence $f(\overline{C}) = J$ and $f(C')$ cross exactly once and it follows from (1) that $f(I_j)$ and $f(I_k)$ lie in different components of $\mathbb{C} \setminus J$ and thus cannot be connected within $\partial G \setminus \{a\}$. $\square$

We now deduce the important *Carathéodory theorem* (Car13a).

Theorem 2.6. *Let f map $\mathbb{D}$ conformally onto the bounded domain G. Then the following three conditions are equivalent:*

(i) *f has a continuous injective extension to $\overline{\mathbb{D}}$;*

(ii) *∂G is a Jordan curve;*

(iii) *∂G is locally connected and has no cut points.*

Proof. The implications (i) $\Rightarrow$ (ii) $\Rightarrow$ (iii) are obvious. Assume now (iii). Then f is continuous in $\overline{\mathbb{D}}$ by Theorem 2.1 and, by Proposition 2.5, every point on ∂G is attained only once on $\mathbb{T}$. Since furthermore $f(\mathbb{D}) = G$ is disjoint from $f(\mathbb{T}) = \partial G$ it follows that (i) holds. $\square$

Corollary 2.7. *Let G and H be Jordan domains and let the points $z_1, z_2, z_3 \in \partial G$ and $w_1, w_2, w_3 \in \partial H$ have the same cyclic order. Then there is a unique conformal map h of G onto H that satisfies*

$$(2) \qquad\qquad h(z_j) = w_j \quad for \quad j = 1, 2, 3.$$

Proof. By the Riemann mapping theorem there are conformal maps f and g of $\mathbb{D}$ onto G and H. By Theorem 2.6, they are continuous and injective on $\mathbb{T}$. We choose $\tau \in \text{Möb}(\mathbb{D})$ such that $\tau(f^{-1}(z_j)) = g^{-1}(w_j)$ for $j = 1, 2, 3$. Then $h = g \circ \tau \circ f^{-1}$ is a conformal map of G onto H satisfying (2). If h^* is another map with the same properties then $\tau^{-1} \circ g^{-1} \circ h^* \circ f$ is a conformal selfmap of $\mathbb{D}$ keeping the three points $f^{-1}(z_j)$ fixed and thus the identity. Hence $h^* = h$. $\qquad\square$

Corollary 2.8. *A conformal map of $\mathbb{D}$ onto a Jordan domain can be extended to a homeomorphism of $\widehat{\mathbb{C}}$ onto $\widehat{\mathbb{C}}$.*

Proof. Let f map $\mathbb{D}$ conformally onto the (bounded) Jordan domain G while f^* maps $\mathbb{D}^* = \{|z| > 1\} \cup \{\infty\}$ conformally onto the outer domain G^* of the Jordan curve ∂G such that $f^*(\infty) = \infty$. Then f and f^* can be extended to bijective continuous maps of $\overline{\mathbb{D}}$ onto $\overline{G}$ and of $\overline{\mathbb{D}}^*$ onto $\overline{G}^*$; the inverse maps are continuous because the sets are compact. If we define $\varphi(\zeta) = f^{*-1}(f(\zeta))$ for $\zeta \in \mathbb{T}$ and

$$(3) \qquad f(r\zeta) = f^*(r\varphi(\zeta)) \quad \text{for} \quad 1 < r < \infty, \ \zeta \in \mathbb{T},$$

then f becomes a homeomorphism of $\widehat{\mathbb{C}} = \overline{\mathbb{D}} \cup \mathbb{D}^*$ onto $\widehat{\mathbb{C}} = \overline{G} \cup G^*$. $\qquad\square$

A consequence is the purely topological *Schoenflies theorem*.

Corollary 2.9. *A bijective continuous map of a Jordan curve onto another can be extended to a homeomorphism of $\mathbb{C}$ onto $\mathbb{C}$.*

Proof. Let f and g map $\mathbb{D}$ conformally onto the inner domains G and H of the two Jordan curves and let φ be the given bijective continuous map of ∂G onto ∂H. We extend f and g according to Corollary 2.8. Then $\psi = g^{-1} \circ \varphi \circ f$ is a bijective continuous map of $\mathbb{T}$ onto $\mathbb{T}$. If we define

$$(4) \qquad \psi(r\zeta) = r\psi(\zeta) \quad \text{for} \quad 0 \le r < \infty, \ z \in \mathbb{T}$$

then $g \circ \psi \circ f^{-1}$ is the required homeomorphic extension of φ. $\qquad\square$

Corollary 2.10. *Let G be a bounded Jordan domain and let f be non-constant and analytic in $\mathbb{D}$. If*

$$(5) \qquad \text{dist}(f(z), \partial G) \to 0 \quad as \quad |z| \to 1$$

then $f(\mathbb{D}) = G$ and f has a continuous extension to $\overline{\mathbb{D}}$. There exists $m = 1, 2, \ldots$ such that f assumes every value of $\overline{G}$ exactly m times in $\overline{\mathbb{D}}$.

The following univalence criterion is a special case: If f is analytic in $\mathbb{D}$ and if $f(z)$ tends to the Jordan curve ∂G as $|z| \to 1$ then f maps $\mathbb{D}$ conformally onto G provided that one value in $\overline{G}$ is assumed only once.

Proof. It follows from Theorem 1.9 with $H = \mathbb{D}$ that $f(\mathbb{D}) = G$. Let φ map $\mathbb{D}$ conformally onto G. We see from Theorem 2.6 that φ is continuous and injective in $\overline{\mathbb{D}}$. Then $g = \varphi^{-1} \circ f$ is an analytic map of $\mathbb{D}$ onto $\mathbb{D}$ and (5) implies that $|g(z)| \to 1$ as $|z| \to 1$. The reflection principle therefore implies that g is a rational function with $g(\mathbb{T}) = \mathbb{T}$, hence a finite Blaschke product

$$(6) \qquad g(z) = c \prod_{k=1}^{m} \frac{z - z_k}{1 - \overline{z}_k z}, \quad |c| = 1, \quad z_k \in \mathbb{D} \quad (k = 1, \ldots, m).$$

Such a function assumes in $\overline{\mathbb{D}}$ every value in $\overline{\mathbb{D}}$ exactly m times (counting multiplicity). Hence $f = \varphi \circ g$ is continuous in $\overline{\mathbb{D}}$ and assumes every value in $\overline{G}$ exactly m times.   □

The next result (Rad 23) deals with sequences of Jordan domains; compare Corollary 2.4.

Theorem 2.11. *For $n = 1, 2, \ldots$ let*

$$J_n : \varphi_n(\zeta), \zeta \in \mathbb{T} \quad \text{and} \quad J : \varphi(\zeta), \zeta \in \mathbb{T}$$

be Jordan curves and let f_n and f map $\mathbb{D}$ conformally onto the inner domains of J_n and J such that $f_n(0) = f(0)$ and $f_n'(0) > 0$, $f'(0) > 0$ for all n. If

$$(7) \qquad \varphi_n \to \varphi \quad \text{as} \quad n \to \infty \quad \text{uniformly on } \mathbb{T}$$

then

$$(8) \qquad f_n \to f \quad \text{as} \quad n \to \infty \quad \text{uniformly in } \overline{\mathbb{D}}.$$

Proof. It follows easily from Theorem 1.8 that we have at least pointwise convergence, i.e. that (2.2.11) holds. In view of Corollary 2.4 it thus suffices to show that the curves J_n are uniformly locally connected.

Suppose this is false. Then there are indices n_ν and points $a_\nu, a_\nu' \in J_{n_\nu}$ with $|a_\nu - a_\nu'| \to 0$ ($\nu \to \infty$) that are not contained in any continuum of diameter $< \varepsilon_0$ in J_{n_ν}. Let $a_\nu = \varphi_{n_\nu}(\zeta_\nu)$ and $a_\nu' = \varphi_{n_\nu}(\zeta_\nu')$. We may assume that $\zeta_\nu \to \zeta$ and $\zeta_\nu' \to \zeta'$ as $\nu \to \infty$. It follows from (7) that $a_\nu \to \varphi(\zeta)$, $a_\nu' \to \varphi(\zeta')$ and thus that $\varphi(\zeta) = \varphi(\zeta')$. Since J is a Jordan curve, we conclude that $\zeta = \zeta'$ and thus $\zeta_\nu \to \zeta$, $\zeta_\nu' \to \zeta$. The function φ_{n_ν} maps the closed arc of $\mathbb{T}$ between ζ_ν and ζ_ν' onto a continuum in J_{n_ν} containing a_ν and a_ν'. Its diameter tends to 0 as $\nu \to \infty$ by (7), and this is a contradiction.   □

Exercises 2.3

1. Let H be a Jordan domain and A a Jordan arc from $a \in H$ to $b \in \partial H$ with $A \subset H \cup \{b\}$. Show that there is a unique conformal map f of $\mathbb{D}$ onto $H \setminus A$ that is continuous in $\overline{\mathbb{D}}$ and satisfies $f(1) = a$ and $f(\pm i) = b$.

2. Let G be a Jordan domain and $z_1, \ldots, z_4$ cyclically ordered points on ∂G. Show that there exists a unique $q > 0$ and a conformal map of G onto $\{u + iv : 0 < u < q, 0 < v < 1\}$ such that $z_1 \mapsto 0$, $z_2 \mapsto q$, $z_3 \mapsto q + i$, $z_4 \mapsto i$.

3. Let J be a Jordan curve and A a Jordan arc that lies inside J except for its endpoints. Show that there is a homeomorphism of $\mathbb{C}$ onto $\mathbb{C}$ that maps $J \cup A$ onto $\mathbb{T} \cup [-1, +1]$.

4. Let H and G be Jordan domains in $\mathbb{C}$ and let f be analytic in H and continuous in $\overline{H}$. Show that f maps H conformally onto G if and only if f maps ∂H bijectively onto ∂G.

5. Under the assumptions of Corollary 2.10, show that f is a conformal map if and only if $f'(z) \neq 0$ for $z \in \mathbb{D}$.

2.4 Crosscuts and Prime Ends

We now consider simply connected domains in $\widehat{\mathbb{C}} = \mathbb{C} \cup \{\infty\}$ and use the spherical metric. All notions that we introduce will be invariant under rotations of the sphere. If necessary we may assume that $\infty \in G$; otherwise we map some interior point of G to ∞ by a rotation of the sphere. This has the technical advantage that ∂G is now a bounded set so that we can use the more convenient euclidean metric in our proofs; the spherical and the euclidean metric are equivalent for bounded sets.

A *crosscut* C of G is an open Jordan arc in G such that $\overline{C} = C \cup \{a, b\}$ with $a, b \in \partial G$; we allow that $a = b$. We first prove two useful topological facts.

Proposition 2.12. *If C is a crosscut of the simply connected domain $G \subset \widehat{\mathbb{C}}$ then $G \setminus C$ has exactly two components G_0 and G_1 and these satisfy*

$$(1) \qquad\qquad G \cap \partial G_0 = G \cap \partial G_1 = C.$$

If G is not simply connected then there are crosscuts C connecting different boundary components and $G \setminus C$ is a single domain (New64, p. 120).

Proof. Let g map G conformally onto $\mathbb{D}$. Then

$$(2) \qquad\qquad h = \frac{g}{1 - |g|} = \frac{|g|}{1 - |g|} \exp(i \arg g)$$

is a non-analytic homeomorphism of G onto $\mathbb{C}$. If z tends to ∂G then $|g(z)| \to 1$ and thus $|h(z)| \to +\infty$. It follows that $J = h(C) \cup \{\infty\}$ is a Jordan curve and, by the Jordan curve theorem, there are exactly two components H_0 and H_1 of $\widehat{\mathbb{C}} \setminus J$ which satisfy $\partial H_0 = \partial H_1 = J$. Then

$$(3) \qquad\qquad G_j = h^{-1}(H_j) = \{z \in G : h(z) \in H_j\} \quad (j = 0, 1)$$

are disjoint domains with $G \cap \partial G_j = h^{-1}(\mathbb{C} \cap \partial H_j) = h^{-1}(\mathbb{C} \cap J) = C$ and $G_0 \cup G_1 = h^{-1}(\mathbb{C} \setminus J) = G \setminus C$. □

Proposition 2.13. *Let G be simply connected and J a Jordan curve with $G \cap J \neq \emptyset$. If $z_0 \in G \setminus J$ then there are countably many crosscuts $C_k \subset J$ of G such that*

$$(4) \qquad G = G_0 \cup \bigcup_k G_k \cup \bigcup_k C_k$$

where G_0 is the component of $G \setminus J$ containing z_0 and where G_k are disjoint domains with

$$(5) \qquad C_k = G \cap \partial G_k \subset G \cap \partial G_0 \quad \text{for} \quad k \geq 1.$$

It follows that if $z \in G$ lies in the other component of $\widehat{\mathbb{C}} \setminus J$ then $z \in G_k$ for exactly one k. Note that G_k for $k \geq 1$ may contain points of J; see Fig. 2.5.

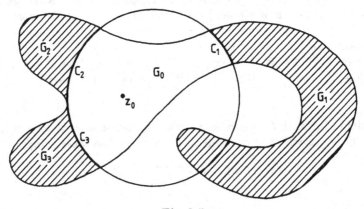

Fig. 2.5

Proof. Let $C_k (k \geq 1)$ denote the (at most) countably many arcs of $J \cap G \cap \partial G_0$. By Proposition 2.12, $G \setminus C_k$ consists of two components and we denote by G_k the component not containing z_0. Then (5) holds by (1) and it follows from the Jordan curve theorem that G_0 lies in the same component of $\widehat{\mathbb{C}} \setminus J$ as z_0.

It is clear that the right-hand side of (4) lies in G. Conversely let $z \in G$ but $z \notin G_k \cup C_k$ for all $k \geq 1$. There is a curve in G from z_0 to z, and since $\operatorname{diam}^{\#} C_k \to 0 \ (k \to \infty)$ it meets only finitely many crosscuts C_k. Hence there exists m such that z_0 and z are not separated by $B \cup \partial G$ where $B = C_{m+1} \cup \dots$. Also z_0 and z are not separated by $C_1 \cup \partial G$. Hence Janiszewski's theorem shows that z_0 and z are not separated by the union $(B \cup C_1) \cup \partial G$ because the intersection $(B \cap C_1) \cap \partial G = \partial G$ is connected. Continuing we see that z_0

and z are not separated by $(B \cup C_1 \cup \cdots \cup C_m) \cup \partial G$ and thus not by $J \cup \partial G$. It follows that $z \in G_0$. This proves (4).   □

We show now that the preimages in the "good" domain $\mathbb{D}$ of curves in the possibly "bad" domain G are again curves. It follows that the preimage of a crosscut of G is a crosscut of $\mathbb{D}$. Note that the converse does not hold because the image may oscillate and thus not end at a definite point of ∂G.

Proposition 2.14. *If f maps $\mathbb{D}$ conformally onto G then the preimage of a curve in G ending at a point of ∂G is a curve in $\mathbb{D}$ ending at a point of $\mathbb{T}$, and curves with distinct endpoints on ∂G have preimages with distinct endpoints on $\mathbb{T}$.*

Thus we are given a curve $\Gamma : w(t)$, $0 < t \le 1$ in G with $w(t) \to \omega \in \partial G$ as $t \to 0$. Then $z(t) = f^{-1}(w(t))$ has a limit $\zeta \in \mathbb{T}$ as $t \to 0$.

Proof. Applying Proposition 2.2 with $h = f^{-1}$, $H = G$ and $c = \omega$, we find $r_n \to 0$ such that

(6) $$\Lambda(f^{-1}(B_n)) \to 0 \quad (n \to \infty), \quad B_n = G \cap \partial D(\omega, r_n).$$

By Proposition 2.13 there is a crosscut C_n of G with $C_n \subset B_n$ that separates $w(1)$ from $w(t)$ ($t > 0$ small); see Fig. 2.6. If U_n denotes the component of $G \setminus C_n$ that does not contain $w(1)$, then $f^{-1}(U_n)$ is a domain in $\mathbb{D}$ not containing $z(1)$ that is bounded by $f^{-1}(C_n)$ and part of $\mathbb{T}$. It follows from (6) that $\Lambda(f^{-1}(C_n)) \to 0$ and thus that diam $f^{-1}(U_n) \to 0$ as $n \to \infty$. We conclude that $\zeta = \lim_{t \to 0} z(t)$ exists and $\zeta \in \mathbb{T}$ because f is a homeomorphism.

To prove the second assertion we may assume that $\infty \in G$ and thus that ∂G is bounded. Let Γ and Γ^* be two curves in G and suppose that $f^{-1}(\Gamma)$ and $f^{-1}(\Gamma^*)$ end at the same point $\zeta \in \mathbb{T}$. We now apply Proposition 2.2 with $h = f$ restricted to a neighbourhood of ζ (in $\mathbb{D}$) where f remains bounded. We find arcs $Q_n = \mathbb{D} \cap \partial D(\zeta, r_n)$ with $\Lambda(f(Q_n)) \to 0$ as $n \to \infty$. Thus there are points on $f^{-1}(\Gamma)$ and $f^{-1}(\Gamma^*)$ arbitrarily close to ζ such that the corresponding points on Γ and Γ^* have arbitrarily small distance. It follows that the endpoints of Γ and Γ^* have to coincide.   □

We say that (C_n) is a *null-chain* of G if

(i) C_n is a crosscut of G ($n = 0, 1, \dots$);
(ii) $\overline{C}_n \cap \overline{C}_{n+1} = \emptyset$ ($n = 0, 1, \dots$);
(iii) C_n separates C_0 and C_{n+1} ($n = 1, 2, \dots$);
(iv) diam$^{\#} C_n \to 0$ as $n \to \infty$.

If ∂G is bounded we can replace the spherical diameter diam$^{\#}$ by the euclidean diameter. By Proposition 2.12 and by (i) there is exactly one component V_n of $G \setminus C_n$ that does not contain C_0. Thus we can write (iii) as

(7) $$V_{n+1} \subset V_n \quad \text{for} \quad n = 1, 2, \dots.$$

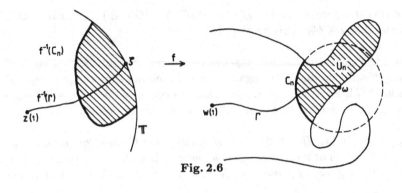

Fig. 2.6

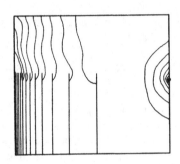

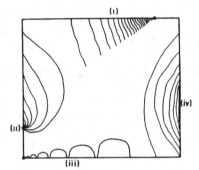

Fig. 2.7. Two null-chains; the second null-chain cuts off a whole boundary segment

Fig. 2.8. None of these four sequences is a null-chain because they violate the corresponding condition

Let now (C_n') be another null-chain of G. We say that (C_n') is *equivalent* to (C_n) if, for every sufficiently large m, there exists n such that

(8) C_m' separates C_n from C_0, C_m separates C_n' from C_0'.

If V_n' is the component of $G \setminus C_n'$ not containing C_0', this can be expressed as

(9) $$V_n \subset V_m', \quad V_n' \subset V_m.$$

It is easy to see that this defines an equivalence relation between null-chains. The equivalence classes are called the *prime ends* of G. Let $P(G)$ denote the set of all prime ends of G. We now prove the main result (Car13b), the *prime end theorem*.

Theorem 2.15. *Let f map $\mathbb{D}$ conformally onto G. Then there is a bijective mapping*

(10) $$\widehat{f} : \mathbb{T} \to P(G)$$

such that, if $\zeta \in \mathbb{T}$ and if (C_n) is a null-chain representing the prime end $\widehat{f}(\zeta)$, then $(f^{-1}(C_n))$ is a null-chain that separates 0 from ζ for large n.

The last statement means more precisely that

(11) $\qquad f^{-1}(C_n) \subset \{z \in \mathbb{D} : \delta_n < |z - \zeta| < \delta_n'\} \quad (n = 1, 2, \ldots)$

for suitable δ_n, δ_n' satisfying $0 < \delta_n < \delta_n' \to 0 \; (n \to \infty)$. In particular we see that the prime ends of $\mathbb{D}$ correspond to the points of $\mathbb{T}$.

Proof. (a) We may assume again that $\infty \in G$ so that ∂G is bounded. Let $\zeta \in \mathbb{T}$ be given and apply Proposition 2.2 with $h = f$ restricted to a neighbourhood of ζ in $\mathbb{D}$. We obtain a sequence of arcs $Q_n = \mathbb{D} \cap \partial D(\zeta, r_n)$ such that

(12) $\qquad r_n > r_{n+1} \to 0, \quad \Lambda(f(Q_n)) \to 0 \quad \text{as} \quad n \to \infty.$

Our null-chain (C_n) will be constructed by slightly modifying the sets $f(Q_n)$. This could be avoided by using Theorem 1.7 or modifying instead condition (ii); see e.g. Pom75, p. 272.

Let $A_n = \{z \in \mathbb{D} : r_n - (r_n - r_{n+1})/3 < |z - \zeta| < r_n + (r_{n-1} - r_n)/3\}$. Suppose that crosscuts C_ν of G with disjoint endpoints and with $f^{-1}(C_\nu) \subset A_\nu$ have already been constructed for $\nu < n$. Since $f(Q_n)$ has finite length, it ends at two points $a_n, b_n \in \partial G$ that may coincide (see Proposition 1.1). We now apply Proposition 2.2 with $h = f^{-1}$ and $c = a_n$ to obtain a circular crosscut of G that intersects $f(Q_n)$. We can choose the radius such that the endpoints of this crosscut do not coincide with the endpoints of any C_ν for $\nu < n$. Furthermore we can choose our crosscut such that its preimage in $\mathbb{D}$ lies in A_n. We do the same for $c = b_n$ choosing in particular a different radius if $b_n = a_n$. Then C_n is obtained from $f(Q_n)$ by replacing the ends near a_n and b_n by parts of the crosscuts just defined (see Fig. 2.9). This completes our construction for $\nu = n$.

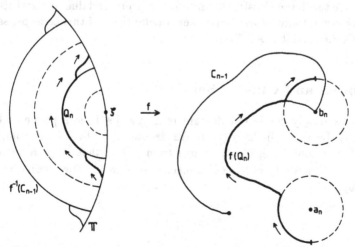

Fig. 2.9. Construction of C_n (indicated by arrows) for the case that C_{n-1} ends at the point b_n

Thus conditions (i) and (ii) are satisfied, and since $f^{-1}(C_n) \subset A_n$ we see that (iii) is also satisfied. Finally (iv) follows from (12). Thus $f^{-1}(C_n)$ is a null-chain separating 0 from ζ. Let now $\widehat{f}(\zeta)$ be the prime end of G represented by (C_n). Thus the mapping (10) is well defined.

(b) Let ζ and ζ' be distinct points of $\mathbb{T}$. If (C_n) and (C'_n) are the null-chains constructed in (a) it is clear that $f^{-1}(C'_m)$ $(m \geq m_0)$ does not separate $f^{-1}(C_n)$ from $f^{-1}(C_0)$ for large n. Since f is a homeomorphism we conclude that (8) does not hold so that (C_n) and (C'_n) are not equivalent, i.e. that $\widehat{f}(\zeta) \neq \widehat{f}(\zeta')$. Hence (10) is injective.

(c) Let now a prime end of G represented by C_n^* be given. Then $B_n = f^{-1}(C_n^*)$ is an open Jordan arc. By Proposition 2.14, it follows from (i) and (ii) that B_n is a crosscut of $\mathbb{D}$ and that $\overline{B}_n \cap \overline{B}_{n+1} = \emptyset$. Let U_n denote the component of $\mathbb{D} \setminus B_n$ not containing 0. Then $U_{n+1} \subset U_n$ for large n and there is thus a point ζ with

(13) $$\zeta \in \mathbb{T} \cap X \cap \partial(\mathbb{D} \setminus X) \quad \text{where} \quad X = \bigcap_n \overline{U}_n \subset \overline{\mathbb{D}}.$$

Suppose now that there is another point $\zeta' \in X$ and consider again the crosscuts C_m constructed in part (a). If m is sufficiently large then $f^{-1}(C_m) \subset D(\zeta, |\zeta - \zeta'|/2)$. Hence B_n intersects $f^{-1}(C_m)$ for large n so that $C_n^* = f(B_n)$ intersects C_m and also C_{m+1}. It would follow that these two sets have a distance $\leq \operatorname{diam} C_n^* \to 0$ $(n \to \infty)$ which is false because $\overline{C}_m \cap \overline{C}_{m+1} = \emptyset$.

Thus there is a unique point ζ satisfying (13). It follows that $\operatorname{diam} B_n \to 0$ as $n \to \infty$. Since $\overline{B}_n \cap \overline{B}_{n+1} = \emptyset$ and $\zeta \in \overline{U}_{n+1}$ we see that each B_n has a positive distance from ζ. Hence (11) holds with suitable $0 < \delta_n < \delta'_n \to 0$. It is now easy to see that (C_n^*) and (C_n) are equivalent. Hence (10) is surjective. $\qquad\square$

We have taken the classical definition of a prime end due to Carathéodory. There are several alternative (equivalent) definitions of this concept, see e.g. Maz36, CoPi64, Oht67, Ahl73.

2.5 Limits and Cluster Sets

Let G be a simply connected domain in $\widehat{\mathbb{C}}$ and let p be a prime end of G represented by the null-chain (C_n). We denote again by V_n the component of $G \setminus C_n$ not containing C_0. It follows from (2.4.7) that $\overline{V}_n$ is a decreasing sequence of non-empty compact connected sets in $\widehat{\mathbb{C}}$. The *impression* of p defined by

(1) $$I(p) = \bigcap_{n=1}^{\infty} \overline{V}_n$$

is therefore a non-empty compact connected set and thus either a single point or a continuum; if $I(p)$ is a single point we call the prime end *degenerate*. It follows from (2.4.9) that equivalent null-chains lead to the same intersection so that our definition is independent of the choice of the null-chain representing the prime end p.

We say that $\omega \in \widehat{\mathbb{C}}$ is a *principal point* of p if this prime end can be represented by a null-chain (C_n) with

$$C_n \subset D(\omega, \varepsilon) \quad \text{for} \quad \varepsilon > 0, \quad n > n_0(\varepsilon).$$

Let $\Pi(p)$ denote the set of all principal points of p.

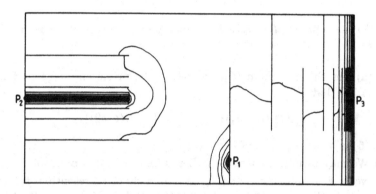

Fig. 2.10. Three prime ends. $I(p_1) = \Pi(p_1)$ is a single point; $I(p_2)$ is a segment, $\Pi(p_2)$ is a point; $I(p_3)$ is the right-hand border, $\Pi(p_3)$ is a smaller segment

Let f be any function in $\mathbb{D}$ with values in $\widehat{\mathbb{C}}$. We define the (unrestricted) *cluster set* $C(f, \zeta)$ of f at $\zeta \in \mathbb{T}$ as the set of all $\omega \in \widehat{\mathbb{C}}$ for which there are sequences (z_n) with

$$(2) \qquad z_n \in \mathbb{D}, \quad z_n \to \zeta, \quad f(z_n) \to \omega \quad \text{as} \quad n \to \infty.$$

It is easy to see that

$$(3) \qquad C(f, \zeta) = \bigcap_{r > 0} \text{clos}\{f(z) : z \in \mathbb{D}, |z - \zeta| < r\}.$$

If f is continuous in $\mathbb{D}$ it follows that $C(f, \zeta)$ is compact and connected and thus either a continuum or a single point.

Let now $E \subset \mathbb{D}$ and $\zeta \in \mathbb{T} \cap \overline{E}$. The *cluster set* $C_E(f, \zeta)$ of f at ζ *along* E consists of all $\omega \in \widehat{\mathbb{C}}$ such that, for some sequence (z_n),

$$(4) \qquad z_n \in E, \quad z_n \to \zeta, \quad f(z_n) \to \omega \quad \text{as} \quad n \to \infty.$$

We shall be interested in the cases that E is a Stolz angle Δ at ζ or that E is a curve Γ in $\mathbb{D}$ ending at ζ. It is easy to see that $C_\Delta(f, \zeta)$ and $C_\Gamma(f, \zeta)$ are connected compact sets if f is continuous in $\mathbb{D}$. See the books Nos60 and CoLo66 for the theory of cluster sets.

The next theorem (Car13b, Lin15) relates the geometrically defined prime ends with the analytically defined cluster sets. Let $\widehat{f}(\zeta)$ again denote the unique prime end of G that corresponds to ζ according to Theorem 2.15.

Theorem 2.16. *Let f map $\mathbb{D}$ conformally onto $G \subset \widehat{\mathbb{C}}$. If $\zeta \in \mathbb{T}$ then*

$$I(\widehat{f}(\zeta)) = C(f, \zeta), \tag{5}$$

$$\Pi(\widehat{f}(\zeta)) = \bigcap_\Gamma C_\Gamma(f, \zeta) = C_{[0,\zeta)}(f, \zeta) = C_\Delta(f, \zeta), \tag{6}$$

where Γ runs through all curves in $\mathbb{D}$ ending at ζ and where Δ is any Stolz angle at ζ.

Proof. (a) Let the null-chain (C_n) represent the prime end $\widehat{f}(\zeta)$. It follows from (2.4.11) that

$$f(\mathbb{D} \cap D(\zeta, \delta_n)) \subset V_n \subset f(\mathbb{D} \cap D(\zeta, \delta'_n))$$

so that (5) is a consequence of (1) and (3) .

(b) We may assume that $\infty \in G$. Let ω be a principal point of $\widehat{f}(\zeta)$ and Γ a curve in $\mathbb{D}$ ending at ζ. If (C_n) is a null-chain representing $\widehat{f}(\zeta)$ that converges to ω then there exists $z_n \in f^{-1}(C_n) \cap \Gamma$ for large n and (4) holds with $E = \Gamma$. Hence $\omega \in C_\Gamma(f, \zeta)$.

It is clear that the intersection of all cluster sets $C_\Gamma(f, \zeta)$ lies in $C_{[0,\zeta)}(f, \zeta) \subset C_\Delta(f, \zeta)$.

Let finally $\omega \in C_\Delta(f, \zeta)$. Then there are $z_n \in \Delta$ with $f(z_n) \to \omega$ and $\rho_n = |z_n - \zeta| \to 0$; we may assume that $\rho_{n+1} < \rho_n/4$. It follows from (2.2.6) that

$$\sum_n \int_{\rho_n}^{2\rho_n} \frac{l(r)^2}{r} dr < \infty, \quad l(r) = \Lambda(f(\mathbb{D} \cap \partial D(\zeta, r))) \, .$$

Hence there exist r_n with $\rho_n < r_n < 2\rho_n$ such that

$$\Lambda(f(Q_n)) \to 0 \quad (n \to \infty), \quad Q_n = \mathbb{D} \cap \partial D(\zeta, r_n) \, . \tag{7}$$

We put $z_n^* = (1 - r_n)\zeta$. Since z_n lies in a Stolz angle Δ we obtain from Corollary 1.6 and Corollary 1.4 that

$$|f(z_n^*) - f(z_n)| \le M(1 - |z_n|^2)|f'(z_n)| \le 4M|f(z_n) - \omega| \, . \tag{8}$$

We now make the same modification of $f(Q_n)$ as in part (a) of the proof of the prime end theorem to construct a null-chain (C_n) with $f(z_n^*) \in C_n$. Since $f(z_n) \to \omega$ as $n \to \infty$ it follows from (7) and (8) that C_n converges to ω so that $\omega \in \Pi(\widehat{f}(\zeta))$. $\qquad \square$

We defined in (1.2.11) that f has the angular limit $f(\zeta)$ at $\zeta \in \mathbb{T}$ if

$$f(z) \to f(\zeta) \quad \text{as} \quad z \to \zeta, \quad z \in \Delta \tag{9}$$

for each Stolz angle Δ. If this is even an unrestricted limit, i.e. if (9) holds with $\mathbb{D}$ instead of Δ, then f is continuous in $\mathbb{D} \cup \{\zeta\}$ (in the spherical metric). Since the cluster set reduces to a point if and only if the corresponding limit exists, we obtain from Theorem 2.16 the following geometric characterization of limits:

Corollary 2.17. *If f maps $\mathbb{D}$ conformally onto $G \subset \widehat{\mathbb{C}}$ then*

(i) *f has the angular limit a at ζ $\Leftrightarrow$ f has the radial limit a at ζ*
    $\Leftrightarrow$ f has the limit a along some curve ending at ζ $\Leftrightarrow$ $\Pi(\widehat{f}(\zeta)) = \{a\}$;
(ii) *f has the unrestricted limit a at ζ $\Leftrightarrow$ $I(\widehat{f}(\zeta)) = \{a\}$.*

Thus f is continuous at ζ if and only if the prime end $\widehat{f}(\zeta)$ is degenerate. Furthermore we see from Theorem 1.7 that f has a finite angular limit almost everywhere.

It is not possible (GaPo67) to replace the Stolz angle Δ in (9) by any fixed domain tangential to $\mathbb{T}$. Much more is however true almost everywhere (NaRuSh82, Two86): For almost all $e^{i\vartheta} \in \mathbb{T}$, the conformal map $f(z)$ tends to a finite limit as

$$(10) \qquad z = re^{it} \to e^{i\vartheta}, \quad r < 1 - e^{-c/|t-\vartheta|}$$

for every $c > 0$. This allows strongly tangential (but not unrestricted) approach.

Exercises 2.5

1. Let $E \neq \emptyset$ be a perfect set (i.e. compact and without isolated points) on $\mathbb{T}$ that contains no arc. Show that

$$G = \mathbb{D} \setminus \{r\zeta : 1/2 \leq r < 1, \, \zeta \in E\}$$

has uncountably many prime ends whose impression is a continuum.

2. Let $G = \mathbb{D} \setminus \{(1 - e^{-t})e^{it} : 0 \leq t < \infty\}$. Show that G has a prime end p with $I(p) = \Pi(p) = \mathbb{T}$.

3. Let H be any simply connected domain. Construct a simply connected subdomain G with a prime end p such that $I(p) = \Pi(p) = \partial H$. Deduce that $I(p)$ and $\Pi(p)$ need not be locally connected.

4. Show that $\Pi(p)$ is a continuum or a single point.

5. Show that f has a radial limit at ζ if and only if the prime end $\widehat{f}(\zeta)$ is *accessible*, i.e. if there is a Jordan arc that lies in G except for one endpoint and that intersects all but finitely many crosscuts of some null-chain of $\widehat{f}(\zeta)$.

6. Let f map $\mathbb{D}$ conformally onto a bounded domain. Use Exercise 2.2.3 to show that the power series of f converges at $\zeta \in \mathbb{T}$ if and only if f has a radial limit. (See (5.2.23) for the absolute convergence and Hay70 for the case of unbounded domains.)

2.6 Countability Theorems

Most of the following results have a strong topological flavour. A *coloured triod* is the union of three Jordan arcs B, R, Y that have all one endpoint, the *junction point*, in common and are otherwise disjoint. These arcs are labeled blue, red and yellow while the junction point is black. We start with a colour version (G. Jensen) of the *Moore triod theorem* (Moo28).

Proposition 2.18. *Let $\mathcal{M}$ be a system of coloured triods in $\mathbb{C}$ with distinct junction points such that sets of different colours do not meet. Then $\mathcal{M}$ is countable.*

Proof. We consider the system $\mathcal{M}^*$ of all "coloured circles" with rational radii and centers that are divided by three points with rational central angles into three arcs which we colour blue, red and yellow; circles with different colour schemes are considered to be different. To each triod in $\mathcal{M}$, we assign a circle in $\mathcal{M}^*$ such that the junction points lies inside and each arc of the triod meets the circle with the first point of intersection being on a circular arc of the same colour. Since $\mathcal{M}^*$ is countable it suffices to show that different coloured circles are assigned to different coloured triods.

Suppose that the same coloured circle C^* with arcs B^*, R^* and Y^* is assigned to the triods T and T'. We replace the arcs of T and T' by the subarcs from the junction points to the circle (see Fig. 2.11). Now $R \cup B \cup Y$ divides the inner domain of C^* into three subdomains; this follows from the Jordan curve theorem. The junction point of T' lies in one of these three domains, say the one bounded by B, R and parts of B^* and R^*. But Y' goes to Y^* and therefore has to meet $B \cup R$ which contradicts our assumption. $\square$

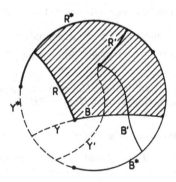

Fig. 2.11

An immediate consequence is: Every system of disjoint (uncoloured) triods in the plane is countable. We deduce now that, if the conformal map f is continuous in $\overline{\mathbb{D}}$ then

(1) $$\operatorname{card}\{\zeta \in \mathbb{T} : f(\zeta) = a\} \le 2$$

except possibly for countably many a. More generally:

Corollary 2.19. *Let f be a homeomorphism of $\mathbb{D}$ into $\mathbb{C}$. Then there are at most countably many points $a \in \mathbb{C}$ such that*

(2) $$f(r\zeta_j) \to a \quad as \quad \rho \to 1- \quad (j = 1, 2, 3)$$

for three distinct points $\zeta_1, \zeta_2, \zeta_3$ on $\mathbb{T}$.

Proof. Since f is a homeomorphism the sets $A_j = \{f(r\zeta_j) : 1/2 \le r < 1\}$ are disjoint, and $A_1 \cup A_2 \cup A_3 \cup \{a\}$ is a triod by (2). Hence Proposition 2.18 shows that there can be only countably many distinct points a. $\qquad\square$

The *Bagemihl ambiguous point theorem* (Bag55) deals with completely arbitrary functions. We say that $\zeta \in \mathbb{T}$ is an *ambiguous point* of f if there are two arcs Γ_1 and Γ_2 ending at ζ such that

(3) $$C_{\Gamma_1}(f, \zeta) \cap C_{\Gamma_2}(f, \zeta) = \emptyset.$$

For example, if $f(z) = \exp[-(1 + z)/(1 - z)]$ then 1 is an ambiguous point because different circles Γ touching $\mathbb{T}$ at 1 have disjoint cluster sets, namely different circles concentric to 0.

Corollary 2.20. *Any function from $\mathbb{D}$ to $\widehat{\mathbb{C}}$ has at most countably many ambiguous points.*

Proof. Let $\mathcal{H}$ denote the system of all open sets in $\widehat{\mathbb{C}}$ that are the union of finitely many disks in $\widehat{\mathbb{C}}$ with rational center and rational radius. Let $\zeta \in \mathbb{T}$ be an ambiguous point of f. Since $C_{\Gamma_1}(f, \zeta)$ and $C_{\Gamma_2}(f, \zeta)$ are disjoint compact sets we can find disjoint $H_j \in \mathcal{H}$ such that $C_{\Gamma_j}(f, \zeta) \subset H_j$ for $j = 1, 2$. Since H_j is open we can replace Γ_j by a subarc ending at ζ such that

(4) $$f(\Gamma_j) \subset H_j \quad for \quad j = 1, 2.$$

Now $\mathcal{H} \times \mathcal{H}$ is countable and we may therefore restrict ourselves to those ζ for which (4) holds with fixed $H_1, H_2 \in \mathcal{H}$ and suitable Γ_1, Γ_2.

We now define a coloured triod with junction point ζ. The blue arc is Γ_1, the red arc is Γ_2 and the yellow arc is $(\zeta, 2\zeta]$. It follows from (4) and $H_1 \cap H_2 = \emptyset$ that blue arcs do not intersect red arcs, and trivially neither meets a yellow arc. Hence, by Proposition 2.18, there are only countably many such triods and thus only countably many points ζ satisfying (4). $\qquad\square$

We come now to the *Collingwood symmetry theorem* (Col60). The *left-hand cluster set* $C^+(f, \zeta)$ at $\zeta \in \mathbb{T}$ consists of all $\omega \in \widehat{\mathbb{C}}$ for which there are (z_n) with

(5) $z_n \in \mathbb{D}, \quad \arg z_n > \arg \zeta, \quad z_n \to \zeta, \quad f(z_n) \to \omega \quad$ as $\quad n \to \infty$.

The *right-hand cluster set* $C^-(f, \zeta)$ is defined similarly with $\arg z_n < \arg \zeta$ in condition (5). It is clear that $C^{\pm}(f, \zeta) \subset C(f, \zeta)$.

Proposition 2.21. *If f is any function from $\mathbb{D}$ to $\widehat{\mathbb{C}}$ then*

(6) $$C^+(f, \zeta) = C^-(f, \zeta) = C(f, \zeta)$$

except for at most countably many $\zeta \in \mathbb{T}$.

Proof. Let D_k and D_k^* be countably many disks in $\widehat{\mathbb{C}}$ with $\overline{D}_k \subset D_k^*$ such that each point of $\widehat{\mathbb{C}}$ lies in infinitely many D_k and that the spherical diameter satisfies $\operatorname{diam}^{\#} D_k^* \to 0$ $(k \to \infty)$. Omitting the symbol f we define

(7) $A_k = \{\zeta \in \mathbb{T} : C(\zeta) \cap D_k \neq \emptyset, \ C^+(\zeta) \cap D_k^* = \emptyset\}, \quad k = 1, 2, \ldots$.

We claim that if $\zeta \in A_k$ then there is an open arc $B_k(\zeta)$ of $\mathbb{T}$ with ζ as its right endpoint such that

(8) $$C(\zeta') \cap D_k = \emptyset \quad \text{for} \quad \zeta' \in B_k(\zeta).$$

Suppose this is false for some k. Then there are points $\zeta_n \in \mathbb{T}$ with $\zeta_n \to \zeta$ and $\arg \zeta_n > \arg \zeta$ such that $\omega_n \in C(\zeta_n) \cap D_k$ for some ω_n. By (2.5.2), we can choose z_n so close to ζ_n that $\arg z_n > \arg \zeta$ and $f(z_n) \in D_k$; we may assume that $f(z_n) \to \omega \in \overline{D}_k$ as $n \to \infty$. Then $\omega \in C^+(\zeta) \cap D_k^*$ contrary to (7).

Let now $\zeta, \zeta' \in A_k$ and $\arg \zeta < \arg \zeta'$. Then $\zeta' \notin B_k(\zeta)$ by (7) and (8) so that $B_k(\zeta) \cap B_k(\zeta') = \emptyset$. It follows that each A_k is countable and therefore also its union A.

Finally let $\zeta \in \mathbb{T} \setminus A$ and $\omega \in C(\zeta)$. We have $\omega \in D_{k_\nu}$ for suitable $k_\nu \to \infty$. Since $\zeta \notin A_{k_\nu}$ we see from (7) that $C^+(\zeta) \cap \overline{D}_{k_\nu}^* \neq \emptyset$ and since $\omega \in D_{k_\nu}^*$, $\operatorname{diam}^{\#} D_{k_\nu}^* \to 0$, we conclude that $\omega \in C^+(\zeta)$. Hence $C^+(\zeta) = C(\zeta)$ except on the countable set A; we handle $C^-(\zeta)$ in a similar way. $\square$

We say that $p = \hat{f}(\zeta)$ is a *symmetric prime end* if (6) holds. In Fig. 2.10, for example, p_1 and p_2 are symmetric whereas p_3 is not symmetric; neither is the "comb" in Fig. 2.2. It seems that our idea of prime ends is strongly influenced by these asymmetric examples whereas all but countably many prime ends are symmetric by Proposition 2.21.

Proposition 2.22. *Let f map $\mathbb{D}$ conformally onto G. Let ζ and ζ' be distinct points on $\mathbb{T}$ where f has radial limits ω and ω'. Suppose that the prime end $p = \hat{f}(\zeta)$ is symmetric and that $\omega' \in I(p)$. Then $I(p)$ is the smallest compact connected subset of ∂G containing ω and ω'. In particular, if $\omega = \omega'$ then $I(p) = \{\omega\}$.*

Together with the Collingwood symmetry theorem this shows that, except possibly for countably many exceptions, the radial limits are either injective or there is an unrestricted limit.

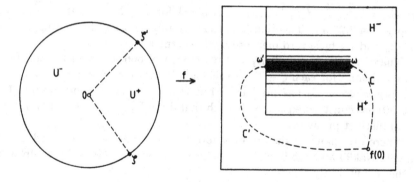

Fig. 2.12

Proof (see Fig. 2.12). If $\omega = \omega'$ let $E = \{\omega\}$, otherwise let E be any continuum with $\omega, \omega' \in E \subset \partial G$. The component H of $\widehat{\mathbb{C}} \setminus E$ containing $f(0)$ is simply connected. The images C and C' of $[0, \zeta)$ and $[0, \zeta')$ end at ω and ω' in E. Hence $H \setminus (C \cup C')$ has exactly two components $H^{\pm}$ by Proposition 2.12. Choosing the notation appropriately we have $f(U^{\pm}) \subset H^{\pm}$ and thus, by (5),

$$I(p) = C(\zeta) = C^+(\zeta) \cap C^-(\zeta) \subset \overline{H}^+ \cap \overline{H}^- \subset E \cup C \cup C'.$$

Since $I(p) \subset \partial G$ we conclude that $I(p) \subset E$.   □

A set $E \subset \mathbb{T}$ is of *first category* if it is the countable union of nowhere dense sets, i.e. if

$$E = \bigcup_{n=1}^{\infty} E_n, \quad \overline{E}_n \text{ contains no proper arc}.$$

This is a topological (as opposed to measure-theoretic) generalization of countability (Oxt80). Non-empty open sets are not of first category by the Baire category theorem.

Proposition 2.23. *If f is continuous in $\mathbb{D}$ then*

$$C(f, \zeta) = C_{[0,\zeta)}(f, \zeta)$$

for $\zeta \in \mathbb{T}$ except for a set of first category.

This is the *Collingwood maximality theorem* (Col57); see e.g. CoLo66, p. 76 for the proof. The prime ends are classified according to the following table:

	$\Pi(p) = I(p)$	$\Pi(p) \neq I(p)$
$\Pi(p)$ singleton	First kind	Second kind
$\Pi(p)$ not singleton	Third kind	Fourth kind

Let $E_j = \{\zeta \in \mathbb{T} : \widehat{f}(\zeta)$ is of the jth kind$\}$ for $j = 1, \ldots, 4$. It follows from Proposition 2.23 and Theorem 2.16 that E_2 and E_4 are of first category. The sets E_3 and E_4 have zero measure by Theorem 1.7.

There is a domain with $E_1 = \emptyset$ (Car13b, CoLo66, p. 184) so that $C(f, \zeta)$ is a continuum for each $\zeta \in \mathbb{T}$. In this case E_2 has measure 2π but is of first category whereas E_3 has measure 0 but is not of first category. Frankl (see CoPi59) has constructed a domain such that diam $I(p) \geq 1$ and $I(p) \cap I(p') = \emptyset$ for all distinct prime ends p, p'.

There are many further results about prime ends; see e.g. Pir58, Pir60, Ham86, Näk90 and the books CoLo66, Oht70. See e.g. McM67, McM69a for further countability theorems.

Exercises 2.6

1. Show that a conformal map has no ambiguous points.

2. In Exercise 2.5.1, show that there are infinitely many non-symmetric prime ends.

3. Draw a symmetric prime end of the fourth kind.

4. Let g be a function in $\mathbb{D}$ such that $(\zeta - z)g(z)$ has an angular limit $a(\zeta)$ for each $\zeta \in \mathbb{T}$. Consider $f(z) = (1 - |z|)g(z)$ to show that $a(\zeta) = 0$ except possibly for countably many ζ.

5. Show that, with at most countably many exceptions, all graph-theoretical components of a planar graph are Jordan curves or Jordan arcs (possibly without endpoints). A planar graph consists of Jordan arcs ("edges") that can intersect only at their endpoints ("nodes"), and two nodes are in the same component if they can be connected by finitely many edges.

Chapter 3. Smoothness and Corners

3.1 Introduction

We now study the behaviour of the derivative f' for the case that the image domain $G = f(\mathbb{D})$ has a reasonably smooth boundary C; the general case will be studied in later chapters.

The most classical case is that C is an *analytic curve*, i.e. there is a parametrization $C : \varphi(\zeta)$, $\zeta \in \mathbb{T}$ where φ is analytic and univalent in $\{1/\rho < |z| < \rho\}$ for some $\rho > 1$ and perhaps undefined otherwise.

Proposition 3.1. *Let f map $\mathbb{D}$ conformally onto the inner domain of the Jordan curve $C \subset \mathbb{C}$. Then C is an analytic curve if and only if f is analytic and univalent in $\{|z| < r\}$ for some $r > 1$.*

Proof (see Fig. 3.1). First let C be an analytic curve. Then there is $r > 1$ such that $h = \varphi^{-1} \circ f$ is analytic and univalent in $\{1/r < |z| < 1\}$ and satisfies $1/\rho < |h(z)| < 1$. Since $|h(z)| \to 1$ as $|z| \to 1$, the reflection principle (Section 1.2) shows that h can be extended to an analytic univalent function in $\{1/r < |z| < r\}$ that satisfies $1/\rho < |h(z)| < \rho$. We conclude that $f = \varphi \circ h$ is analytic and univalent in $\{1/r < |z| < r\}$ and thus in $\{|z| < r\}$. The converse is trivial. □

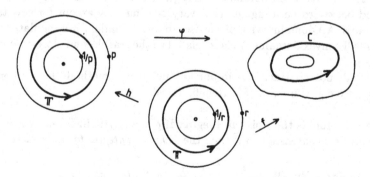

Fig. 3.1

In the next two sections we restrict ourselves to Jordan curves. A theorem of Lindelöf states that

$$C \text{ smooth} \quad \Leftrightarrow \quad \arg f' \text{ is continuous in } \overline{\mathbb{D}}.$$

This is a very satisfactory result because it gives a one-to-one correspondence between a geometric and an analytic condition. The fact that $\operatorname{Im} \log f' = \arg f'$ is continuous in $\mathbb{D} \cup \mathbb{T}$ does not imply that $\operatorname{Re} \log f' = \log |f'|$ is continuous there. No geometric characterization for the continuity of $\log f'$ in $\mathbb{D} \cup \mathbb{T}$ is known. We shall show

$$C \text{ Dini-smooth} \quad \Rightarrow \quad f' \text{ and } \log f' \text{ are continuous in } \overline{\mathbb{D}}.$$

Corresponding results hold for higher derivatives. For example, the *Kellogg-Warschawski theorem* states that

$$C \in C^{\infty} \quad \Leftrightarrow \quad \text{all derivatives are continuous in } \overline{\mathbb{D}}$$

where C^{∞} is the class of all smooth curves with some infinitely differentiable parametrization.

Many domains that occur in applications have corners. If there is a corner of opening $\pi\alpha$ at $f(\zeta)$ then $f(z) - f(\zeta)$ behaves like $(z - \zeta)^{\alpha}$ as $z \to \zeta$. The detailed behaviour is more complicated than might be expected (Section 3.4).

The *Schwarz-Christoffel formula* states that

$$(1) \qquad f'(z) = f'(0) \prod_{k=1}^{n} (1 - \overline{z}_k z)^{\alpha_k - 1} \quad \text{for} \quad z \in \mathbb{D}$$

if f maps $\mathbb{D}$ conformally onto a polygonal domain with inner angles $\pi\alpha_k \, (k = 1, \ldots, n)$ at the corners $w_k = f(z_k)$. Unfortunately it is not easy to determine the points $z_k \in \mathbb{T}$ if w_k is given; see e.g. Neh52, KoSt59 or LaSc67. Domains bounded by finitely many circular arcs can be handled via the Schwarzian derivative S_f; see Neh52, p. 201.

We shall generalize the Schwarz-Christoffel formula to the wide class of *regulated domains*: The boundary of a regulated domain has everywhere forward and backward halftangents that vary continuously except for countably many jumps. All domains with piecewise smooth boundary are regulated, furthermore all convex domains. This is related to the starlike and close-to-convex domains.

We shall not consider the important numerical methods for conformal mapping; see e.g. Gai64, Gai83, Hen86, Tre86.

The main tool is the Poisson integral. If g is analytic in $\mathbb{D}$ and if $v = \operatorname{Im} g$ has a continuous extension to $\overline{\mathbb{D}}$ then the *Schwarz integral formula* states that

$$(2) \qquad g(z) = \operatorname{Re} g(0) + \frac{i}{2\pi} \int_0^{2\pi} \frac{e^{it} + z}{e^{it} - z} v(e^{it}) \, dt \quad \text{for} \quad z \in \mathbb{D}.$$

Taking the imaginary part leads to the *Poisson integral formula*

$$(3) \qquad \operatorname{Im} g(z) = \frac{1}{2\pi} \int_0^{2\pi} \frac{1 - |z|^2}{|e^{it} - z|^2} v(e^{it}) \, dt = \int_{\mathbb{T}} p(z, \zeta) v(\zeta) |d\zeta|$$

where $p(z, \zeta)$ is the Poisson kernel

(4) $\qquad p(z, \zeta) = \dfrac{1}{2\pi} \dfrac{1 - |z|^2}{|\zeta - z|^2} = \dfrac{1}{2\pi} \operatorname{Re} \dfrac{\zeta + z}{\zeta - z} > 0 \quad (|z| < |\zeta| = 1).$

Conversely if g is given by (2) then (Ahl66a, p. 168)

(5) $\qquad v$ continuous at $\zeta \in \mathbb{T} \quad \Rightarrow \quad \operatorname{Im} g(z) \to v(\zeta)$ as $z \to \zeta$, $z \in \mathbb{D}$.

It is easy to prove geometrically that

(6) $\qquad\qquad |e^{it} - re^{i\vartheta}| \geq \max\left(\dfrac{|t - \vartheta|}{\pi}, 1 - r \right)$

for $0 \leq r \leq 1$ and $|t - \vartheta| \leq \pi$.

Exercises 3.1

We assume that f maps $\mathbb{D}$ conformally onto the inner domain of the Jordan curve $C \subset \mathbb{C}$.

1. Suppose that the open arc A of $\mathbb{T}$ is mapped onto an analytic arc on C. Show that f is analytic in some neighbourhood of A.

2. Let f be analytic in some neighbourhood of $\zeta \in \mathbb{T}$ and let $f'(\zeta) = 0$. Show that C consists near $f(\zeta)$ of two analytic arcs that form an inward pointing (zero-angle) cusp.

3. If f is analytic near $z = e^{it}$ and $f'(z) \neq 0$, show that the curvature $\kappa(z)$ of C at $f(z)$ satisfies

$$\kappa(z) = |f'(z)|^{-1} \operatorname{Re}[1 + z f''(z)/f'(z)].$$

$$\frac{\partial}{\partial t} \kappa(z) = -|f'(z)|^{-1} \operatorname{Im}[z^2 S_f(z)].$$

3.2 Smooth Jordan Curves

Let C be a Jordan curve in $\mathbb{C}$. We say that C is *smooth* if there is a parametrization $C : w(\tau)$, $0 \leq \tau \leq 2\pi$ such that $w'(\tau)$ is continuous and $\neq 0$; we have chosen the parameter range $[0, 2\pi]$ for convenience. We extend $w(\tau)$ to $-\infty < \tau < +\infty$ as a 2π-periodic function.

The curve C is smooth if and only if it has a continuously varying tangent, i.e. if there is a continuous function β such that, for all t,

(1) $\quad \arg[w(\tau) - w(t)] \to \beta(t) \quad$ as $\quad \tau \to t+, \quad \to \beta(t) + \pi \quad$ as $\quad \tau \to t - .$

We call $\beta(\tau)$ the *tangent direction angle* of C at $w(t)$.

Now let f map $\mathbb{D}$ conformally onto the inner domain of C. Since the characterization of smoothness in terms of tangents does not depend on the parametrization, we may choose the *conformal parametrization*

(2) $$C : w(t) = f(e^{it}), \quad 0 \le t \le 2\pi.$$

We first give an analytic characterization of smoothness (Lin16).

Theorem 3.2. *Let f map $\mathbb{D}$ conformally onto the inner domain of the Jordan curve C. Then C is smooth if and only if $\arg f'$ has a continuous extension to $\overline{\mathbb{D}}$. If C is smooth then*

(3) $$\arg f'(e^{it}) = \beta(t) - t - \frac{\pi}{2} \quad (t \in \mathbb{R})$$

for the conformal parametrization and

(4) $$\log f'(z) = \log |f'(0)| + \frac{i}{2\pi} \int_0^{2\pi} \frac{e^{it} + z}{e^{it} - z} \left(\beta(t) - t - \frac{\pi}{2} \right) dt \quad (z \in \mathbb{D}).$$

Proof. (a) First let C be smooth. The functions

(5) $$g_n(z) = \log \frac{f(e^{i/n} z) - f(z)}{(e^{i/n} - 1)z} \quad (n = 1, 2, \dots)$$

are analytic in $\mathbb{D}$ and continuous in $\overline{\mathbb{D}}$ by Theorem 2.6. Hence we see from (3.1.2) that

(6) $$g_n(z) = \operatorname{Re} g_n(0) + \frac{i}{2\pi} \int_0^{2\pi} \frac{e^{it} + z}{e^{it} - z} \operatorname{Im} g_n(e^{it}) dt \quad (z \in \mathbb{D}).$$

It follows from (5) and (1) that

$$\operatorname{Im} g_n(e^{it}) = \arg[f(e^{it + i/n}) - f(e^{it})] - t - \arg(e^{i/n} - 1)$$
$$\to \beta(t) - t - \pi/2$$

as $n \to \infty$. The convergence is uniform in t because $\beta(t)$ is continuous; see e.g. Proposition 3.12 below. Hence (4) follows from (6) for $n \to \infty$ because $g_n(z) \to \log f'(z)$ for each $z \in \mathbb{D}$ and we conclude from (3.1.5) that

(7) $$\arg f'(z) \to \beta(t) - t - \pi/2 \quad \text{as} \quad z \to e^{it}, \quad z \in \mathbb{D}.$$

(b) Conversely, let $v = \arg f'$ be continuous in $\overline{\mathbb{D}}$ and let first $\zeta, z \in \mathbb{D}$ and $\zeta \ne z$. We write

$$q(z) \equiv \frac{f(z) - f(\zeta)}{z - \zeta} e^{-iv(\zeta)} = \int_0^1 |f'(\zeta + (z - \zeta)s)| e^{iv(\zeta + (z - \zeta)s) - iv(\zeta)} ds.$$

If $0 < \varepsilon < \pi/2$ and if $\delta > 0$ is chosen so small that $|v(z) - v(\zeta)| < \varepsilon$ for $|z - \zeta| < \delta$, it follows that

$$|\operatorname{Im} q| \le \sin \varepsilon \int_0^1 |f'| ds, \quad \operatorname{Re} q \ge \cos \varepsilon \int_0^1 |f'| ds$$

for $|z - \zeta| < \delta$ and therefore $|\arg q(z)| \leq \varepsilon$. Hence, by continuity,

(8) $\quad \arg \dfrac{f(z) - f(\zeta)}{z - \zeta} = v(\zeta) + \arg q(z) \rightarrow v(\zeta) \quad$ as $\quad z \rightarrow \zeta \in \mathbb{T}, \quad z \in \overline{\mathbb{D}}$

and it follows that C has at $f(\zeta)$ a tangent of direction angle $v(\zeta) + \arg \zeta + \pi/2$ which varies continuously. Thus C is smooth. $\square$

The smoothness of C does not imply that f' has a continuous extension to $\overline{\mathbb{D}}$. To see this let h be analytic in $\mathbb{D}$, $\operatorname{Im} h$ continuous in $\overline{\mathbb{D}}$ and

$$|\operatorname{Re} h(z)| < 1, \quad |\operatorname{Im} h(z)| < \pi/2 \quad \text{for} \quad z \in \mathbb{D}.$$

If $\log f' = h$ then $e^{-1} < |f'| = e^{\operatorname{Re} h} < e$ and $|\arg f'| = |\operatorname{Im} h| < \pi/2$. Hence f maps $\mathbb{D}$ conformally onto a Jordan domain G by Proposition 1.10, and ∂G is smooth by Theorem 3.2. We can choose h such that $\operatorname{Re} h$ is not continuous in $\overline{\mathbb{D}}$. This is for instance the case if h maps $\mathbb{D}$ conformally onto the domain of Fig. 2.2 as Corollary 2.17 (ii) shows. Thus $f' = e^h$ has no continuous extension to $\overline{\mathbb{D}}$.

The next result (War30) provides a substitute that is sufficient for our purposes.

Proposition 3.3. *Let C be smooth. Then C is rectifiable and*

(9) $\qquad \Lambda(\{f(e^{i\vartheta}) : t_1 \leq \vartheta \leq t_2\}) \leq M(t_2 - t_1)^{1/2}$

for $t_1 \leq t_2 \leq t_1 + 2\pi$ where M is a constant.

Proof. By Theorem 3.2, the function $\arg f'(e^{it})$ is continuous and 2π-periodic. By the Weierstraß approximation theorem (Zyg68, I, p. 90), it can therefore be approximated by a trigonometric polynomial. Hence we can find a polynomial p such that $|\arg f'(z) - \operatorname{Im} p(z)| < \pi/6$ holds for $z \in \mathbb{T}$ and thus also for $z \in \mathbb{D}$. Let M_1 denote the maximum of $\operatorname{Re} p$ on $\overline{\mathbb{D}}$.

We write $z = re^{it}$ with $0 < r < 1$ and obtain

$$e^{-2M_1} \int_0^{2\pi} |f'(z)|^2 \, dt \leq \int_0^{2\pi} |f' e^{-p}|^2 2 \cos[2(\arg f' - \operatorname{Im} p)] \, dt$$

$$= \operatorname{Re} \left[\frac{2}{i} \int_{|z|=r} (f' e^{-p})^2 z^{-1} \, dz \right] = M_2$$

where M_2 is constant by the residue theorem. Let $n \in \mathbb{N}$ and $t_1 = \vartheta_0 < \vartheta_1 < \cdots < \vartheta_n = t_2$. Applying the Schwarz inequality twice we see that

$$\sum_{\nu=1}^{n} |f(re^{i\vartheta_\nu}) - f(re^{i\vartheta_{\nu-1}})| \leq \sum_{\nu=1}^{n} \int_{\vartheta_{\nu-1}}^{\vartheta_\nu} |f'(re^{it})| \, dt$$

$$\leq \left(\sum_{\nu} (\vartheta_\nu - \vartheta_{\nu-1}) \sum_{\nu} \int_{\vartheta_{\nu-1}}^{\vartheta_\nu} |f'|^2 \, dt \right)^{1/2}.$$

If we use the above estimate and let $r \to 1$ we deduce that

$$\sum_{\nu=1}^{n} |f(e^{i\vartheta_\nu}) - f(e^{i\vartheta_{\nu-1}})| \le M_3 (t_2 - t_1)^{1/2}$$

which implies (9).   □

Exercises 3.2

1. Show that the tangent direction angle changes by 2π if a smooth Jordan curve is run through once in the positive sense.

2. Let f map $\mathbb{D}$ onto the inner domain G of a smooth Jordan curve. Show that f can be extended to an angle-preserving homeomorphism of $\overline{\mathbb{D}}$ onto $\overline{G}$.

3. Show that $f(z) = 2z + (1-z)\log(1-z)$ maps $\mathbb{D}$ onto the inner domain of a smooth Jordan curve but that $f'(x) \to +\infty$ as $x \to 1-$.

3.3 The Continuity of the Derivatives

We need a few facts about the modulus of continuity. Let the function φ be uniformly continuous on the connected set $A \subset \mathbb{C}$. Its *modulus of continuity* is defined by

$$(1) \quad w(\delta) \equiv w(\delta, \varphi, A) = \sup\{|\varphi(z_1) - \varphi(z_2)| : z_1, z_2 \in A, |z_1 - z_2| \le \delta\}$$

for $\delta \le 0$. This is an increasing continuous function with $w(0) = 0$. If A is convex it is easy to see that

$$(2) \quad w(n\delta) \le n w(\delta) \quad \text{for} \quad \delta \ge 0, \quad n = 1, 2, \dots .$$

If φ is analytic in $\mathbb{D}$ and continuous in $\overline{\mathbb{D}}$ then the moduli of continuity in $\overline{\mathbb{D}}$ and $\mathbb{T}$ are essentially the same because (RuShTa75)

$$(3) \quad w(\delta, \varphi, \mathbb{T}) \le w(\delta, \varphi, \overline{\mathbb{D}}) \le 3w(\delta, \varphi, \mathbb{T}) \quad \text{for} \quad \delta \le \pi/2;$$

see e.g. Tamr73 and Hin89 for further results in this direction.

The function φ is called *Dini-continuous* if

$$(4) \quad \int_0^\pi t^{-1} w(t)\, dt < \infty;$$

the limit π could be replaced by any positive constant. For Dini-continuous φ and $0 < \delta < \pi$, we define

$$(5) \quad w^*(\delta) \equiv w^*(\delta, \varphi, A) = \int_0^\delta \frac{w(t)}{t}\, dt + \delta \int_\delta^\pi \frac{w(t)}{t^2}\, dt .$$

The following estimates are closely connected with conjugate functions (see e.g. Gar81, p. 106).

Proposition 3.4. *Let φ be 2π-periodic and Dini-continuous in $\mathbb{R}$. Then*

$$(6) \qquad g(z) = \frac{i}{2\pi} \int_0^{2\pi} \frac{e^{it} + z}{e^{it} - z} \varphi(t)\, dt \quad (z \in \mathbb{D})$$

has a continuous extension to $\overline{\mathbb{D}}$. Furthermore

$$(7) \qquad |g'(z)| \leq \frac{2}{\pi} \frac{\omega(1-r)}{1-r} + 2\pi \int_{1-r}^{\pi} \frac{\omega(t)}{t^2}\, dt \leq 2\pi \frac{\omega^*(1-r)}{1-r}$$

for $|z| \leq r < 1$, and if $z_1, z_2 \in \overline{\mathbb{D}}$ then

$$(8) \qquad |g(z_1) - g(z_2)| \leq 20\,\omega^*(\delta) \quad \text{for} \quad |z_1 - z_2| \leq \delta < 1.$$

The function φ is *Hölder-continuous* with exponent $\alpha (0 < \alpha \leq 1)$ in A if

$$(9) \qquad |\varphi(z_1) - \varphi(z_2)| \leq M|z_1 - z_2|^\alpha \quad \text{for} \quad z_1, z_2 \in A,$$

i.e. if $\omega(\delta) \leq M\delta^\alpha$. Then φ is Dini-continuous and $\omega^*(\delta) = O(\delta^\alpha)$ if $0 < \alpha < 1$ but only $\omega^*(\delta) = O(\delta \log 1/\delta)$ if $\alpha = 1$. Thus Proposition 3.4 implies that g is Hölder-continuous in $\overline{\mathbb{D}}$ if $0 < \alpha < 1$.

Proof. It follows from (6) that

$$(10) \qquad g'(z) = \frac{i}{\pi} \int_0^{2\pi} \frac{e^{it}}{(e^{it} - z)^2} \varphi(t)\, dt \quad \text{for} \quad z \in \mathbb{D}.$$

Since the integral vanishes for constant φ, the substitution $t = \vartheta + \tau$ shows that

$$g'(re^{i\vartheta}) = \frac{i}{\pi} \int_{-\pi}^{\pi} \frac{e^{i\tau - i\vartheta}}{(e^{i\tau} - r)^2} [\varphi(\vartheta + \tau) - \varphi(\vartheta)]\, d\tau$$

for $0 \leq r < 1$ and thus, by (1),

$$|g'(re^{i\vartheta})| \leq \frac{2}{\pi} \int_0^{\pi} \frac{\omega(\tau)}{|e^{i\tau} - r|^2}\, d\tau.$$

If we consider the integrals over $[0, 1-r]$ and $[1-r, \pi]$ separately and use (3.1.6), we obtain the first inequality (7), and the second one then follows from (see (2))

$$\frac{1}{3}\omega(\delta) \leq \omega\left(\frac{\delta}{3}\right) \leq \int_{\delta/3}^{\delta} \frac{\omega(t)}{t}\, dt \leq \int_0^{\delta} \frac{\omega(t)}{t}\, dt.$$

The estimates hold also for $|z| < r$ by the maximum principle.

Let $z_j = r_j \zeta_j$, $r_j < 1$, $\zeta_j \in \mathbb{T}$ $(j = 1, 2)$ with $|z_1 - z_2| \leq \delta$ and put $r = 1 - \delta$. Integrating the second estimate (7) over $[z_1, z_2]$ we obtain

$$(11) \qquad |g(z_1) - g(z_2)| \leq 2\pi\omega^*(\delta) \quad \text{for} \quad r_1 \leq r, \quad r_2 \leq r.$$

Suppose now that $r_j > r$ for some j. From the first estimate (7) we see that

$$|g(r\zeta_j) - g(r_j\zeta_j)| \leq \int_r^{r_j} |g'(\xi\zeta_j)|\, d\xi \leq \frac{2}{\pi} \int_0^\delta \frac{\omega(x)}{x}\, dx + 2\pi \int_0^\delta \left(\int_x^\pi \frac{\omega(t)}{t^2}\, dt \right) dx \,.$$

Exchanging the order of integration we obtain that the last term is

$$= 2\pi \int_0^\delta \frac{\omega(t)}{t}\, dt + 2\pi\delta \int_\delta^\pi \frac{\omega(t)}{t^2}\, dt$$

so that $|g(r\zeta_j) - g(r_j\zeta_j)| \leq 7\omega^*(\delta)$ by (5). Furthermore $|g(r\zeta_1) - g(r\zeta_2)| \leq 2\pi\omega^*(\delta)$ by (11). It follows that (8) holds in all cases. $\qquad\square$

We say that the curve C is *Dini-smooth* if it has a parametrization $C : w(\tau)$, $0 \leq \tau \leq 2\pi$ such that $w'(\tau)$ is Dini-continuous and $\neq 0$. Let $M_1, M_2, \ldots$ denote suitable positive constants.

Theorem 3.5. *Let f map $\mathbb{D}$ conformally onto the inner domain of the Dini-smooth Jordan curve C. Then f' has a continuous extension to $\overline{\mathbb{D}}$ and*

$$(12) \qquad \frac{f(\zeta) - f(z)}{\zeta - z} \to f'(z) \neq 0 \quad for \quad \zeta \to z, \quad \zeta, z \in \overline{\mathbb{D}},$$

$$(13) \qquad |f'(z_1) - f'(z_2)| \leq M_1\omega^*(\delta) \quad for \quad z_1, z_2 \in \overline{\mathbb{D}}, \quad |z_1 - z_2| \leq \delta \,.$$

See Kel12, War32, War61. Our assumption is that

$$(14) \qquad |w'(\tau_1) - w'(\tau_2)| \leq \omega(\tau_2 - \tau_1) \quad for \quad \tau_1 < \tau_2$$

where (4) holds and ω^* is defined by (5). We can write the assertion (13) as $\omega(\delta, f', \overline{\mathbb{D}}) \leq M_1\omega^*(\delta, w', \mathbb{R})$.

Proof. Let $0 < t_2 - t_1 \leq \delta \leq \pi$ and define τ_j $(j = 1, 2)$ by $w(\tau_j) = f(e^{it_j})$. Since $|w'(\tau)| \geq 1/M_2$ we obtain from Proposition 3.3 that

$$(15) \qquad \tau_2 - \tau_1 \leq M_2 \int_{\tau_1}^{\tau_2} |w'(\tau)|\, d\tau \leq M_3(t_2 - t_1)^{1/2} \leq M_3\delta^{1/2} \,.$$

Let $\beta(t)$ again denote the tangent direction angle at $f(e^{it})$. It follows from (3.1.6) that

$$|r_1 e^{i\beta_1} - r_2 e^{i\beta_2}| \geq \frac{r}{\pi}|\beta_2 - \beta_1| \quad for \quad r_1 \geq r, \quad r_2 \geq r, \quad |\beta_1 - \beta_2| \leq \pi \,.$$

Since $|w'(\tau)| \leq 1/M_2$ and since ω is increasing we therefore get from (14) and (15) that

$$|\beta(t_1) - \beta(t_2)| \leq \pi M_2\omega(\tau_2 - \tau_1) \leq M_4\omega\left(M_3\sqrt{\delta}\right) \,.$$

Since $t \leq M_5\omega(t)$ by (2), we conclude that the modulus of continuity of $\gamma(t) = \beta(t) - t - \pi/2$ satisfies $\omega(\delta, \gamma) \leq M_6\omega(\sqrt{\delta})$ and therefore

$$\int_0^\pi \frac{\omega(\delta,\gamma)}{\delta}\,d\delta \le M_6 \int_0^\pi \frac{\omega(\sqrt{\delta})}{\delta}\,d\delta = 2M_6 \int_0^{\sqrt{\pi}} \frac{\omega(x)}{x}\,dx < \infty$$

so that γ is Dini-continuous. Hence $\log f'$ has a continuous extension to $\overline{\mathbb{D}}$ by (3.2.4) and Proposition 3.4 with $\varphi = \gamma$.

Hence f' is continuous and $\ne 0$ in $\overline{\mathbb{D}}$ and thus

$$(16) \qquad \frac{f(\zeta) - f(z)}{\zeta - z} = \int_0^1 f'(z + s(\zeta - z))\,ds \qquad \text{for} \quad z,\zeta \in \overline{\mathbb{D}}$$

which implies (12). Since we know now that f' is bounded we see as in (15) that $\tau_2 - \tau_1 \le M_7(t_2 - t_1)$ and therefore that $\omega(\delta,\gamma) \le M_8\omega(\delta)$. Hence $|\log f'(z_1) - \log f'(z_2)| \le M_9\omega^*(\delta)$ by Proposition 3.4 which implies (13). $\qquad\square$

We now turn to higher derivatives restricting ourselves for simplicity to Hölder conditions.

The Jordan curve C is of class C^n $(n = 1, 2, \dots)$ if it has a parametrization $C : w(\tau)$, $0 \le \tau \le 2\pi$ that is n times continuously differentiable and satisfies $w'(\tau) \ne 0$. It is of class $C^{n,\alpha}$ $(0 < \alpha \le 1)$ if furthermore

$$(17) \qquad |w^{(n)}(\tau_1) - w^{(n)}(\tau_2)| \le M_{10}|\tau_1 - \tau_2|^\alpha\,.$$

The following *Kellogg-Warschawski theorem* (War32) shows that, for $0 < \alpha < 1$, we can always take the conformal parametrization; this is not always true if $\alpha = 1$.

Theorem 3.6. *Let f map $\mathbb{D}$ conformally onto the inner domain of the Jordan curve C of class $C^{n,\alpha}$ where $n = 1, 2, \dots$ and $0 < \alpha < 1$. Then $f^{(n)}$ has a continuous extension to $\overline{\mathbb{D}}$ and*

$$(18) \qquad |f^{(n)}(z_1) - f^{(n)}(z_2)| \le M_{11}|z_1 - z_2|^\alpha \qquad \text{for} \quad z_1, z_2 \in \overline{\mathbb{D}}.$$

Proof. We prove by induction on $n = 1, 2, \dots$ that the assertions of the theorem hold and that furthermore, with $\gamma(t) = \arg f'(e^{it})$,

$$(19) \qquad \left(z\frac{d}{dz}\right)^n \log f'(z) = \frac{i^{1-n}}{2\pi}\int_0^{2\pi} \frac{e^{it} + z}{e^{it} - z}\gamma^{(n)}(t)\,dt \qquad (z \in \mathbb{D}),$$

$$(20) \qquad \left(|f'(e^{it})|^{-1}\frac{d}{dt}\right)^{n-1}\left[\gamma(t) + t + \frac{\pi}{2}\right] = \left(|w'(\tau)|^{-1}\frac{d}{d\tau}\right)^{n-1}\arg w'(\tau)$$

where t and τ are related by $f(e^{it}) = w(\tau)$.

Let first $n = 1$. We know from Theorem 3.5 that $\log f'$ is continuous in $\overline{\mathbb{D}}$ and that (18) is true. It follows from (3.2.4) by differentiation that

$$z\frac{d}{dz}\log f'(z) = \frac{i}{2\pi}\int_0^{2\pi} \frac{2ze^{it}}{(e^{it} - z)^2}\gamma(t)\,dt\,,$$

and we obtain (19) for $n = 1$ integrating by parts and using that γ is 2π-periodic. Finally (20) for $n = 1$ is equivalent to (3.2.3).

Suppose now that our claim holds for n and that C is of class $C^{n+1,\alpha}$. Then $|w'(\tau)| \neq 0$ and $\arg w'(\tau)$ have nth Hölder-continuous derivatives (with exponent α) so that the right-hand side of (20) has a Hölder-continuous derivative with respect to τ. Since

$$(21) \qquad \frac{d\tau}{dt} = |f'(e^{it})|/|w'(\tau)|$$

is Hölder-continuous, the left-hand side of (20) has a Hölder-continuous derivative with respect to t and has the form $|f'|^{-(n-1)}\gamma^{(n-1)} + \ldots$ where the remaining terms are already known to have a Hölder-continuous derivative. Hence $\gamma^{(n)}$ exists and is Hölder continuous.

Applying Proposition 3.4 with $\varphi = \gamma^{(n)}$ we see from (19) that $\left(z\frac{d}{dz}\right)^n \log f'(z)$ is Hölder-continuous in $\overline{\mathbb{D}}$. Since $f^{(\nu)}$ is Hölder-continuous for $\nu \leq n$, it follows that $f^{(n+1)}$ is Hölder-continuous. Finally, if we differentiate (19) for n and then integrate by parts we obtain (19) for $n + 1$; if we differentiate (20) for n and use (21) we obtain (20) for $n + 1$.   □

It follows at once from Theorem 3.6 that all derivatives $f^{(n)}$ are continuous in $\overline{\mathbb{D}}$ if the Jordan curve C is of class C^∞, i.e. of class C^n for all n. The converse is trivial because we can choose the conformal parametrization.

Exercises 3.3

1. If φ is uniformly continuous on the convex set A, show that

$$\omega(\delta_1 + \delta_2) \leq \omega(\delta_1) + \omega(\delta_2).$$

2. Let $\omega(\delta) = O((\log 1/\delta)^{-c})$ with $c > 1$. Show that $\omega^*(\delta) = O((\log 1/\delta)^{-c+1})$.

3. Let g be analytic in $\mathbb{D}$ and let $0 < \alpha < 1$. Show that g is Hölder-continuous in $\overline{\mathbb{D}}$ with exponent α if and only if $g'(z) = O((1 - |z|)^{\alpha-1})$ as $|z| \to 1-$ (Hardy-Littlewood, see e.g. Dur70, p. 74).

4. Let G be the inner domain of the Dini-smooth Jordan curve C and let ψ be a real-valued continuous function on C with $\int_C \psi|dz| = 0$. Show that there is a harmonic function u in G with grad u continuous in $\overline{G}$ whose normal derivative at $z \in C$ is $\psi(z)$. If $G = \mathbb{D}$ show that

$$u(z) = \text{const} - \frac{1}{\pi} \int_0^{2\pi} \psi(e^{it}) \log |e^{it} - z|\, dt.$$

5. Let the Jordan curve C be of class $C^{1,\alpha}$ with $0 < \alpha < 1$. Show that the conformal map f can be extended to a (non-analytic) homeomorphism of $\mathbb{C}$ onto $\mathbb{C}$ with non-vanishing Hölder-continuous gradient; compare Corollary 2.9.

6. Let C be of class $C^{2,\alpha}$ with $0 < \alpha < 1$. Show that, uniformly in $z, \zeta \in \overline{\mathbb{D}}$,

$$\frac{f'(\zeta)}{f(\zeta) - f(z)} = \frac{1}{\zeta - z} + \frac{1}{2}\frac{f''(z)}{f'(z)} + O(|\zeta - z|^\alpha) \qquad \text{as} \quad |z - \zeta| \to 0.$$

3.4 Tangents and Corners

Let now G be any simply connected domain in $\mathbb{C}$ with locally connected boundary. By Theorem 2.1 this means that the conformal map f of $\mathbb{D}$ onto G is continuous in $\overline{\mathbb{D}}$. We shall study the local behaviour at a point $\zeta = e^{i\vartheta} \in \mathbb{T}$.

We say that ∂G has a *corner* of opening $\pi\alpha(0 \leq \alpha \leq 2\pi)$ at $f(\zeta) \neq \infty$ if

$$(1) \qquad \arg[f(e^{it}) - f(e^{i\vartheta})] \to \begin{cases} \beta & \text{as } t \to \vartheta+, \\ \beta + \pi\alpha & \text{as } t \to \vartheta - . \end{cases}$$

It would be more precise to say "at the prime end corresponding to ζ" instead of "at $f(\zeta)$" because there may exist other $\zeta' \in \mathbb{T}$ with $f(\zeta') = f(\zeta)$ where there may be a corner of opening $\pi\alpha'$ or none at all, see Fig. 3.2.

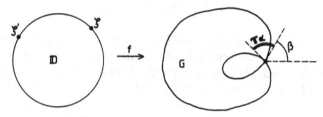

Fig. 3.2

If $\alpha = 1$ then we have a *tangent* of direction angle β; compare (3.2.1). If $\alpha = 0$ we have an *outward-pointing cusp*; if $\alpha = 2$ we have an *inward-pointing cusp*.

As an example, the function

$$f(z) = \left(\frac{1-z}{1+z}\right)^\alpha = e^{-i\pi\alpha}\left(\frac{z-1}{z+1}\right)^\alpha \qquad (0 < \alpha \leq 2)$$

maps $\mathbb{D}$ onto the sector $G = \{|\arg w| < \pi\alpha/2\}$, and ∂G has a corner of opening $\pi\alpha$ with $\beta = -\pi\alpha/2$ at 0. If $\alpha = 2$ then $G = \mathbb{C} \setminus (-\infty, 0]$ and there is an inward-pointing cusp at 0. We always use the branch of the logarithm determined by

$$(2) \qquad \vartheta + \pi/2 < \operatorname{Im}\log(z - e^{i\vartheta}) = \arg(z - e^{i\vartheta}) < \vartheta + 3\pi/2.$$

Theorem 3.7. *Let f map $\mathbb{D}$ conformally onto G where ∂G is locally connected. Let $\zeta = e^{i\vartheta} \in \mathbb{T}$ and $f(\zeta) \neq \infty$. Then ∂G has a corner of opening $\pi\alpha(0 \leq \alpha \leq 2)$ at $f(\zeta)$ if and only if*

$$(3) \qquad \arg\frac{f(z) - f(\zeta)}{(z - \zeta)^\alpha} \to \beta - \alpha(\vartheta + \pi/2) \qquad as \quad z \to \zeta, \quad z \in \mathbb{D}.$$

To prove this theorem (Lin15) we need the following representation formula.

Proposition 3.8. *If f maps $\mathbb{D}$ conformally onto $G \subset \mathbb{C}$ and if $a \neq \infty$, $a \notin G$ then*

$$(4) \qquad \log(f(z) - a) = \log|f(0) - a| + \frac{i}{2\pi} \int_0^{2\pi} \frac{e^{it} + z}{e^{it} - z} \arg(f(e^{it}) - a)\, dt.$$

Proof. We may assume that $a = 0$. We have

$$\frac{1}{2\pi} \int_0^{2\pi} |\log f(re^{it})|^2\, dt = \text{const} + \frac{1}{\pi} \int_0^{2\pi} (\log|f(re^{it})|)^2\, dt$$

for $0 \leq r < 1$. Since $x^2 \leq e^x + e^{-x}$ we have

$$\left(\frac{2}{5} \log|f|\right)^2 \leq |f|^{2/5} + \left|\frac{1}{f}\right|^{2/5}$$

and since both f and $1/f$ are analytic and univalent in $\mathbb{D}$, it follows from Theorem 1.7 that the integral is bounded. Hence $g = \log f$ belongs to the Hardy space H^2, and (4) follows from the Schwarz integral formula (3.1.2) which is valid for $g \in H^1$ and thus for $g \in H^2$. $\qquad \square$

Proof of Theorem 3.7. Let ∂G have a corner at $f(\zeta)$. Taking the imaginary part in (4) we easily see that

$$(5) \qquad \arg\frac{f(z) - f(\zeta)}{(z - \zeta)^\alpha} = \frac{1}{2\pi} \int_0^{2\pi} \frac{1 - |z|^2}{|e^{it} - z|^2} \arg\frac{f(e^{it}) - f(\zeta)}{(e^{it} - \zeta)^\alpha}\, dt.$$

Since $\arg(e^{it} - e^{i\vartheta}) \to \vartheta + \pi/2$ as $t \to \vartheta+$, $\to \vartheta + 3\pi/2$ as $t \to \vartheta-$, we conclude from (1) that

$$(6) \qquad \arg\frac{f(e^{it}) - f(\zeta)}{(e^{it} - \zeta)^\alpha} \to \beta - \alpha\left(\vartheta + \frac{\pi}{2}\right) \qquad \text{as} \quad t \to \vartheta \pm.$$

Hence (3) follows from (5) by (3.1.5). Conversely, if (3) holds it follows by continuity that (6) is true which at once implies (1). $\qquad \square$

We say that ∂G has a *Dini-smooth corner* at $f(\zeta)$ if there are closed arcs $A^\pm \subset \mathbb{T}$ ending at $\zeta \in \mathbb{T}$ and lying on opposite sides of ζ that are mapped onto Dini-smooth Jordan arcs C^+ and C^- forming the angle $\pi\alpha$ at $f(\zeta)$. Then Theorem 3.7 can be strengthened (War32).

Theorem 3.9. *If ∂G has a Dini-smooth corner of opening $\pi\alpha$ $(0 < \alpha \leq 2)$ at $f(\zeta) \neq \infty$ then the functions*

$$\frac{f(z) - f(\zeta)}{(z - \zeta)^\alpha} \qquad \text{and} \qquad \frac{f'(z)}{(z - \zeta)^{\alpha - 1}}$$

are continuous and $\neq 0, \infty$ in $\overline{\mathbb{D}} \cap D(\zeta, \rho)$ for some $\rho > 0$.

Proof. We shall reduce the present situation to the case of a Dini-smooth Jordan curve by localizing and then straightening out the angle. We may assume that $f(\zeta) = 0$.

If $\rho_1 > 0$ is sufficiently small we can find circles that touch $\partial D(0, \rho_1)$ and $C^{\pm}$. Let G' be the subdomain of $D(0, \rho_1) \cap G$ indicated in Fig. 3.3. We next apply the transformation $w \mapsto w^{1/\alpha}$ to G' and obtain a domain H that is bounded by a Jordan curve with a tangent at 0.

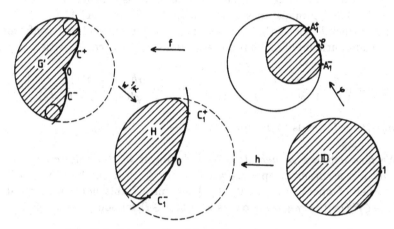

Fig. 3.3. Localization (shaded) and straightening

We now show that ∂H is Dini-smooth. Let $w_{\pm}(\tau)$, $0 \leq \tau \leq b_1$ be parametrizations of the arcs $C^{\pm} \cap \partial G'$ such that $w_{\pm}(0) = 0$ and $w'_{\pm}(\tau) \neq 0$ is Dini-continuous. The corresponding arc $C_1^{\pm}$ on ∂H can then be parametrized as

$$(7) \qquad v(t) = w(ct^{\alpha})^{1/\alpha}, \quad 0 \leq t \leq b_2$$

where $c = 1/|w'(0)|$; we have dropped the subscripts $\pm$. It follows that

$$(8) \qquad v'(t) = c^{1/\alpha} w'(ct^{\alpha}) u(ct^{\alpha}), \quad u(\tau) \equiv (\tau^{-1} w(\tau))^{-1+1/\alpha}$$

with $v'(0) = [cw'(0)]^{1/\alpha}$ and thus $|v'(0)| = 1$. If $0 < \tau_1 < \tau_2 \leq b_1$ then

$$\left| \frac{w(\tau_1)}{\tau_1} - \frac{w(\tau_2)}{\tau_2} \right| = \left| \int_0^1 (w'(\tau_2 x) - w'(\tau_1 x)) \, dx \right| \leq \omega(\tau_2 - \tau_1, w')$$

where ω is the modulus of continuity (3.3.1). Since $|\tau^{-1} w(\tau)|$ is bounded from below (because $w'(0) \neq 0$) it easily follows, by (8), that $\omega(\delta, u) \leq M_1 \omega(\delta, w')$ and thus $\omega(\delta, v') \leq M_2 \omega(c\delta^{\alpha}, w')$. Hence

$$(9) \qquad \int_0^{b_2} \delta^{-1} \omega(\delta, v'_{\pm}) \, d\delta \leq M_2 \alpha^{-1} \int_0^{b_1} t^{-1} \omega(t, w') \, dt < \infty.$$

Now $|v'_+(0)| = 1$ and $\arg v'_+(0) = \arg v'_-(0)$ because ∂H has a tangent at 0. It follows that $v'_+(0) = v'_-(0)$. Hence (9) shows that the curve arc $C_1^- \cup C_1^+$: $v_\pm(|t|)$, $0 \le \pm t \le b_2$ is Dini-smooth and therefore also ∂H.

Let h map $\mathbb{D}$ conformally onto H such that $h(1) = 0$. Since $G' \subset G$ there is a conformal map φ of $\mathbb{D}$ into $\mathbb{D}$ such that $\varphi(1) = \zeta$ and

$$(10) \qquad h(s) = f(\varphi(s))^{1/\alpha} \qquad \text{for} \quad s \in \mathbb{D};$$

see Fig. 3.3. We see from Theorem 3.5 that $h'(s)$ is continuous and $\ne 0, \infty$ for $s \in \overline{\mathbb{D}}$. Since φ maps an arc of $\mathbb{T}$ with $1 \in \mathbb{T}$ onto $A_1^- \cup A_1^+ \subset \mathbb{T}$, the reflection principle shows that φ maps some neighbourhood of 1 conformally onto a neighbourhood of ζ. Thus our assertions follow because, with $z = \varphi(s)$,

$$\frac{f(z)}{(z-\zeta)^\alpha} = \left(\frac{h(s)}{s-1} \frac{s-1}{\varphi(s)-\zeta} \right)^\alpha , \qquad \frac{f'(z)}{(z-\zeta)^{\alpha-1}} = \frac{\alpha h'(s)}{\varphi'(s)} \left(\frac{h(s)}{s-1} \frac{s-1}{\varphi(s)-\zeta} \right)^{\alpha-1}$$

and $h(s)/(s-1)$ tends to a limit $\ne 0$ as $s \to 1$ by (3.3.12).   □

We now define a *corner at* ∞ by reducing it to the finite case. We say that ∂G has a corner of opening $\pi\alpha$ $(0 \le \alpha \le 2)$ at ∞ if $w \mapsto 1/w$ leads to a corner of opening $\pi\alpha$ at 0. By Theorem 3.7 this holds if and only if $\arg[(z-\zeta)^\alpha f(z)] \to \gamma + \alpha\pi + \alpha\vartheta$ as $z \to \zeta$, $z \in \overline{\mathbb{D}}$ for some γ and thus

$$(11) \qquad \arg f(e^{it}) \to \gamma \pm \alpha\pi/2 \qquad \text{as} \quad t \to \vartheta\pm$$

where $\zeta = e^{i\vartheta}$, see (2). This means that the domain approximately contains an infinite sector of opening $\alpha\pi$ and midline inclination angle γ. We call the corner at ∞ Dini-smooth if the corresponding corner at 0 is Dini-smooth.

It follows at once from Theorem 3.9 that the functions $(z-\zeta)^\alpha f(z)$ and $(z-\zeta)^{\alpha+1} f'(z)$ are continuous and $\ne 0, \infty$ in $\overline{\mathbb{D}} \cap D(\rho, \zeta)$ if ∂G has a Dini-smooth corner of opening $\pi\alpha > 0$ at ∞.

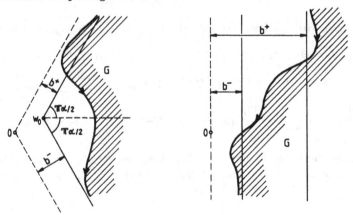

Fig. 3.4. Corners at ∞ of openings $\pi\alpha$ and π with $\gamma = 0$

Proposition 3.10. *Let ∂G have a Dini-smooth corner of opening $\pi\alpha$ with $0 < \alpha < 2$ at $f(\zeta) = \infty$ where $\zeta = e^{i\vartheta}$. Suppose that*

$$\text{(12)} \qquad \text{Im}[e^{-i\gamma \mp i\pi\alpha/2} f(e^{it})] \to \mp b^{\pm} \qquad as \quad t \to \vartheta \pm .$$

If $0 < \alpha < 1$ then

$$\text{(13)} \qquad f(z) = \frac{a}{(z - \zeta)^\alpha} + w_0 + o(1) \qquad as \quad z = r\zeta, \quad r \to 1$$

with some constant $a \neq 0$, $\arg a = \gamma + \alpha\pi + \alpha\vartheta$. If $\alpha = 1$ then

(14)

$$f(z) = \frac{a}{z - \zeta} + ie^{i\gamma} \frac{b^+ - b^-}{\pi} \log(z - \zeta) + o\left(\log \frac{1}{1 - r}\right) \qquad as \quad z = r\zeta, \quad r \to 1.$$

If $1 < \alpha < 2$ then (13) holds with an additional term $a'/(z - \zeta)^{\alpha-1}$ where $a' \in \mathbb{C}$.

The assumption (12) says that the curves forming the angle $\pi\alpha$ at ∞ are asymptotic to the lines $w = e^{i\gamma \pm i\pi\alpha/2}(s \mp ib^\pm)$, $s \in \mathbb{R}$. If $\alpha \neq 1$ they intersect at

$$\text{(15)} \qquad w_0 = e^{i\gamma}(b^- e^{i\pi\alpha/2} + b^+ e^{-i\pi\alpha/2})/\sin(\pi\alpha)$$

and this is the constant in (13). For example, the function $[(z + 1)/(z - 1)]^\alpha$ maps $\mathbb{D}$ onto $\{-3\pi\alpha/2 < \arg w < -\pi\alpha/2\}$ according to (2); we have $\gamma = -\pi\alpha$ and $b^+ = b^- = 0$.

If $\alpha = 1$ then the two asymptotes are parallel. For example, the function

$$\text{(16)} \qquad f(z) = \frac{1 + z}{1 - z} - i \log \frac{1 + z}{1 - z} + \frac{\pi}{2}$$

maps $\mathbb{D}$ conformally onto the domain of Fig. 3.5; we have

$$f(e^{it}) = \begin{cases} i(\cot t/2 - \log(\cot t/2)) + \pi & \text{for } 0 < t < \pi, \\ -i(|\cot t/2| + \log|\cot t/2|) & \text{for } -\pi < t < 0. \end{cases}$$

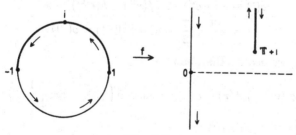

Fig. 3.5

Proof. (a) Let first $0 < \alpha < 1$. Then w_0 exists and we may assume that $\gamma = 0$, $b^+ = b^- = 0$ and $\zeta = 1$. Then $w_0 = 0$ and (12) shows that

$$\arg f(e^{it}) = \pm\frac{\pi\alpha}{2} + o\left(\frac{1}{|f(e^{it})|}\right) = \pm\frac{\pi\alpha}{2} + o(|t|^\alpha) \quad \text{as} \quad t \to 0\pm$$

because the corner is Dini-smooth. It follows that

$$(17) \quad \arg[(e^{it} - 1)^\alpha f(e^{it})] = \pi\alpha + \varphi(t), \quad \varphi(t) = o(|t|^\alpha) \quad \text{as} \quad t \to 0\pm \,.$$

Given $\varepsilon > 0$ we can therefore find δ such that $|\varphi(t)| < \varepsilon|t|^\alpha$ for $|t| \leq \delta$.

Assuming for simplicity that $w_0 = 0 \notin G$ we have as in Proposition 3.8

$$(18) \qquad g(z) \equiv \log[(z - 1)^\alpha f(z)] = a_1 + \frac{i}{2\pi}\int_{-\pi}^{\pi}\frac{e^{it} + z}{e^{it} - z}\varphi(t)\,dt$$

for $z \in \mathbb{D}$ and thus

$$|g'(z)| = \left|\frac{i}{\pi}\int_{-\pi}^{\pi}\frac{e^{it}\varphi(t)}{(e^{it} - z)^2}\,dt\right| \leq \frac{1}{\pi}\int_{-\pi}^{\pi}\frac{|\varphi(t)|}{|e^{it} - z|^2}\,dt\,.$$

We see from (3.1.6) that, for $0 < r < 1 - \delta$,

$$|g'(r)| \leq \frac{2\varepsilon}{\pi}\int_0^{1-r}\frac{(1 - r)^\alpha}{(1 - r)^2}\,dt + \frac{2\varepsilon}{\pi}\int_{1-r}^{\delta}\frac{\pi^2 t^\alpha}{t^2}\,dt + M_1\int_\delta^{\pi}\frac{t^\alpha}{t^2}\,dt$$

$$= \left(\frac{2\varepsilon}{\pi} + \frac{2\varepsilon\pi}{1 - \alpha}\right)(1 - r)^{\alpha-1} + M_2(\varepsilon)\,.$$

Since $0 < \alpha < 1$ it follows by integration that

$$|g(1) - g(r)| \leq \varepsilon M_3(1 - r)^\alpha + M_2(\varepsilon)(1 - r)\,.$$

Hence $g(r) = g(1) + o((1-r)^\alpha)$ as $r \to 1-$, and (17), (18) show that $\operatorname{Im} g(1) = \pi\alpha$. Thus (13) follows by exponentiation.

(b) Now let $\alpha = 1$. We may assume that $\gamma = 0$ and $\zeta = 1$. The function

$$(19) \qquad h(z) = i\pi^{-1}(b^+ - b^-)\log(z - 1) + 3b^+/2 - b^-/2$$

satisfies $\operatorname{Re} h(e^{it}) \to b^\pm$ as $t \to 0\pm$. Since also $\operatorname{Re} f(e^{it}) \to b^\pm$ by (12) and since $|f(e^{it})|^{-1} = O(|t|)$ as $t \to 0\pm$, we conclude from (12) that

$$\varphi(t) \equiv \arg[(e^{it} - 1)(f(e^{it}) - h(e^{it}))] - \pi$$

$$= \mp\frac{\pi}{2} \pm \frac{\pi}{2} + o(|f(e^{it})|^{-1}) = o(|t|)\,.$$

As in part (a), we deduce that, as $r \to 1-$,

$$\log[(r - 1)(f(r) - h(r))] = a_0 + o\left((1 - r)\log\frac{1}{1 - r}\right)$$

from which (14) follows by (19).

(c) Finally let $1 < \alpha < 2$. We define g again by (18) and show now that $g''(r) = o((1 - r)^{\alpha-2})$ as $r \to 1-$. It follows that $g(r) = g(1) + (r - 1)g'(1) + o((1 - r)^\alpha)$ which leads to (13) with an additional term $g'(1)e^{g(1)}(r - 1)^{1-\alpha}$. $\qquad\square$

We return to the case of a corner at a finite point by $w \mapsto 1/w$. The asymptotes now become circles of curvature.

Theorem 3.11. *Let ∂G have a Dini-smooth corner of opening $\pi\alpha(0 < \alpha < 2)$ at $f(\zeta) \neq \infty$ and suppose that the arcs $C^\pm$ forming the corner have curvature $\kappa^\pm$ at $f(\zeta)$. If $0 < \alpha < 1$ then, as $z = r\zeta$, $r \to 1-$,*

$$(20) \qquad f(z) = f(\zeta) + b_1(z - \zeta)^\alpha + b_2(z - \zeta)^{2\alpha} + o((1 - r)^{2\alpha})$$

where $\arg b_1 = \beta - \pi\alpha/2 - \alpha\vartheta$ and

$$(21) \qquad b_2 = -e^{-i\beta}b_1^2(\kappa^- + \kappa^+ e^{-i\pi\alpha})/(2\sin\pi\alpha).$$

If $\alpha = 1$ then

$$(22)\ f(z) = f(\zeta) + b_1(z - \zeta) + b_2(z - \zeta)^2 \log(z - \zeta) + o\left((1 - r)^2 \log\frac{1}{1 - r}\right)$$

where $b_2 = e^{-i\beta}b_1^2(\kappa^+ - \kappa^-)/(2\pi)$. If $1 < \alpha < 2$ then (20) holds with an additional term $b'(z - \zeta)^{\alpha+1}$.

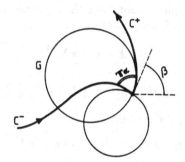

Fig. 3.6. Corner and circles of curvature with $\kappa^+ > 0$, $\kappa^- < 0$

For example, if $0 < \alpha < 1$ and $a = e^{i\pi\alpha/2}$ then

$$(23) \qquad f(z) = 1/\left(a\left(\frac{z+1}{z-1}\right)^\alpha + 1\right) = \frac{\bar{a}}{2^\alpha}(z-1)^\alpha - \frac{\bar{a}^2}{2^{2\alpha}}(z-1)^{2\alpha} + \cdots$$

maps $\mathbb{D}$ onto the domain bounded by $[0, 1]$ and a circular arc; we have $\kappa^+ = 0$, $\kappa^- = 2\sin\pi\alpha$ and $\beta = 0$ at $f(1) = 0$. If $\alpha = 1$ we see that there is always a logarithmic term except if $\kappa^+ = \kappa^-$. The case $\alpha = 0$ of an outward pointing cusp will be considered in Chapter 11.

Proof. We may assume that $f(\zeta) = 0$. Now $w \mapsto 1/w$ transforms the circles of curvature of $C^\pm$ at 0 into the lines $\operatorname{Im}[e^{-i\gamma \mp i\pi\alpha/2}w] = \mp\kappa^\pm/2$ where $\gamma = -\beta - \pi\alpha/2$. The existence of the curvature is equivalent to (12) with $b^\pm = \kappa^\pm/2$. Hence we can apply Proposition 3.10 to $1/f$. If $0 < \alpha < 1$ we conclude from (13) that

$$f(z) = \left(\frac{a}{(z-\zeta)^\alpha} + w_0 + o(1)\right)^{-1} = \frac{(z-\zeta)^\alpha}{a} - \frac{w_0}{a^2}(z-\zeta)^{2\alpha} + o(|z-\zeta|^{2\alpha})$$

and (20) with (21) follows from (15). The cases $1 \leq \alpha < 2$ are similar. □

In most applications we have the case of an *analytic corner*, i.e. a corner of opening $\pi\alpha$ formed by two arcs $C^\pm : w_\pm(t)$, $0 \leq t \leq \delta$ where $w_\pm$ is analytic in $|t| < \delta$ and $w_\pm(0) = f(\zeta)$, $w'_\pm(0) \neq 0$.

Lewy and Lehman (Lehm57) have given an asymptotic expansion: If $\alpha > 0$ is irrational then, as $z \to \zeta$, $z \in \overline{\mathbb{D}}$,

$$(24) \qquad f(z) \sim \sum_{k=0}^\infty \sum_{j=1}^\infty a_{kj}(z-\zeta)^{k+\alpha j}, \qquad a_{01} \neq 0;$$

if $\alpha = p/q$ with relatively prime positive p, q then

$$(25) \qquad f(z) \sim \sum_{k=0}^\infty \sum_{j=1}^q \sum_{m=0}^{[k/p]} a_{kjm}(z-\zeta)^{k+pj/q}(\log(z-\zeta))^m, \qquad a_{010} \neq 0.$$

We now give an example to show that the logarithmic terms may appear. Let $q = 1, 2, \ldots$. It follows from (16) that

$$(26) \qquad \begin{aligned} f(z) &= \left(\frac{1+z}{1-z} - i\log\frac{1+z}{1-z} + \frac{\pi}{2}\right)^{-1/q} \\ &= b_1(z-1)^{1/q} + b_2(z-1)^{1+1/q}\log(z-1) + \ldots \end{aligned}$$

maps $\mathbb{D}$ conformally onto the domain

$$\{|\arg w| < \pi/(2q)\} \setminus C, \quad C : t(i + \pi t^q)^{-1/q}, \quad 0 \leq t \leq 1.$$

Thus 0 is a corner of opening π/q formed by a line and the analytic curve C.

These asymptotic expansions have been generalized to corners with weaker conditions (Wig65). See e.g. PaWaHo86 for a discussion of the numerical computation of conformal maps at corners. Many explicit examples can be found in KoSt59.

Exercises 3.4

1. Let ∂G have a corner of opening $\alpha\pi > 0$ at $f(\zeta) \neq \infty$ formed by two Hölder-smooth curves with exponent γ ($0 < \gamma < 1$). Show that, with $a \neq 0$,

$$f(z) = f(\zeta) + a(z-\zeta)^\alpha + O(|z-\zeta|^{\alpha+\alpha\gamma}) \quad \text{as} \quad z \to \zeta, \quad z \in \overline{\mathbb{D}}.$$

2. Let $f(z) = z + b(z-1)^2 \log(z-1)$ where $b > 0$ is sufficiently small. Show that f maps $\mathbb{D}$ conformally onto a smooth Jordan domain and determine $\kappa^\pm$.

3. Let f map $\mathbb{D}$ conformally onto $\{u + iv : u > 0, v > 1/u\}$ such that $f(1) = \infty$. Use "squaring" to show that $f(x) = a(1-x)^{-1/2} + O((1-x)^{1/2})$ as $x \to 1-$.

4. Show that a Jordan curve can have only countably many corners of angle $\neq \pi$. (Apply the Bagemihl ambiguous point theorem to $f(e^{-t}\zeta) = \arg[w(e^{it}\zeta) - w(\zeta)]$ for $t > 0$, $\zeta \in \mathbb{T}$.)

3.5 Regulated Domains

The typical domain in the applications of conformal mapping is bounded by finitely many smooth arcs that may form corners or may go to infinity; parts of the boundary may be run through twice. We shall introduce (Ost35) a class of domains, the regulated domains, that is wide enough to allow all these possibilities but also narrow enough to have reasonable properties.

We call β a *regulated function* on the interval $[a, b]$ if the one-sided limits

$$(1) \qquad \beta(t-) = \lim_{\tau \to t-} \beta(\tau), \quad \beta(t+) = \lim_{\tau \to t+} \beta(\tau)$$

exist for $a \leq t \leq b$. The function β is regulated if and only if (Die69, p. 145) it can be uniformly approximated by step functions, i.e. if for every $\varepsilon > 0$ there exist $a = t_0 < t_1 < \cdots < t_n = b$ and constants $\gamma_1, \ldots, \gamma_n$ such that

$$(2) \qquad |\beta(t) - \gamma_\nu| < \varepsilon \quad \text{for} \quad t_{\nu-1} < t < t_\nu, \quad \nu = 1, \ldots, n.$$

We have $\beta(t+) = \beta(t-)$ except possibly for countably many t.

We assume again that G is a simply connected domain in $\mathbb{C}$ with locally connected boundary C and that f maps $\mathbb{D}$ conformally onto G, so that f is continuous in $\overline{\mathbb{D}}$ by Theorem 2.1. We shall use the conformal parametrization

$$(3) \qquad C = \partial G : w(t) = f(e^{it}), \quad 0 \leq t \leq 2\pi;$$

this curve may have many multiple points and may pass through ∞.

We call G a *regulated domain* if each point on C is attained only finitely often by f and if

$$(4) \qquad \beta(t) = \begin{cases} \lim\limits_{\tau \to t+} \arg[w(\tau) - w(t)] & \text{for } w(t) \neq \infty, \\ \lim\limits_{\tau \to t+} \arg w(\tau) + \pi & \text{for } w(t) = \infty \end{cases}$$

exists for all t and defines a regulated function. The limit $\beta(t)$ is the direction angle of the *forward tangent* of C at $w(t)$, more precisely of the forward half-tangent. Thus a domain is regulated if it is bounded by a (possibly non-Jordan) curve C with regulated forward tangent.

It follows from Proposition 3.13 below that $\beta(t+) = \beta(t)$ and that $\beta(t-)$ is the direction of the *backward tangent* at $w(t)$. Since β is continuous except for countably many jumps it follows that C has a tangent except for at most countably many corners or cusps. If $w(t) \neq \infty$ is a corner we determine the argument by

$$(5) \qquad \beta(t+) - \beta(t-) = \pi(1 - \alpha), \quad \pi\alpha \text{ opening angle}$$

in accordance with (3.4.1). If $w(t) = \infty$ we define

(6) $$\beta(t+) - \beta(t-) = \pi(1 + \alpha), \quad \pi\alpha \text{ sector angle};$$

see (4) and (3.4.11).

If C is piecewise smooth then G is regulated and β is continuous except for the finitely many values that correspond to the corners. As another example consider the function (3.4.16), see Fig. 3.5. Then $\beta(t) = 3\pi/2$ for $0 \le t < \pi/2$ and for $\pi \le t < 2\pi$ while $\beta(t) = \pi/2$ for $\pi/2 \le t < \pi$.

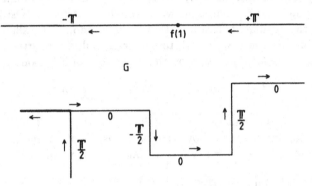

Fig. 3.7. A regulated domain with the values of $\beta(t)$ written to the corresponding parts of the boundary; in this example $\beta(t)$ is constant on each line; the apparent jump at $f(1)$ is caused by the periodicity

Suppose now that we have another parametrization $C : w^*(t^*)$, $a \le t^* \le b$ of the boundary. Then $w^* = w \circ \varphi$ where φ is a strictly increasing continuous map from $[a, b]$ to $[0, 2\pi]$ and

$$\arg[\omega^*(\tau^*) - w^*(t^*)] = \arg[w(\tau) - w(t)] \to \beta(t)$$

as $\tau = \varphi(\tau^*) \to t = \varphi(t^*)$, $\tau^* > t^*$ if $w(t) \ne \infty$. Hence $w \circ \varphi$ is again a regulated function and vice versa. Our definition is therefore independent of the choice of boundary parametrization. This is important because in general the conformal map f is not explicitly known.

We first develop some geometric properties.

Proposition 3.12. *Let G be regulated. If $0 < \varepsilon < \pi$ and*

(7) $$|\beta(t) - \gamma| < \varepsilon, \quad w(t) \ne \infty \quad \text{for} \quad t \in I$$

where I is an interval on $\mathbb{R}$, then

(8) $$|\arg[w(\tau) - w(t)] - \gamma| < \varepsilon \quad \text{for} \quad \tau, t \in I, \quad t < \tau.$$

Proof. Suppose that (8) is false and consider (see Fig. 3.8) the sector $A = \{z : |\arg(z - w(t)) - \gamma| < \varepsilon\}$. Since $w(\tau') \in A$ for small $\tau' > t$ but $w(\tau) \notin A$ there is a last point $w(t') \in \partial A$ with $t < t' < \tau$. Since

$$|\arg[w(\tau') - w(t')] - \gamma| \geq \varepsilon \quad \text{for} \quad t' < \tau' < \tau$$

we conclude from (4) that $|\arg\beta(t') - \gamma| \geq \varepsilon$ in contradiction to (7). $\square$

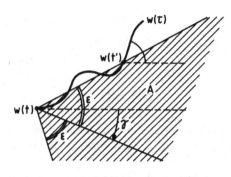

Fig. 3.8

Proposition 3.13. *Let G be regulated. Then $\beta(t+) = \beta(t)$ for all t, and if $w(t) \neq \infty$ then*

$$(9) \qquad \beta(t-) = \lim_{\tau \to t-} \arg[w(t) - w(\tau)].$$

Proof. We assume first that $w(t) \neq \infty$. Since $\beta(t-)$ exists by the definition of a regulated function, there is $\delta > 0$ such that $|\beta(\tau) - \beta(t-)| < \varepsilon$ for $t - \delta < \tau < t$. Hence it follows from Proposition 3.12 by continuity that

$$|\arg[w(t) - w(\tau)] - \beta(t-)| \leq \varepsilon \quad \text{for} \quad t - \delta < \tau < t.$$

This implies (9), and $\beta(t+) = \beta(t)$ for $w(t) \neq \infty$ is proved similarly.

Assume now that $w(t) = \infty$. We can choose τ' with $\tau' > t$ such that $|\beta(\tau) - \beta(t+)| < \varepsilon$ for $t < \tau \leq \tau'$. Hence

$$|\arg[w(\tau') - w(\tau)] - \beta(t+)| < \varepsilon \quad \text{for} \quad t < \tau < \tau'$$

by Proposition 3.12 and thus

$$|\arg w(\tau) + \pi - \beta(t+)| < \left|\arg \frac{w(\tau) - w(\tau')}{w(\tau)}\right| + \varepsilon < 2\varepsilon$$

if τ is sufficiently close to t. It follows that $\arg w(\tau) + \pi \to \beta(t+)$ as $\tau \to t+$ and therefore $\beta(t) = \beta(t+)$ by (4). The proof of (9) is similar. $\square$

Theorem 3.14. *The domain $G \subset \mathbb{C}$ is regulated if and only if, for every $\varepsilon > 0$, there are finitely many points $0 = t_0 < t_1 < \cdots < t_n = 2\pi$ such that*

$$(10) \qquad w(t) \neq \infty, \quad |\arg[w(\tau) - w(t)] - \gamma_\nu| < \varepsilon \quad (t_{\nu-1} < t < \tau < t_\nu)$$

for $\nu = 1, \ldots, n$ with real constants γ_ν.

Condition (10) says that $C_\nu : w(t)$, $t_{\nu-1} < t < t_\nu$ are rather flat open Jordan arcs in $\mathbb{C}$. These arc may be unbounded and are obtained by rotating the graph of a suitable real Lipschitz function with small constant.

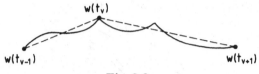

Fig. 3.9

Proof. First let G be a regulated domain. Since β is a regulated function, given $\varepsilon > 0$ we can find t_ν and γ_ν such that (2) holds. Thus (10) holds by Proposition 3.12.

Assume now conversely that, for every $\varepsilon > 0$, condition (10) holds with suitable t_ν and γ_ν. Let s be given; we restrict ourselves to the case $w(s) \neq \infty$. There exists ν such that $t_{\nu-1} \leq s < t_\nu$. By continuity we have

$$(11) \qquad |\arg[w(\tau) - w(t)] - \gamma_\nu| \leq \varepsilon \quad \text{for} \quad s \leq t < \tau < t_\nu.$$

Hence the limit points of $\arg[w(\tau) - w(s)]$ as $\tau \to s+$ lie in some interval of length 2ε for every $\varepsilon > 0$ and it follows that the limit $\beta(s)$ exists and satisfies $|\beta(s) - \gamma_\nu| \leq \varepsilon$. We also see from (11) that $|\beta(t) - \gamma_\nu| \leq \varepsilon$ for $s \leq t < t_\nu$. Hence $\beta(t) \to \beta(s)$ as $t \to s+$.

Furthermore we have $t_{\nu-1} < s \leq t_\nu$ for some ν and an analogue of (11) is valid for $t_{\nu-1} < t < \tau \leq s$. This can be used to show that $\beta(s-)$ exists. Hence β is regulated. $\qquad\square$

We now turn to a representation formula.

Theorem 3.15. *Let f map $\mathbb{D}$ conformally onto a regulated domain. Then, for $z \in \mathbb{D}$,*

$$(12) \qquad \log f'(z) = \log |f'(0)| + \frac{i}{2\pi} \int_0^{2\pi} \frac{e^{it} + z}{e^{it} - z} \left(\beta(t) - t - \frac{\pi}{2} \right) dt$$

where $\beta(t)$ is the direction angle of the forward tangent at $f(e^{it})$, and furthermore

$$(13) \qquad \arg f'(z) + \frac{\beta(t+) - \beta(t-)}{\pi} \left(\arg(z - e^{it}) - t - \frac{\pi}{2} \right) \to \beta(t) - t - \frac{\pi}{2}$$

as $z \to e^{it}$, $z \in \mathbb{D}$ for each t.

It follows that $\beta(t + 2\pi) = \beta(t) + 2\pi$, furthermore that

$$(14) \qquad \arg f'(re^{it}) \to \frac{\beta(t+) + \beta(t-)}{2} - t - \frac{\pi}{2} \quad \text{as} \quad r \to 1-\ .$$

We mention without proof that the converse is also true: If a conformal map has an integral representation (12) where β is regulated, then the image domain is regulated.

Proof. By Theorem 3.14 there is a constant M' such that $|\arg(w - \omega)| < M'$ for $w, \omega \in C$. It follows from Proposition 3.8 that this also holds for $w \in G$. The function

(15) $v(z, \zeta) = \arg[(f(z) - f(\zeta))/(z - \zeta)]$ for $z \in \mathbb{D}, \quad \zeta \in \overline{\mathbb{D}}, \quad z \neq \zeta$

is defined except for finitely many ζ and satisfies $|v(z, \zeta)| < M = M' + 2\pi$ for $z \in \mathbb{D}$ and $\zeta \in \mathbb{T}$. Since $v(z, \zeta)$ is harmonic in $\zeta \in \mathbb{D}$ (for fixed $z \in \mathbb{D}$) the maximum principle shows that $|v(z, \zeta)| \leq M$ holds for $z, \zeta \in \mathbb{D}$. Hence the analytic functions g_n defined as in (3.2.5) satisfy

(16) $|\operatorname{Im} g_n(z)| = |v(e^{i/n}z, z)| \leq M$ for $z \in \mathbb{D}, \quad n = 1, 2, \ldots$.

We apply the Schwarz integral formula (3.1.2) to $g_n(rz)$. If we let $r \to 1-$ we obtain from (16) and Lebesgue's bounded convergence theorem that

$$g_n(z) = \operatorname{Re} g_n(0) + \frac{i}{2\pi} \int_0^{2\pi} \frac{e^{it} + z}{e^{it} - z} \operatorname{Im} g_n(e^{it}) \, dt$$

for $z \in \mathbb{D}$. We now let $n \to \infty$. Since $g_n(0) = \log f'(0)$ and $\operatorname{Im} g_n(e^{it}) \to \beta(t) - t - \pi/2$ by (4), we obtain that (12) holds for each $z \in \mathbb{D}$.

To prove (13) we may assume that $t = 0$. We use that

$$\arg(z - 1) - \frac{\pi}{2} = \int_0^{2\pi} p(z, e^{it}) \left[\arg(e^{it} - 1) - \frac{\pi}{2}\right] dt$$

where p is the Poisson kernel (3.1.4). Taking the imaginary part in (12) we see that, with $\beta(0) - \beta(0-) = \pi\sigma$,

$$\arg f'(z) + \sigma \left(\arg(z - 1) - \frac{\pi}{2}\right)$$
$$= \int_0^{2\pi} p(z, e^{it}) \left[\beta(t) - t - \frac{\pi}{2} + \sigma \left(\arg(e^{it} - 1) - \frac{\pi}{2}\right)\right] dt.$$

The function in the square bracket has the same limit $\beta(0) - \pi/2$ as $t \to 0+$ and as $t \to 0-$. Hence (13) is a consequence of (3.1.5).   □

Now we introduce a stronger condition (Paa31; Dur83, p. 269). The regulated domain G is of *bounded boundary rotation* if $\beta(t)$ has bounded variation, i.e. if

(17) $\displaystyle \int_0^{2\pi} |d\beta(t)| = \sup_{(t_\nu)} \sum_{\nu=1}^n |\beta(t_\nu) - \beta(t_{\nu-1})| < \infty$

for all partitions $0 = t_0 < t_1 < \cdots < t_n = 2\pi$. This definition does not depend on the parametrization of $C = \partial G$ because the value of (17) is not changed if we replace β by $\beta \circ \varphi$ with continuous increasing φ.

Corollary 3.16. *If f maps $\mathbb{D}$ conformally onto a domain G of bounded boundary rotation then*

$$(18) \qquad \log f'(z) = \log f'(0) - \frac{1}{\pi} \int_0^{2\pi} \log(1 - e^{-it}z)d\beta(t) \quad (z \in \mathbb{D}).$$

This is a Stieltjes integral. The branch of the logarithm is determined such that $|\arg(1 - e^{-it}z)| < \pi/2$ for $z \in \mathbb{D}$.

Proof. Since $(\zeta + z)/(\zeta - z) = 1 + 2\bar{\zeta}z/(1 - \bar{\zeta}z)$ for $|\zeta| = 1$ we see from (12) that

$$\log f'(z) = c + \frac{i}{\pi} \int_0^{2\pi} \frac{e^{-it}z}{1 - e^{-it}z} \left(\beta(t) - t - \frac{\pi}{2}\right) dt$$

with a complex constant c. Since $\beta(t) - t - \pi/2$ is 2π-periodic an integration by parts gives

$$\log f'(z) = c - \frac{1}{\pi} \int_0^{2\pi} \log(1 - e^{-it}z)d(\beta(t) - t)$$

and (18) follows because $\int \log(1 - e^{-it}z) dt = 0$; the constant c is determined by putting $z = 0$. $\qquad\qquad\square$

Every function of bounded variation can be written as

$$(19) \qquad\qquad \beta = \beta_{\text{jump}} + \beta_{\text{sing}} + \beta_{\text{abs}}$$

where β_{jump} is constant except for countably many jumps, β_{sing} is continuous and $\beta'_{\text{sing}} = 0$ for almost all t, and β_{abs} is absolutely continuous, i.e. the indefinite integral of its derivative. We shall now consider the case that there are only finitely many jumps and that $\beta_{\text{sing}} = 0$.

Corollary 3.17. *If β is absolutely continuous except for the jumps $\pi\sigma_k$ at t_k $(k = 1, \ldots, n)$ then*

$$(20) \quad f'(z) = f'(0) \prod_{k=1}^n (1 - e^{-it_k}z)^{-\sigma_k} \exp\left(-\frac{1}{\pi} \int_0^{2\pi} \log(1 - e^{-it}z)\beta'(t)\, dt\right).$$

Proof. In the decomposition (19) we have $d\beta_{\text{abs}}(t) = \beta'(t)dt$ except at the jumps t_k. Since

$$\frac{1}{\pi} \int_0^{2\pi} \log(1 - e^{-it}z)d\beta_{\text{jump}} = \sum_{k=1}^m \sigma_k \log(1 - e^{-it_k}z),$$

we obtain (20) from (18) by exponentiation. $\qquad\qquad\square$

This is the *generalized Schwarz-Christoffel formula*. The classical form
(3.1.1) holds for the special case that the domain is bounded by line segments
and thus $\beta'(t) = 0$ except at the corners. The difficulty in applying (20) is
that $\beta(t)$ is the forward tangent direction at $f(e^{it})$ and therefore refers to the
conformal parametrization which is not explicitly known if the boundary is
given geometrically.

If C has a finite corner of opening $\pi\alpha_k$ at $f(e^{it_k})$ then $\sigma_k = 1 - \alpha_k$ by (5);
if e^{it_k} corresponds to an infinite sector of opening $\pi\alpha_k$ then $\sigma_k = 1 + \alpha_k$ by
(6). The difficulty remains that the t_k are not explicitly known.

Exercises 3.5

1. Let G be a bounded regulated domain. Show that ∂G has finite length and that
   its arc length parametrization is differentiable with at most countably many exceptions.

2. If $f'(z) = (1 - z^6)^{1/2}(1 - z^2)^{-5/2}$ determine $f(\mathbb{D})$.

3. Let G be a bounded regulated domain without outward-pointing cusps. Show
   that f is Hölder-continuous in $\mathbb{D}$.

4. Let f map $\mathbb{D}$ conformally onto a domain containing ∞ such that $f(0) = \infty$.
   Generalize the concept "regulated" to this case and show that

$$\log[z^2 f'(z)] = \text{const} + \frac{i}{2\pi} \int_0^{2\pi} \frac{e^{it} + z}{e^{it} - z}(\beta(t) + t)\, dt.$$

5. Construct a Jordan curve that has forward and backward tangents everywhere
   but is not regulated.

3.6 Starlike and Close-to-Convex Domains

The domain $G \subset \mathbb{C}$ is called *convex* if $w, w' \in G \Rightarrow [w, w'] \subset G$, and the analytic function f is *convex* in $\mathbb{D}$ if it maps $\mathbb{D}$ conformally onto a convex domain.
Every convex domain is regulated and furthermore of bounded boundary rotation. The direction angle $\beta(t)$ of the forward (half-)tangent is increasing and
satisfies $\beta(2\pi) - \beta(0) = 2\pi$.

It follows from Corollary 3.16 that, if f is convex, then

$$(1) \quad 1 + z\frac{f''(z)}{f'(z)} = 1 + \frac{1}{2\pi}\int_0^{2\pi} \frac{2e^{-it}z}{1 - e^{-it}z}\, d\beta(t) = \frac{1}{2\pi}\int_0^{2\pi} \frac{e^{it} + z}{e^{it} - z}\, d\beta(t)$$

for $z \in \mathbb{D}$. Since $d\beta(t) \geq 0$ we conclude from (3.1.4) that

$$(2) \qquad \text{Re}\left[1 + z\frac{f''(z)}{f'(z)}\right] > 0 \quad \text{for} \quad z \in \mathbb{D}.$$

Conversely, if $f'(z) \neq 0$ and (2) holds then f is convex (Pom75, p. 44; Dur83,
p. 42) and (1) holds, furthermore (ShS69, Suf70)

$$(3) \qquad \operatorname{Re}\left(\frac{2zf'(z)}{f(z)-f(\zeta)} - \frac{z+\zeta}{z-\zeta}\right) \geq 0 \quad \text{for} \quad z,\zeta \in \mathbb{D}.$$

A deep property (RuShS73; Dur83, p. 247) is that the *Hadamard convolution*

$$(4) \qquad f * g(z) = \sum_{n=0}^{\infty} a_n b_n z^n \quad (z \in \mathbb{D})$$

of two convex (univalent) functions $f(z) = \sum a_n z^n$ and $g(z) = \sum b_n z^n$ is again convex. See e.g. Rus75, Rus78, ShS78 for further developments.

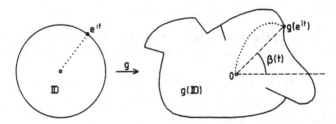

Fig. 3.10. A starlike domain

The domain G is called *starlike* with respect to 0 if $w \in G \Rightarrow [0, w] \subset G$, and the analytic function g is *starlike* in $\mathbb{D}$ if it maps $\mathbb{D}$ conformally onto a starlike domain such that $g(0) = 0$. This holds (Pom75, p. 42; Dur83, p. 41) if and only if $g'(0) \neq 0$ and

$$(5) \qquad \operatorname{Re}[zg'(z)/g(z)] > 0 \quad \text{for} \quad z \in \mathbb{D}.$$

Hence we see from (2) that

$$(6) \qquad f(z) \text{ convex} \quad \Leftrightarrow \quad g(z) = zf'(z) \text{ starlike}.$$

Theorem 3.18. *If g is starlike then the limits*

$$(7) \qquad \beta(t) = \lim_{r \to 1-} \arg g(re^{it}), \quad g(e^{it}) = \lim_{r \to 1-} g(re^{it}) \in \widehat{\mathbb{C}}$$

exist for all t and

$$(8) \qquad g(z) = zg'(0) \exp\left(-\frac{1}{\pi} \int_0^{2\pi} \log(1 - e^{-it}z)d\beta(t)\right) \quad (z \in \mathbb{D}).$$

Conversely, if $\beta(t)$ is increasing and $\beta(2\pi) - \beta(0) = 2\pi$ then (8) represents a starlike function.

The radial limit $g(e^{it})$ is an angular limit by Corollary 2.17. For the present case of starlike functions the approach may be rather tangential (Two86).

Proof. We apply Corollary 3.16 to the convex function f defined as in (6). We replace $\beta(t)$ by $\pi/2 + (\beta(t+) + \beta(t-))/2$ which does not change the value of the integral. Then (8) follows from (3.5.18), and (3.5.14) gives the first limit (7). The second limit then also exists because $|g(re^{it})|$ is increasing in r; indeed (5) shows that

(9) $\qquad \dfrac{\partial}{\partial r} \log|g(re^{it})| = \operatorname{Re}[e^{it} g'(re^{it})/g(re^{it})] > 0 \qquad \text{for} \quad 0 < r < 1.$

Conversely if g is given by (8) then, as in (1),

(10) $\qquad\qquad z\dfrac{g'(z)}{g(z)} = \dfrac{1}{2\pi} \int_0^{2\pi} \dfrac{e^{it} + z}{e^{it} - z} d\beta(t).$

Since $d\beta(t) \geq 0$ it follows from (3.1.4) that (5) holds so that g starlike. $\qquad \square$

As an example, let $z_k \in \mathbb{T}$, $\sigma_k > 0$ $(k = 1, \ldots, n)$ and $\sigma_1 + \cdots + \sigma_n = 2$. Then

(11) $\qquad\qquad g(z) = z \displaystyle\prod_{k=1}^m (1 - \bar{z}_k z)^{-\sigma_k} \quad (z \in \mathbb{D})$

is starlike and $g(\mathbb{D})$ is the plane slit along n rays $\{se^{i\theta_k} : \rho_k \leq s < \infty\}$ $(k = 1, \ldots, n)$ where $\theta_k - \theta_{k-1} = \pi\sigma_k$. This is a consequence of the following result (ShS70).

Proposition 3.19. *Let g be starlike and $0 \leq \vartheta \leq 2\pi$. If $\beta(\vartheta+) - \beta(\vartheta-) = \pi\sigma > 0$ then*

(12) $\qquad |g(re^{i\vartheta})| \geq r|g'(0)| 2^{-2+\sigma} (1 - r)^{-\sigma} \qquad \text{for} \quad 0 \leq r < 1,$

and $g(\mathbb{D})$ contains the infinite sector $\{\beta(\vartheta-) < \arg w < \beta(\vartheta+)\}$ which cannot be replaced by a larger infinite sector.

Proof. We can write $\beta = \beta_0 + \beta_1$, where β_0 is constant except for a jump of height $\pi\sigma$ at ϑ and where β_1 is increasing. Hence we see from (8) that

$$|g(re^{i\vartheta})| = r|g'(0)|(1 - r)^{-\sigma} \exp\left(-\frac{1}{\pi} \int_0^{2\pi} \log|1 - e^{-it+i\vartheta} r| d\beta_1(t)\right)$$

which implies (12) because $\log|\ldots| \leq \log 2$ and $\beta_1(2\pi) - \beta_1(0) = 2\pi - \pi\sigma$.

Now suppose that the above sector does not lie in $g(\mathbb{D})$. Then there exists $w \in \mathbb{C}$ with $\beta(\vartheta-) < \arg w < \beta(\vartheta+)$ such that $[0, w) \subset g(\mathbb{D})$ but $w \in \partial g(\mathbb{D})$. Hence we see from Corollary 2.17 that $w = g(e^{it})$ for some t which is $\neq \vartheta$ by (12). But $\beta(t) = \arg g(e^{it})$ by (7), and this is $\leq \beta(\vartheta-)$ for $t < \vartheta$ and $\geq \beta(\vartheta+)$ for $t > \vartheta$ which is a contradiction.

On the other hand there are points $t_n^\pm \to \vartheta\pm$ $(n \to \infty)$ such that $f(e^{it_n^\pm}) \neq \infty$. Since $\arg f(e^{it_n^\pm}) \to \beta(\vartheta\pm)$ it follows that $g(\mathbb{D})$ contains no larger sector $\{\alpha^- < \arg w < \alpha^+\}$ with either $\alpha^- < \beta(\vartheta-)$ or $\alpha^+ > \beta(\vartheta+)$. $\qquad \square$

The height $\pi\alpha$ of the largest jump of $\beta(t)$ determines the growth of g. If $M(r) = \max\{|g(z)| : |z| = r\}$ then (Pom62, Pom63, ShS70)

$$(13) \qquad \frac{\log M(r)}{\log 1/(1-r)} \to \alpha, \quad (1-r)\frac{M'(r)}{M(r)} \to \alpha \quad \text{as} \quad r \to 1 - .$$

It follows from Proposition 3.19 that β is continuous, i.e. $\alpha = 0$, if and only if $g(\mathbb{D})$ contains no infinite sector. In Theorem 6.27 we shall give a geometric characterization for the absolute continuity of β.

The function f is called *close-to-convex* (Kap52) in $\mathbb{D}$ if it is analytic and if there exists a convex (univalent) function h such that

$$(14) \qquad \operatorname{Re}[f'(z)/h'(z)] > 0 \quad \text{for} \quad z \in \mathbb{D};$$

by (6) this is equivalent to the existence of a starlike function g such that $\operatorname{Re}[zf'(z)/g(z)] > 0$. Hence (5) shows that every starlike function is close-to-convex.

As an example, consider the convex function $h(z) = \log[(1+z)/(1-z)]$ that maps $\mathbb{D}$ onto a horizontal strip. Then (14) is equivalent to

$$(15) \qquad \operatorname{Re}[(1-z^2)f'(z)] > 0 \quad \text{for} \quad z \in \mathbb{D}.$$

Proposition 3.20. *Every close-to-convex function f is univalent in $\mathbb{D}$, and if (14) holds with the convex function h then*

$$(16) \qquad \operatorname{Re}\frac{f(z_1) - f(z_2)}{h(z_1) - h(z_2)} > 0 \quad \text{for} \quad z_1, z_2 \in \mathbb{D}.$$

Proof. The function $\varphi = f \circ h^{-1}$ is analytic in the convex domain $H = h(\mathbb{D})$. Hence φ is univalent in H by Proposition 1.10 and it follows that $f = \varphi \circ h$ is univalent in $\mathbb{D}$. The inequality (16) follows from (1.5.4) applied to φ. □

Let f be close-to-convex in $\mathbb{D}$ and let (14) be satisfied where h is convex. We denote by $\beta(t)$ the tangent direction angle of the convex curve $\partial h(\mathbb{D})$; see (3.5.4). We need the concept of a prime end and its impression defined by (2.5.1). Let $I(t)$ denote the impression of the prime end $\widehat{f}(e^{it})$; see (2.4.10).

Theorem 3.21. *Let $G = f(\mathbb{D})$ and $0 \le t < 2\pi$. If $I(t)$ contains finite points then there exists $b(t) \in I(t)$ such that*

$$(17) \quad S(t) = \left\{ w : \left| \arg(w - b(t)) - \frac{\beta(t+) + \beta(t-)}{2} + \frac{\pi}{2} \right| \le \frac{\beta(t+) - \beta(t-)}{2} \right\}$$

lies in $\mathbb{C} \setminus G$ and contains $I(t)$. The radial limit $f(e^{it}) \in \widehat{\mathbb{C}}$ exists unless $\beta(t+) - \beta(t-) = \pi$.

If the increasing function β is continuous at t then $S(t)$ is the halfline $b(t) - ise^{i\beta(t)}$, $0 \leq s < \infty$; if β has a jump at t then $S(t)$ is an infinite sector.

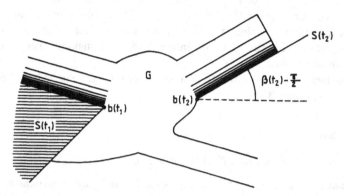

Fig. 3.11. A close-to-convex domain with two non-degenerate prime ends. The function β has a jump at t_1 but not at t_2

Proof. Let $b(t)$ be a principal point of $\widehat{f}(e^{it})$. By definition there are crosscuts Q_n of $\mathbb{D}$ separating 0 from e^{it} such that $\operatorname{diam} Q_n \to 0$ and $f(Q_n) \to b(t)$ as $n \to \infty$. We assume that $\beta(t+) > \beta(t-)$; the other case is simpler. If $\beta(t-) \leq \theta \leq \beta(t+) = \beta(t)$ then we can find $z_n \in Q_n$ with

$$\arg(z_n - e^{it}) \to \pi(\beta(t) - \theta)/(\beta(t) - \beta(t-)) + t + \pi/2 \quad \text{as} \quad n \to \infty$$

and thus $\arg z_n h'(z_n) \to \theta - \pi/2$ by Theorem 3.15. Furthermore it follows from (16) and (3) that

$$\left| \arg \frac{f(z) - f(z_n)}{h(z) - h(z_n)} \right| < \frac{\pi}{2}, \quad -\frac{3\pi}{2} < \arg \frac{z_n h'(z_n)}{h(z) - h(z_n)} < -\frac{\pi}{2}$$

for $|z| < |z_n| < 1$ and thus $0 < \arg[f(z) - f(z_n)] - \arg z_n h'(z_n) < 2\pi$. Since $f(z_n) \in f(Q_n)$ and therefore $f(z_n) \to b(t)$ as $n \to \infty$, we conclude that, for all $z \in \mathbb{D}$,

$$0 \leq \arg(f(z) - b(t)) - \theta + \pi/2 \leq 2\pi \quad \text{for} \quad \beta(t-) \leq \theta \leq \beta(t+).$$

Equality cannot hold by the maximum principle and it follows from (17) that $S(t) \subset \mathbb{C} \setminus G$.

Let $e^{it_n^{\pm}}$ be the endpoints of Q_n. Since $S(t_n^{\pm}) \cap G = \emptyset$ we see that $f(\mathbb{D} \setminus Q_n)$ does not meet $S(t_n^+) \cup S(t_n^-) \cup f(Q_n)$. Since $\beta(t_n^{\pm}) \to \beta(t\pm)$ and $f(Q_n) \to b(t)$, it follows from (2.5.1) that $I(t) \subset S(t)$.

Finally if $\widehat{f}(e^{it})$ has a principal point $b^*(t) \neq b(t)$ then $b^*(t) \in I(t) \subset S(t)$ and thus $S^*(t) \subset S(t)$ by (17), hence $S^*(t) = S(t)$ by symmetry. This is possible only if $S(t)$ is a halfplane, i.e. if $\beta(t+) - \beta(t-) = \pi$. Hence the radial limit exists except possibly in this case; see Corollary 2.17. $\square$

It follows from Theorem 3.21 that at every point of ∂G there is a halfline that lies in $\mathbb{C} \setminus G$, and two such halflines do not properly cross each other. It is easy to fill in any rest of $\mathbb{C} \setminus G$ by non-intersecting halflines.

This gives one part of the following geometric characterization (Lew58, Lew60): A univalent function is close-to-convex if and only if the complement of the image domain is the union of halflines that are disjoint except possibly for their initial points.

For further results on close-to-convex functions, see e.g. Tho67, Scho75, Mak87 and Section 5.3.

Exercises 3.6

1. If f is convex show that $\mathrm{Re}[zf'(z)/f(z)] > \frac{1}{2}$ for $z \in \mathbb{D}$.

2. Show that $z(1 - 2az^2 + z^4)^{-1/2}$ $(a > 0)$ is starlike in $\mathbb{D}$ and determine the image domain.

3. Let g be starlike. Show that (Keo59)

$$\int_0^{2\pi} |g'(re^{it})|\, dt = O\left(M(r) \log \frac{1}{1-r}\right) \qquad \text{as} \quad r \to 1- \, .$$

4. Let a close-to-convex function map $\mathbb{D}$ onto the inner domain of the smooth Jordan curve C. Show that the tangent direction angle of C never turns back by more than π (Kap52).

5. Let f satisfy (15). Show that f is convex in the direction of the imaginary axis (Rob36), i.e. every vertical line intersects $f(\mathbb{D})$ in an interval or not at all. (The converse holds if $f(x)$ is real for real x.)

Chapter 4. Distortion

4.1 An Overview

Let f map $\mathbb{D}$ conformally onto $G \subset \mathbb{C}$. We shall study the behaviour of f' for general domains G. The tangent angle is related to $\arg f'$ whereas $|f'|$ describes how sets are compressed or expanded. The results of this chapter will be often used later on.

The functions $g = \log f'$ and $g = \log(f - a)$ $(a \notin G)$ are Bloch functions, i.e. satisfy

$$\|g\|_{\mathcal{B}} = \sup_{|z|<1} \left(1 - |z|^2\right)|g'(z)| < \infty.$$

Bloch functions satisfy a non-trivial maximum (and minimum) principle; see Theorem 4.2. This is actually true for the wider class of *normal functions* (LeVi57, see e.g. Pom75), i.e. meromorphic functions g in $\mathbb{D}$ with

$$(1) \qquad \sup_{|z|<1} \left(1 - |z|^2\right)g^{\#}(z) = \sup_{|z|<1} \frac{\left(1 - |z|^2\right)|g'(z)|}{1 + |g(z)|^2} < \infty.$$

Lehto-Virtanen Theorem. *For normal functions, every asymptotic value is an angular limit, i.e. if $g(z) \to a$ as $z \to \zeta \in \mathbb{T}$ along some curve then $g(z) \to a$ as $z \to \zeta$ in every Stolz angle. In particular, radial and angular limits are the same.*

While a conformal map has an angular limit $f(\zeta)$ at almost all $\zeta \in \mathbb{T}$ the angular derivative

$$(2) \qquad f'(\zeta) = \lim_{z \to \zeta, z \in \Delta} \frac{f(z) - f(\zeta)}{z - \zeta}, \quad \Delta \text{ Stolz angle at } \zeta \in \mathbb{T}$$

may exist almost nowhere. The angular derivative is particularly useful if it is finite. Then it is equal to the angular limit of f'. If it is moreover nonzero then f is angle-preserving at ζ.

Now let φ be analytic in $\mathbb{D}$ and $\varphi(0) = 0$, $\varphi(\mathbb{D}) \subset \mathbb{D}$. Then φ is expanding on the parts of $\mathbb{T}$ that are mapped to $\mathbb{T}$:

Julia-Wolff-Lemma. *If $\zeta \in \mathbb{T}$ and $\varphi(\zeta) \in \mathbb{T}$ then $\varphi'(\zeta)$ exists and satisfies $1 \leq |\varphi'(\zeta)| \leq \infty$.*

Löwner's Lemma. *If $A \subset \mathbb{T}$ and $\varphi(A) \subset \mathbb{T}$ then $\Lambda(A) \leq \Lambda(\varphi(A))$.*

These results lead to monotonicity properties of the angular derivative (Theorem 4.14) and the harmonic measure (Corollary 4.16).

The harmonic measure $\omega(z, A)$ of a set $A \subset \mathbb{T}$ at $z \in \mathbb{D}$ is the total non-euclidean angle under which A is seen from z normalized such that $\omega(z, \mathbb{T}) = 1$; see Fig. 4.7. While we use it mainly in order to have an elegant conformally invariant formulation, the true power of harmonic measure appears only for multiply connected domains.

A useful fact about conformal maps f is (Corollary 4.18) that every point $z \in \mathbb{D}$ can be connected to the arc $I \subset \mathbb{T}$ by a curve S such that

$$(3) \qquad \Lambda(f(S)) \leq K(\omega)\,\mathrm{dist}(f(z), \partial G) \leq K(\omega)\big(1 - |z|^2\big)|f'(z)|\,,$$

where the constant $K(\omega)$ depends only on the harmonic measure $\omega(z, I)$ and Λ denotes again the length.

Gehring-Hayman Theorem. *If C is a curve in $\mathbb{D}$ and if S is the non-euclidean segment connecting the endpoints of C then*

$$\Lambda(f(S)) \leq K\Lambda(f(C))\,, \quad K \text{ absolute constant}.$$

Thus the hyperbolic (non-euclidean) geodesic between two points of G is not much longer in the euclidean sense than any curve in G between these two points. We shall discuss these geodesics in Section 4.6 and compare the (analytical) hyperbolic metric with the (geometrical) quasihyperbolic metric.

4.2 Bloch Functions

The function g is called a *Bloch function* if it is analytic in $\mathbb{D}$ and if

$$(1) \qquad \|g\|_{\mathcal{B}} = \sup_{z \in \mathbb{D}}\big(1 - |z|^2\big)|g'(z)| < \infty\,.$$

This defines a seminorm, and the Bloch functions form a complex Banach space $\mathcal{B}$ with the norm $|g(0)| + \|g\|_{\mathcal{B}}$, see AnClPo74.

It follows from (1.2.4) that

$$(2) \qquad \left|\big(1 - |z|^2\big)\frac{d}{dz}g(\tau(z))\right| = \big(1 - |\tau(z)|^2\big)|g'(\tau(z))| \quad (z \in \mathbb{D})$$

for $\tau \in \mathrm{Möb}(\mathbb{D})$. Hence $\|g\|_{\mathcal{B}}$ is conformally invariant, i.e.

$$(3) \qquad \|g \circ \tau\|_{\mathcal{B}} = \|g\|_{\mathcal{B}} \quad \text{for} \quad \tau \in \mathrm{Möb}(\mathbb{D})\,.$$

Furthermore (1) shows that

$$|g(z) - g(0)| = \left| z \int_0^1 g'(rz)\, dr \right| \le \|g\|_B \int_0^1 \frac{|z|}{1 - r^2 |z|^2}\, dr$$

for $z \in \mathbb{D}$ and thus

(4) $$|g(z) - g(0)| \le \frac{1}{2}\|g\|_B \log \frac{1 + |z|}{1 - |z|} = \|g\|_B \lambda(z, 0)$$

where λ is the non-euclidean distance (1.2.8). Using the identity (3) with $\tau(z) = (z + z_1)/(1 + \bar{z}_1 z)$ we can write (4) in the invariant form

(5) $$|g(z_1) - g(z_2)| \le \|g\|_B \lambda(z_1, z_2) \quad \text{for} \quad z_1, z_2 \in \mathbb{D}.$$

Every bounded analytic function belongs to $\mathcal{B}$. Indeed if $|g(z)| \le 1$ for $z \in \mathbb{D}$ then $(1 - |z|^2)|g'(z)| \le 1 - |g(z)|^2 \le 1$. An example of an unbounded Bloch function is $g(z) = \log[(1+z)/(1-z)]$. It satisfies $\|g\|_B = 2$ and equality holds in (4) for $z > 0$.

All Bloch functions are normal, i. e. satisfy (4.1.1). Furthermore all analytic normal functions belong to the MacLane class $\mathcal{A}$, see MacL63; Hay89, p. 785.

There is a close connection between Bloch functions and univalent functions, in particular their derivatives.

Proposition 4.1. *If f maps $\mathbb{D}$ conformally into $\mathbb{C}$ then*

(6) $$\| \log(f - a) \|_B \le 4 \quad \text{for} \quad a \notin f(\mathbb{D}),$$

(7) $$\| \log f' \|_B \le 6.$$

Conversely if $\|g\|_B \le 1$ then $g = \log f'$ for some conformal map f.

Proof. It follows from Proposition 1.2 that, for $z \in \mathbb{D}$,

$$\left(1 - |z|^2\right) \left| \frac{d}{dz} \log f'(z) \right| = \left(1 - |z|^2\right) \left| \frac{f''(z)}{f'(z)} \right| \le 4 + 2|z| < 6$$

and since $|f(z) - a| \ge d_f(z)$ we see from Corollary 1.4 that

$$\left(1 - |z|^2\right) \left| \frac{d}{dz} \log(f(z) - a) \right| = \frac{(1 - |z|^2)|f'(z)|}{|f(z) - a|} \le 4.$$

In the opposite direction, define f by $f(0) = 0$ and $g = \log f'$. If $\|g\|_B \le 1$ then

$$\left(1 - |z|^2\right) \left| z \frac{f''(z)}{f'(z)} \right| \le \left(1 - |z|^2\right)|g'(z)| \le 1 \quad \text{for} \quad z \in \mathbb{D}$$

so that f is univalent by Theorem 1.11. $\qquad\square$

Theorem 4.2. *Let C be a circle or line with $C \cap \mathbb{T} \ne \emptyset$ and let G be a domain with*

(8) $$G \subset \mathbb{D} \setminus C, \quad A = \partial G \setminus C \subset \mathbb{D}.$$

If $g \in \mathcal{B}$ and

(9) $$c = \frac{\gamma}{2 \sin \gamma} \|g\|_{\mathcal{B}} \quad (0 \le \gamma < \pi)$$

where γ is the angle between C and $\mathbb{T}$ towards G, then

(i) $|g(z)| \le a \le c$ for $z \in A$ $\Rightarrow$ $|g(z)| \le \dfrac{2c}{\log(c/a) + 1}$ for $z \in G$,

(ii) $|g(z)| \ge a \ge ec$ for $z \in A$ $\Rightarrow$ $|g(z)| \ge e^{-1}a$ for $z \in G$,

(iii) $\operatorname{dist}(g(z), g(A)) \le ec$ for $z \in G$.

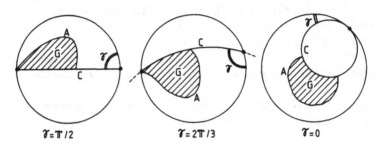

$\gamma = \mathbb{T}/2$ $\gamma = 2\mathbb{T}/3$ $\gamma = 0$

Fig. 4.1. The situation of Theorem 4.2; the domain G lies on one side of the circle C

The important point is that we have to know a bound for $|g(z)|$ only on $A = \partial G \setminus C$ and not on C. Thus (i) is a non-linear *maximum principle* for Bloch functions while (ii) is a *minimum principle*. Both are consequences of the Lehto-Virtanen maximum principle (LeVi57; Pom75, p. 263) for normal functions; see AnClPo74 for (iii).

We must have some restriction on a in (ii) because $|g(z)| \ge e^{-1}a$ excludes any zeros in G. The proof will show that, in (9), we may replace $\|g\|$ by $\sup\{(1 - |z|^2)|g'(z)| : z \in G\}$.

If $\gamma = 0$ then C touches $\mathbb{T}$ and $c = (1/2)\|g\|$; the case $\gamma = \pi$ is not allowed. Since $1 \le \pi \gamma^{-1}(\pi - \gamma)^{-1} \sin \gamma \le 4/\pi$ we see from (9) that

(10) $$c \le \frac{\pi}{2(\pi - \gamma)} \|g\|_{\mathcal{B}} \le \frac{4}{\pi} c.$$

Proof. (a) We first study the case that $\gamma > 0$ and $\partial G \subset \mathbb{D}$. In order to establish (i) and (ii) we consider the increasing function

(11) $$\varphi(x) = x \log \frac{x}{a} \quad \text{for} \quad e^{-1}a \le x < +\infty$$

which satisfies $\varphi(e^{-1}a) = -e^{-1}a$, $\varphi(a) = 0$ and $\varphi(+\infty) = +\infty$. Hence, for each $a > 0$, there exists $b > a$ such that $\varphi(b) = c$, and if $c \le e^{-1}a$ then there exists b^* with $e^{-1}a \le b^* < a$ such that $\varphi(b^*) = -c$.

We may assume that $\pm i \in C$ and that G lies to the right of C; otherwise we replace g by $g \circ \tau$ with suitable $\tau \in \text{Möb}(\mathbb{D})$ which does not change the angles, the bounds and the seminorm, by (3). For $0 < \gamma' < \gamma$ let G' denote the open subset of G cut off by the circle C' through $\pm i$ that forms the angle γ' with $\mathbb{T}$; see Fig. 4.2.

(i) Let $|g(z)| \leq a \leq c$ for $z \in A = \partial G \setminus C$. We claim that $|g(z)| \leq b$ for $z \in G$. Suppose this is false. Since $a < b$ and since g is continuous in $\overline{G} \subset \mathbb{D}$ there exists $\gamma' < \gamma$ such that

$$(12) \qquad |g(z)| \leq b = |g(x_0)| \quad (z \in G') \quad \text{with} \quad x_0 \in C' \cap \partial G',$$

and we may assume that $x_0 \in \mathbb{R}$; otherwise we consider $g \circ \sigma$ with suitable $\sigma \in \text{Möb}(\mathbb{D})$. We have $(x_0, x_0 + \delta) \subset G'$ for some $\delta > 0$ because otherwise $|g(x_0)| \leq a < b$. Let now $u(z) = \arg(i + z) - \arg(1 + iz)$ for $z \in \mathbb{D}$. Then
(13)

$$u(z) > 0 \; (z \in \mathbb{D}), \quad u(z) = \gamma' \; (z \in C'), \quad u(x) = \frac{\pi}{2} - 2\arctan x \; (x \in \mathbb{R}).$$

Thus $v = cu/(b\gamma') - \log|g|$ satisfies $v(z) \geq -\log a$ for $z \in \partial G' \setminus C' \subset \partial G \setminus C$ and furthermore, by (12),

$$v(z) \geq \frac{c}{b} - \log b = -\log a \quad \text{for} \quad z \in \partial G' \cap C'$$

with equality for $z = x_0$. Since v is harmonic except for positive logarithmic poles, we conclude from the minimum principle that $v(z) \geq -\log a$ for $z \in G'$. Hence

$$v(x) \geq -\log a = v(x_0) \quad \text{for} \quad x_0 < x < x_0 + \delta$$

and therefore, by (1), (12) and (13),

$$0 \leq v'(x_0) = -\frac{2c/(b\gamma')}{1 + x_0^2} - \text{Re}\,\frac{g'(x_0)}{g(x_0)} \leq -\frac{2c/(b\gamma')}{1 + x_0^2} + \frac{\|g\|}{(1 - x_0^2)b}.$$

Hence we see from (9) and (13) that

$$\frac{\gamma\|g\|}{2\sin\gamma} = c \leq \frac{\gamma'\|g\|}{2}\frac{1 + x_0^2}{1 - x_0^2} = \frac{\gamma'\|g\|}{2\sin\gamma}$$

which contradicts $\gamma' < \gamma$.

Hence we have shown that $|g(z)| \leq b$ for $z \in G$. Since $c = \varphi(b) = b\log(b/a)$ by (11), we have

$$\frac{1}{b} = \frac{1}{c}\log\frac{b}{a} \geq \frac{1}{2c}\left[\log\frac{b}{a} + \log\log\frac{b}{a} + 1\right] = \frac{1}{2c}\left[\log\frac{c}{a} + 1\right]$$

and this proves (i); the right-hand side is > 0 because $a \leq c$.

(ii) Now let $|g(z)| \geq a \geq ec$ for $z \in \partial G \setminus C$. We claim that $|g(z)| \geq b^*$ for $z \in G$. Since $b^* < a$ there would otherwise exist $\gamma' < \gamma$ such that

$$|g(z)| \geq b^* = |g(x_0)| \quad (z \in G') \quad \text{with} \quad x_0 \in C' \cap \partial G'$$

where we may again assume that $x_0 \in \mathbb{R}$. Since $\varphi(b^*) = -c$ the function $v = cu/(b^*\gamma') + \log|g|$ satisfies $v(z) \geq \log a$ both for $z \in \partial G' \setminus C'$ and for $z \in C'$, hence for $z \in G'$ by the minimum principle. We conclude that $v'(x_0) \geq 0$ which leads to a contradiction as above. Hence we have proved that $|g(z)| \geq b^* \geq e^{-1}a$ for $z \in G$.

(iii) Let $z \in G$ and suppose that $\mathrm{dist}(g(z), g(A)) \geq ec$. Then $|g(\zeta) - g(z)| \geq ec$ for $\zeta \in A$ and thus $|g(\zeta) - g(z)| \geq c > 0$ for $\zeta \in G$ by (ii) which contradicts $z \in G$.

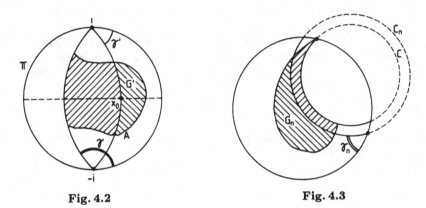

Fig. 4.2   **Fig. 4.3**

(b) We now turn to the general case; see Fig. 4.3. Let C_n be circles with $C_n \cap G \neq \emptyset$ and $C_n \cap C = \emptyset$ such that $C_n \to C(n \to \infty)$ and let G_n be the open subsets of G cut off by C_n. Then $\partial G_n \subset \mathbb{D}$ and $\gamma_n > 0$ so that we can apply part (a) to G_n, and our assertions follow for $n \to \infty$ because $\gamma_n \to \gamma.\square$

Theorem 4.3. *Let Γ be a Jordan arc in $\mathbb{D}$ ending at $\zeta \in \mathbb{T}$. If $g \in \mathcal{B}$ and if*

$$(14) \qquad g(z) \to b \in \widehat{\mathbb{C}} \quad as \quad z \to \zeta, \quad z \in \Gamma$$

then g has the angular limit b at ζ.

This result (LeVi57) can be expressed as follows. For Bloch functions (and more generally for normal functions), every asymptotic value is an angular limit; we say that g has the *asymptotic value* b at ζ if (14) holds for some curve Γ. The proof will show that we have only to assume that Γ is a halfopen Jordan arc and that $\zeta \in \overline{\Gamma} \setminus \Gamma \subset \mathbb{T}$.

Proof. Let Δ be a Stolz angle at ζ. We first consider the case that $b \neq \infty$. Let $3\pi/4 < \gamma < \pi$ and define c by (9). Let Γ^* be a subarc of Γ ending at ζ such that

$$(15) \qquad |g(z) - b| < \varepsilon < c \quad for \quad z \in \Gamma^*.$$

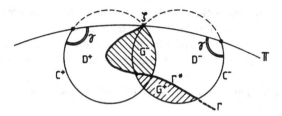

Fig. 4.4

We construct disks $D^\pm$ as in Fig. 4.4 such that $C^\pm = \partial D^\pm$ intersects $\mathbb{T}$ at ζ under the angle γ and $\Gamma^* \setminus (D^+ \cup D^-) \neq \emptyset$. If $\delta > 0$ and $\pi - \gamma$ are sufficiently small then $\Delta^* = \{z \in \Delta : |z - \zeta| < \delta\} \subset D^+ \cap D^-$.

Let $V^\pm$ be the component of $D^\pm \setminus \Gamma^*$ that contains points of $\mathbb{T}$ and let $G^\mp = D^\pm \setminus (V^\pm \cup \Gamma^*)$. Since Γ^* connects $\{\zeta\}$ and $\mathbb{D} \cap (C^+ \cup C^-)$ without meeting $V^+ \cup V^-$, it follows that V^+ and V^- are disjoint. Hence

$$(16) \qquad\qquad \Delta^* \subset D^+ \cap D^- \subset G^+ \cup G^- \cup \Gamma^*.$$

We see from (15) and Theorem 4.2 (i) that

$$|g(z) - b| < 2c/[\log(c/\varepsilon) + 1]$$

for $z \in G^\pm$ and thus for $z \in \Delta^*$ by (16). Since the right-hand side tends $\to 0$ as $\varepsilon \to 0$ we conclude that g has the angular limit b at ζ. The case $b = \infty$ is handled the same way using now (ii) instead of (i). □

In a similar way, we obtain (AnClPo74) from Theorem 4.2 (iii):

Proposition 4.4. *Let Γ be a Jordan arc in $\mathbb{D}$ ending at $\zeta \in \mathbb{T}$ and let $g \in \mathcal{B}$. For $0 < \alpha < \pi/2$ there exists $K = K(\alpha)$ such that*

$$(17) \qquad\qquad \operatorname{dist}(g(z), g(\Gamma)) \leq K\|g\|_{\mathcal{B}} \quad for \quad z \in \Delta,$$

where $\Delta = \{|\arg(1 - \bar\zeta z)| < \alpha, |z - \zeta| < \rho\}$ with suitable $\rho > 0$.

In particular it follows for $g \in \mathcal{B}$ that

$$(18) \qquad\qquad |g| \text{ bounded on } \Gamma \quad \Rightarrow \quad |g| \text{ bounded on } \Delta;$$

$$(19) \qquad \operatorname{Re} g \to \pm\infty \text{ on } \Gamma \quad \Rightarrow \quad \operatorname{Re} g \text{ has the angular limit } \pm\infty \text{ at } \zeta.$$

Corollary 4.5. *Let h be analytic and nonzero in $\mathbb{D}$. If $\log h \in \mathcal{B}$ and if h has the asymptotic value $b \in \widehat{\mathbb{C}}$ at $\zeta \in \mathbb{T}$, then h has the angular limit b at ζ.*

Proof. This follows at once from Theorem 4.3 if $b \neq 0, \infty$. If $b = 0$ then $\operatorname{Re} \log h$ has the asymptotic value $-\infty$ at ζ and thus the angular limit $-\infty$ by (19). It follows that h has the angular limit 0 at ζ. The case $b = \infty$ is handled similarly. □

Another way to prove this would be to show that h is normal and then appeal to the normal function version of Theorem 4.3. The next result implies that every Bloch function is radially bounded on a dense subset of $\mathbb{T}$; see StSt81. In fact much more is true as Makarov has proved; see (10.4.12).

Proposition 4.6. *Let I be an arc of $\mathbb{T}$ and let $g \in \mathcal{B}$. Then there exists $\zeta \in I$ such that*

(20) $$|g(r\zeta) - g(0)| \le 22\|g\|_\mathcal{B}/\Lambda(I) \quad \text{for} \quad 0 \le r < 1.$$

Proof (see Fig. 4.5). Let $\vartheta = \Lambda(I)/2$. We may assume that $g(0) = 0$, $\|g\|_\mathcal{B} = 1$ and that I is the arc of $\mathbb{T}$ between $e^{-i\vartheta}$ and $e^{i\vartheta}$. The circle C through $e^{i\vartheta}$, $e^{-i\vartheta}$ and 0 forms the inner angle $\gamma = \pi - \vartheta$ with $\mathbb{T}$ at $e^{\pm i\vartheta}$. Let G be the component of

$$\{z \text{ inside } C : |g(z)| < a\}, \quad a = (e\pi)/(2\vartheta)$$

with $0 \in \partial G$. We claim that $I \cap \partial G \ne \emptyset$. Since $a \ge ec$ by (10), it would otherwise follow from Theorem 4.2 (ii) that $0 = |g(0)| \ge e^{-1}a > 0$. Thus there exist $z_n \in G$ with $z_n \to \zeta \in I \cap \partial G$ as $n \to \infty$. Since 0 and z_n can be connected by a curve $P_n \subset G$ it follows from Theorem 4.2 (iii) with $\gamma = \pi/2$ (hence $c = \pi/4$ by (9)) that

$$|g(rz_n)| \le \frac{e\pi}{2\vartheta} + \frac{e\pi}{4} < \frac{11}{\vartheta} \quad \text{for} \quad 0 \le r \le 1$$

and (20) follows for $n \to \infty$.   □

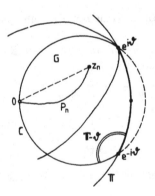

Fig. 4.5

The name "Bloch function" derives from the close connection to the Bloch constant problem. Let $d_g(z)$ denote the radius of the largest unramified disk around $g(z)$ on the Riemann image surface $g(\mathbb{D})$. Then (Ahl38; Gol57, p. 320)

(21) $$d_g(z) \le (1 - |z|^2)|g'(z)| \le 2(3M - d_g(z))\sqrt{\frac{d_g(z)}{3M}} \le \frac{4M}{3}\sqrt{3} \quad \text{for } z \in \mathbb{D}$$

if $M = \sup\{d_g(\zeta) : \zeta \in \mathbb{D}\} < \infty$; see also Bon90.

The space $\mathcal{B}_0$ consists of all Bloch functions g such that

(22) $$\left(1 - |z|^2\right)|g'(z)| \to 0 \quad \text{as} \quad |z| \to 1-.$$

For example all functions that are analytic in $\mathbb{D}$ and continuous in $\mathbb{D} \cup \mathbb{T}$ belong to $\mathcal{B}_0$. See e.g. AnClPo74, ArFiPe85, Bis90 and Section 11.2 for a discussion of the "little Bloch space" $\mathcal{B}_0$.

Exercises 4.2

1. Let $f \in S$. Show that $g(z) = \log(f(z)/z)$ is a Bloch function.

2. Let f map $\mathbb{D}$ conformally into $\mathbb{C}$. Show that $f \in \mathcal{B}$ if and only if $f(\mathbb{D})$ contains no arbitrarily large disks.

3. Let $g \in \mathcal{B}$ and $g(0) = 0$, $\|g\|_{\mathcal{B}} = 1$. Show that $\mathbb{T} \cap \partial H \neq \emptyset$ where H is the component of $\{z \in \mathbb{D} : |g(z)| < r\}$ ($r > e/2$) that contains 0 (StSt81, GnHPo86).

4. Let g be a nonconstant Bloch function and $g(z) \to b$ as $|z| \to 1$, $z \in \Gamma$ where Γ is a halfopen Jordan arc in $\mathbb{D}$ with $\overline{\Gamma} \setminus \Gamma \subset \mathbb{T}$. Show that $\overline{\Gamma} = \Gamma \cup \{\zeta\}$ for some $\zeta \in \mathbb{T}$.

5. Let f map $\mathbb{D}$ conformally into $\mathbb{C}$ and suppose that $|\arg f'|$ is bounded on some curve ending at $\zeta \in \mathbb{T}$. Show that $|\arg f'(r\zeta)|$ is bounded for $0 < r < 1$.

6. Let g be analytic and let $(1 - |z|)|\operatorname{Im}[zg'(z)]|$ be bounded in $\mathbb{D}$. Show that $\left(1 - |z|\right)^2 |g''(z)|$ is bounded and deduce that $g \in \mathcal{B}$.

4.3 The Angular Derivative

Let f be analytic in $\mathbb{D}$. We say that f has the *angular derivative* $f'(\zeta)$ at $\zeta \in \mathbb{T}$ if the angular limit $f(\zeta) \neq \infty$ exists and if

(1) $$\frac{f(z) - f(\zeta)}{z - \zeta} \to f'(\zeta) \quad \text{as} \quad z \to \zeta, \quad z \in \Delta$$

for each Stolz angle Δ at ζ. We allow that $f'(\zeta) = \infty$ though this case has properties different from those of the finite case. For the wide class of John domains we shall show (Theorem 5.5) that if $f'(\zeta) \neq \infty$ exists then (1) holds with $\mathbb{D}$ instead of Δ.

Proposition 4.7. *The analytic function f has a finite angular derivative $f'(\zeta)$ if and only if f' has the finite angular limit $f'(\zeta)$ at ζ.*

The proof uses the following auxiliary result that we formulate with a weaker hypothesis in view of a later application. We omit the proof which either uses the Schwarz integral formula (3.1.2) or normal families as in Pom75, p. 306.

Proposition 4.8. *Let h be analytic in $\mathbb{D}$. If $\operatorname{Im} h(z)$ has a finite angular limit at $\zeta \in \mathbb{T}$ then $(z - \zeta)h'(z)$ has the angular limit 0 at ζ.*

Proof of Proposition 4.7. Let (1) hold. Proposition 4.8 applied to $h(z) = (f(z) - f(\zeta))/(z - \zeta)$ shows that

$$f'(z) - \frac{f(z) - f(\zeta)}{z - \zeta} = (z - \zeta)h'(z) \to 0 \quad \text{as} \quad z \to \zeta, \quad z \in \Delta$$

so that $f'(z) \to f'(\zeta)$ again by (1).

Conversely suppose that f' has the finite angular limit $f'(\zeta)$. It follows that

$$f(z) = f(0) + \int_0^z f'(s)\,ds$$

has a finite angular limit $f(\zeta)$ and that

$$(2) \quad \frac{f(z) - f(\zeta)}{z - \zeta} = \int_0^1 f'(t\zeta + (1 - t)z)\,dt \to f'(\zeta) \quad \text{as} \quad z \to \zeta, \quad z \in \Delta. \;\square$$

We now turn to univalent functions. The next result follows at once from Proposition 4.1 and Corollary 4.5.

Proposition 4.9. *Let f map $\mathbb{D}$ conformally into $\mathbb{C}$ and let f have an angular limit $f(\zeta) \neq \infty$ at $\zeta \in \mathbb{T}$. If*

$$(3) \quad \frac{f(z) - f(\zeta)}{z - \zeta} \to a \quad \text{as} \quad z \to \zeta, \quad z \in \Gamma$$

for some curve Γ in $\mathbb{D}$ ending at ζ, or if

$$(4) \quad f'(z) \to a \neq \infty \quad \text{as} \quad z \to \zeta, \quad z \in \Gamma,$$

then $f'(\zeta) = a$ exists.

We say that f is *conformal* at $\zeta \in \mathbb{T}$ if the angular derivative $f'(\zeta)$ exists and is $\neq 0, \infty$. We say that f is *isogonal* at ζ if there is a finite angular limit $f(\zeta)$ and if

$$(5) \quad \arg \frac{f(z) - f(\zeta)}{z - \zeta} \to \gamma \quad \text{as} \quad z \to \zeta, \quad z \in \Delta$$

for some finite γ and every Stolz angle Δ. It is clear that

$$(6) \quad f \text{ conformal at } \zeta \quad \Rightarrow \quad f \text{ isogonal at } \zeta.$$

We have seen in Section 3.2 that the converse does not hold. Ostrowski (Ost36, Fer42) has given a geometric characterization of isogonality; see Theorem 11.6. We shall study conditions for conformality in Section 11.4.

If $\partial f(\mathbb{D})$ is locally connected then, by Theorem 3.7 with $\alpha = 1$,

(7) $\qquad\qquad \partial G$ has a tangent at $f(\zeta) \quad \Rightarrow \quad f$ isogonal at ζ.

The converse holds for John domains (Theorem 5.5) and thus for quasidisks (see Section 5.4).

The following property justifies the name isogonal (= angle-preserving).

Proposition 4.10. *If the conformal map f is isogonal at $\zeta \in \mathbb{T}$ then smooth curves in a Stolz angle at ζ are mapped onto smooth curves, and the angles between curves are preserved.*

Proof. Let $\Gamma_j : z_j(\tau)$, $0 \le \tau \le 1$ for $j = 1, 2$ be two smooth curves in Δ with $z_1(1) = z_2(1) = \zeta$ and consider the image curves $f(\Gamma_j) : w_j(\tau) = f(z_j(\tau))$, $0 \le \tau \le 1$. It follows from (5) that

$$\arg[w_j(1) - w_j(\tau)] = \arg \frac{f(\zeta) - f(z_j(\tau))}{\zeta - z_j(\tau)} + \arg[\zeta - z_j(\tau)]$$
$$\to \gamma + \arg z_j'(1) \qquad \text{as} \quad \tau \to 1 - .$$

Hence $f(\Gamma_j)$ has a tangent of direction angle $\gamma + \arg z_j'(1)$ at $f(\zeta)$ and it follows from (8) below that the direction angle of $f(\Gamma_j)$ depends continuously on $\tau \in [0, 1]$ so that $f(\Gamma_j)$ is smooth. Finally the angle between Γ_1 and Γ_2 at ζ is $\arg z_1'(1) - \arg z_2'(1)$ which is also the angle between $f(\Gamma_1)$ and $f(\Gamma_2)$. $\square$

Proposition 4.11. *The conformal map f is isogonal at $\zeta \in \mathbb{T}$ if and only if f and $\arg f'$ have finite angular limits at ζ. In this case*

(8) $\qquad \dfrac{(z - \zeta)f'(z)}{f(z) - f(\zeta)} \to 1 \quad$ *as* $\quad z \to \zeta \quad$ *in every Stolz angle.*

The Visser-Ostrowski condition (8) will be studied in Section 11.3; see Vis33 and WaGa55.

Proof. If f is isogonal at ζ it follows from (5) and from Proposition 4.8 applied to

(9) $\qquad\qquad h(z) = \log \dfrac{f(z) - f(\zeta)}{z - \zeta} \quad (z \in \mathbb{D})$

that (8) holds, and using again (5) we conclude that $\arg f'(z) \to \gamma$ in every Stolz angle.

Conversely suppose that

$$f(z) \to f(\zeta) \neq \infty, \quad \arg f'(z) \to \gamma \qquad \text{as} \quad z \to \zeta, \ z \in \Delta$$

for every Stolz angle Δ. Since

$$\frac{f(z) - f(\zeta)}{z - \zeta} e^{-i\gamma} = \int_0^1 |f'(t\zeta + (1-t)z)| e^{i(\arg f' - \gamma)}\, dt$$

and since $\arg f'(t\zeta + (1-t)z) \to \gamma$ uniformly in t as $z \to \zeta$, $z \in \Delta$ it follows that

$$\tan\left[\arg \frac{f(z) - f(\zeta)}{z - \zeta} - \gamma\right] = \int_0^1 |f'| \sin(\arg f' - \gamma)\, dt \Big/ \int_0^1 |f'| \cos(\arg f' - \gamma)\, dt$$

tends $\to 0$. Hence (5) holds. $\qquad\qquad\square$

Proposition 4.12. *There is a conformal map onto a Jordan domain (even a quasidisk) that is nowhere isogonal and thus nowhere conformal on* $\mathbb{T}$.

It follows from (6) that this conformal map has nowhere on $\mathbb{T}$ a finite nonzero angular derivative. It does however have infinite and zero angular derivatives at very many points (see Theorem 10.13). Furthermore (7) shows that the bounding Jordan curve has nowhere a tangent even though it may be a quasicircle (see Section 5.4). Most Julia sets (see e.g. Bla84, Lju86, Bea91) have this property. We postpone the proof to the end of Section 8.6.

The *Julia-Wolff lemma* deals with selfmaps of $\mathbb{D}$ that need not be univalent.

Proposition 4.13. *Let* φ *be analytic in* $\mathbb{D}$ *with an angular limit* $\varphi(\zeta)$ *at* $\zeta \in \mathbb{T}$. *If*

$$\varphi(\mathbb{D}) \subset \mathbb{D}, \quad \varphi(\zeta) \in \mathbb{T}$$

then the angular derivative $\varphi'(\zeta)$ *exists and*

$$(10) \qquad 0 < \zeta \frac{\varphi'(\zeta)}{\varphi(\zeta)} = \sup_{z \in \mathbb{D}} \frac{1 - |z|^2}{|\zeta - z|^2} \frac{|\varphi(\zeta) - \varphi(z)|^2}{1 - |\varphi(z)|^2} \le +\infty.$$

Proof. We may assume that $\zeta = \varphi(\zeta) = 1$; otherwise we consider $\varphi(\zeta z)/\varphi(\zeta)$. The function $h = (1 + \varphi)/(1 - \varphi)$ is analytic in $\mathbb{D}$ and satisfies

$$(11) \qquad \operatorname{Re} h(z) = \frac{1 - |\varphi(z)|^2}{|1 - \varphi(z)|^2} > 0 \quad \text{for} \quad z \in \mathbb{D}.$$

Let $z_n \in \mathbb{D}$ and $\tau_n(s) = (s + z_n)/(1 + \bar{z}_n s)$. Then

$$(12) \qquad \left| \frac{h(\tau_n(s)) - i \operatorname{Im} h(z_n)}{\operatorname{Re} h(z_n)} \right| \le \frac{1 + |s|}{1 - |s|} \le \frac{4}{1 - |s|^2} \quad \text{for} \quad s \in \mathbb{D}$$

because the left-hand function is analytic and of positive real part for $s \in \mathbb{D}$ and has the value 1 at $s = 0$.

We define

$$(13) \qquad c = \inf_{z \in \mathbb{D}} \frac{|1 - z|^2}{1 - |z|^2} \operatorname{Re} h(z)$$

and choose (z_n) such that this infimum is approached as $n \to \infty$. Let $0 < x < 1$. It follows from (12) with $s = \tau_n^{-1}(x)$ that

$$|h(x) - i \operatorname{Im} h(z_n)| \le \frac{4|1 - \bar{z}_n x|^2}{(1 - |z_n|^2)(1 - x^2)} \operatorname{Re} h(z_n).$$

Since $h(x) \to \infty$ as $x \to 1-$ we conclude that

$$(14) \qquad \limsup_{x \to 1-} \frac{1-x}{1+x} |h(x)| \le \frac{|1 - z_n|^2}{1 - |z_n|^2} \operatorname{Re} h(z_n), \quad \text{thus} \le c$$

by our choice of (z_n). Since $[(1 - x)/(1 + x)] \operatorname{Re} h(x) \ge c$ by (13), it follows from (14) that

$$\frac{1-x}{1+x} \frac{1 + \varphi(x)}{1 - \varphi(x)} = \frac{1-x}{1+x} h(x) \to c \quad \text{as} \quad x \to 1-.$$

Hence $(1 - \varphi(x))/(1 - x) \to 1/c$ because $\varphi(x) \to 1$.

It is easy to see that $\log[(1 - \varphi(z))/(1 - z)]$ is a Bloch function (with norm ≤ 4). Hence we see from Theorem 4.3 that $(1 - \varphi(z))/(1 - z)$ has the angular limit $1/c$ at 1 so that $\varphi'(1) = 1/c$, and (10) now follows from (11) and (13).$\Box$

We now prove a comparison theorem that goes back to Carathéodory and Lelong-Ferrand; see Fig. 4.6.

Theorem 4.14. *For $j = 1, 2$ let f_j map $\mathbb{D}$ conformally onto G_j where $G_1 \subset G_2$. Let Γ be a Jordan arc in G_1 ending at $\omega \in \partial G_1 \cap \partial G_2$ and let $\Gamma_j = f_j^{-1}(\Gamma)$ end at $\zeta_j \in \mathbb{T}$. If the angular derivative $f_1'(\zeta_1) \ne \infty$ exists then $f_2'(\zeta_2) \ne \infty$ exists and $|f_2'(\zeta_2)| = c|f_1'(\zeta_1)|$ with $0 \le c < \infty$.*

It follows from Proposition 2.14 that Γ_j is a Jordan arc ending on $\mathbb{T}$. If G_1 and G_2 are Jordan domains then the assumptions involving Γ can be replaced simply by $f_1(\zeta_1) = f_2(\zeta_2) = \omega$.

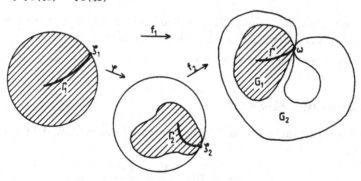

Fig. 4.6

Proof. The function $\varphi = f_2^{-1} \circ f_1$ is analytic in $\mathbb{D}$ and satisfies $\varphi(\mathbb{D}) \subset \mathbb{D}$ and $\varphi(\Gamma_1) = \Gamma_2$. Hence φ has the asymptotic value ζ_2 along Γ_1 by Proposition 2.14

and therefore the angular limit ζ_2 at ζ_1. Hence $\varphi'(\zeta_1)$ exists by the Julia-Wolff lemma (Proposition 4.13) and thus

$$\frac{f_2(\zeta_2) - f_2(\varphi(r\zeta_1))}{\zeta_2 - \varphi(r\zeta_1)} = \frac{\zeta_1 - r\zeta_1}{\varphi(\zeta_1) - \varphi(r\zeta_1)} \frac{f_1(\zeta_1) - f_1(r\zeta_1)}{\zeta_1 - r\zeta_1} \rightarrow \frac{f_1'(\zeta_1)}{\varphi'(\zeta_1)}$$

as $r \rightarrow 1-$. This means that $(f_2(\zeta_2) - f_2(z))/(\zeta_2 - z)$ has the asymptotic value $f_1'(\zeta_1)/\varphi'(\zeta_1)$ along the curve $\varphi(r\zeta_1)$, $0 \le r \le 1$ ending at ζ_2 and it follows from Proposition 4.1 and Corollary 4.5 that $f_2'(\zeta_2)$ exists and that $|f_2'(\zeta_2)| = c|f_1'(\zeta_1)|$ with $0 \le c = 1/|\varphi'(\zeta_1)| < +\infty$. $\square$

As an application, let J_1, J and J_2 be Jordan curves such that J_1 lies inside J and J lies inside J_2 except for the point $\omega \in J_1 \cap J \cap J_2$. Let f_1, f and f_2 map $\mathbb{D}$ onto the corresponding inner domains such that $f_1(\zeta) = f(\zeta) = f_2(\zeta) = \omega$. We suppose that the comparison curves J_1 and J_2 are Dini-smooth; see Section 3.3. Then it follows from Theorem 3.5 that $f_1'(\zeta)$ and $f_2'(\zeta)$ exist and are finite and nonzero. Hence we conclude from Theorem 4.14 that $f'(\zeta) \ne 0, \infty$ exist so that f is conformal at ζ. More general comparison curves can be obtained from the results in Section 11.4.

Exercises 4.3

Let f map $\mathbb{D}$ conformally onto G.

1. Let G be bounded by the (non-Jordan) curve C and let ζ, ζ' be distinct points on $\mathbb{T}$ where f is isogonal. If $f(\zeta) = f(\zeta') = \omega$ show that C has a tangent at ω.

2. Let $z_0 \in \mathbb{D}$ and let $f(\zeta)$ be the boundary point nearest to $f(z_0)$. Show that $f'(\zeta)$ exists and is finite but may be zero.

3. If G is bounded and convex show that $f'(\zeta) \ne 0$ exists for all $\zeta \in \mathbb{T}$.

4. Find $\zeta_n \in \mathbb{T}$ and $0 < r_n < 1$ such that $f'(\zeta)$ exists and is infinite for all ζ with $f(\zeta) \in \mathbb{T}$ if

$$G = \mathbb{D} \setminus \bigcup_n [r_n \zeta_n, \zeta_n].$$

5. Let φ be analytic in $\mathbb{D}$ and $\varphi(\mathbb{D}) \subset \mathbb{D}$. Show that φ has at most one attractive fixed point in $\overline{\mathbb{D}}$, i.e. a point $z_0 \in \overline{\mathbb{D}}$ such that $\varphi(z_0) = z_0$ and $|\varphi'(z_0)| < 1$.

6. Prove the converse of Proposition 4.10.

4.4 Harmonic Measure

Let A be a measurable set on $\mathbb{T}$. The *harmonic measure* of A at $z \in \mathbb{D}$ is defined by

$$(1) \qquad \omega(z, A) \equiv \omega(z, A, \mathbb{D}) = \frac{1}{2\pi} \int_A \frac{1 - |z|^2}{|\zeta - z|^2} |d\zeta|,$$

i.e. by the Poisson integral of the characteristic function of A. It is clear that this is a harmonic function satisfying $0 \le \omega(z, A) \le 1$ for $z \in \mathbb{D}$. Furthermore (e.g. Dur70, p. 5), as $r \rightarrow 1-$,

(2)
$$w(r\zeta, A) \rightarrow \begin{cases} 1 & \text{for almost all } \zeta \in A, \\ 0 & \text{for almost all } \zeta \in \mathbb{T} \setminus A. \end{cases}$$

If $\tau \in \text{Möb}(\mathbb{D})$, i.e. if τ is a Möbius transformation of $\mathbb{D}$ onto $\mathbb{D}$, then

(3)
$$\frac{1 - |\tau(z)|^2}{|\tau(\zeta) - \tau(z)|^2} |\tau'(\zeta)| = \frac{1 - |z|^2}{|\zeta - z|^2} .$$

Hence (1) shows that

(4)
$$w(\tau(z), \tau(A)) = w(z, A) \quad \text{for} \quad \tau \in \text{Möb}(\mathbb{D}) .$$

Furthermore

(5)
$$w(0, A) = \frac{1}{2\pi} \int_A |d\zeta| = \frac{\Lambda(A)}{2\pi}$$

is the normalized total angle under which the set A is seen from 0. Hence it follows from (4) and the conformal invariance of angles that $2\pi w(z, A)$ is the total angle under which A can be seen from z along non-euclidean lines; see Fig. 4.7.

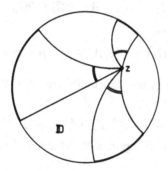

Fig. 4.7. The sum of the angles is $2\pi w(z, A)$

Proposition 4.15. *Let* φ *be analytic in* $\mathbb{D}$ *and let* $A, B \subset \mathbb{T}$ *be measurable. If the angular limit* $\varphi(\zeta)$ *exists for all* $\zeta \in A$ *and if*

(6)
$$\varphi(\mathbb{D}) \subset \mathbb{D}, \quad \varphi(A) \subset B \subset \mathbb{T}$$

then

(7)
$$w(\varphi(z), B) \geq w(z, A) \quad \text{for} \quad z \in \mathbb{D} .$$

If $\varphi(0) = 0$ then (5) and (7) show that

(8)
$$\Lambda(B) \geq \Lambda(A)$$

if (6) is satisfied. This is *Löwner's lemma*. The problem of the measurability of the image set will be discussed in Section 6.2.

Proof. Let V be any open subset of $\mathbb{T}$ with $B \subset V$. Then $\omega(z, V)$ is continuous on $\mathbb{D} \cup V$. The function $u(z) = \omega(\varphi(z), V)$ is harmonic in $\mathbb{D}$ and satisfies $0 \le u \le 1$. If $\zeta \in A$ then $\varphi(\zeta) \in B \subset V$ so that u has the radial limit 1 at ζ. Hence

$$\omega(\varphi(z), V) = u(z) = \frac{1}{2\pi} \int_0^{2\pi} \frac{1 - |z|^2}{|\zeta - z|^2} u(\zeta) |d\zeta| \ge \omega(z, A)$$

by (1), and (7) follows from (1) because B is measurable.   □

Now let G be a simply connected domain with locally connected boundary and let f map $\mathbb{D}$ conformally onto G. Then f is continuous in $\overline{\mathbb{D}}$ by Theorem 2.1. If A is a Borel set of ∂G then $f^{-1}(A)$ is a Borel set on $\mathbb{T}$ and therefore measurable. We define the *harmonic measure* of A at $z \in G$ relative to G by

$$(9) \qquad \omega(z, A, G) = \omega(f^{-1}(z), f^{-1}(A), \mathbb{D}).$$

It follows from (4) that this definition does not depend on the choice of the function f mapping $\mathbb{D}$ onto G.

Corollary 4.16. *Let G_1, G_2 be Jordan domains. If*

$$(10) \qquad G_1 \subset G_2, \quad A \subset \partial G_1 \cap \partial G_2$$

where A is a Borel set then

$$(11) \qquad \omega(z, A, G_1) \le \omega(z, A, G_2) \quad \text{for} \quad z \in G_1.$$

Proof. Let $f_j (j = 1, 2)$ map $\mathbb{D}$ conformally onto G_j. We see from (9) and (7) with $\varphi = f_2^{-1} \circ f_1$ and thus $\varphi \circ f_1^{-1} = f_2^{-1}$ that

$$\omega(z, A, G_1) = \omega(f_1^{-1}(z), f_1^{-1}(A)) \le \omega(f_2^{-1}(z), f_2^{-1}(A))$$

because $f_j^{-1}(A) \subset \mathbb{T}$; this is again a Borel set and thus measurable. Hence (11) follows from (9).   □

This is Carleman's *principle of domain extension* for harmonic measure (Nev70, p. 68). For reasonable sets A (say union of arcs) it follows easily from the maximum principle for harmonic functions applied to $\omega(z, A, G_1) - \omega(z, A, G_2)$. This proof has the advantage that it carries over to multiply connected domains.

Harmonic measure can be generalized (see e.g. Nev70, Ahl73) to domains of any connectivity where it is of great importance; conformal mapping fails here and its generalization, the universal covering map of $\mathbb{D}$ onto G, is not easy to handle.

Exercises 4.4

1. If $\lambda(z, z')$ denotes the non-euclidean distance then $\omega(z', A) \le e^{2\lambda(z, z')} \omega(z, A)$.

2. Let the functions $\psi^\pm$ be continuous in $[a, b]$ and positive except for $\psi^\pm(a) = \psi^\pm(b) = 0$. Let G be the domain between the graphs $C^\pm : \pm \psi^\pm(\xi)$, $a \le \xi \le b$. If $\psi^- \le \psi^+$ show that $\omega(x, C^-, G) \ge \omega(x, C^+, G)$ for $a < x < b$; see Küh67.

3. Let g be analytic in the Jordan domain G and continuous in $\overline{G}$. If $|g(z)| \le a$ for $z \in A$, $|g(z)| \le 1$ for $z \in \partial G \setminus A$ where $A \subset \partial G$ is a Borel set, show that

$$|g(z)| \le a^{\omega(z,A,G)} \quad \text{for} \quad z \in G.$$

This is the *two-constants theorem*, see Nev70, p. 41.

4.5 Length Distortion

Let $K_1, K_2, \ldots$ be positive absolute constants and let $\omega(z, A)$ again denote the harmonic measure of $A \subset \mathbb{T}$ at $z \in \mathbb{D}$ relative to $\mathbb{D}$.

Proposition 4.17. *Let f map $\mathbb{D}$ conformally into $\mathbb{C}$. If $z_0 \in \mathbb{D}$ and if I is an arc of $\mathbb{T}$, then there is $\zeta \in I$ such that*

(1) $$|\log f'(z) - \log f'(z_0)| < K_1/\omega(z_0, I) \quad \text{for} \quad z \in S$$

where S is the non-euclidean segment from z_0 to ζ.

Proof. We consider the Bloch function

$$g(z) = \log f'(\tau(z)), \quad \tau(z) = \frac{z + z_0}{1 + \overline{z}_0 z} \quad (z \in \mathbb{D}).$$

We see from (4.2.3) and (4.2.7) that $\|g\|_B \le 6$. Hence, by Proposition 4.6, there exists $\zeta^* \in \tau^{-1}(I)$ such that

$$|\log f'(\tau(r\zeta^*)) - \log f'(z_0)| \le 132/\Lambda(\tau^{-1}(I)) \quad \text{for} \quad 0 \le r < 1$$

which implies (1) with $\zeta = \tau(\zeta^*) \in I$. $\qquad\square$

Let $d_f(z)$ again denote the distance of $f(z)$ to the boundary. It follows from Corollary 1.4 that, for $z \in \mathbb{D}$,

(2) $$\frac{1}{4}(1 - |z|^2)|f'(z)| \le d_f(z) \equiv \text{dist}(f(z), \partial f(\mathbb{D})) \le (1 - |z|^2)|f'(z)|.$$

Corollary 4.18. *If $z_0 \in \mathbb{D}$ and if I is an arc of $\mathbb{T}$ then*

(3) $$\Lambda(f(S)) < d_f(z_0) \exp[K_2/\omega(z_0, I)]$$

where S is the non-euclidean segment from z_0 to a suitable point on I.

Proof. We may assume that $z_0 = 0$; otherwise we consider $f \circ \tau$ with suitable $\tau \in \text{Möb}(\mathbb{D})$ and use $d_{f\circ\tau}(0) = d_f(z_0)$ and (4.4.4). Now we see from (1) that

$$|f'(r\zeta)| < |f'(0)| \exp[K_1/\omega(0, I)] \quad (0 \le r < 1)$$

for some $\zeta \in I$, and (3) follows from (2) because

$$\Lambda(f(S)) = \int_0^1 |f'(r\zeta)|\, dr, \quad S = [0, \zeta]. \qquad\square$$

Later we shall prove the much stronger result (see Corollary 9.20) that $\Lambda(f(S))$ is not much larger than $d_f(z_0)$ except for the non-euclidean segments connecting z_0 to a set of small capacity and thus of small measure. The next estimate goes in the opposite direction.

Proposition 4.19. *Let f map $\mathbb{D}$ conformally into $\mathbb{C}$. If $z_0 \in \mathbb{D}$ and I is an arc on $\mathbb{T}$ then*

$$(4) \qquad d_f(z_0) \leq \mathrm{dist}(f(z_0), f(I)) \leq K_3 \frac{\mathrm{diam}\, f(I)}{\omega(z_0, I)^{5/2}}$$

where $f(I)$ is the set of angular limits on I whenever they exist.

Proof. We may assume that $z_0 = 0$ and $f(0) = 0$. Choose $a \in f(I)$. Since $f/(f - a)$ is analytic and univalent in $\mathbb{D}$ it follows from Theorem 1.7 (for $r \to 1-$) that

$$\left(\int_I \left| \frac{f(\zeta)}{f(\zeta) - a} \right|^{2/5} |d\zeta| \right)^{5/2} \leq K_4 \left| \frac{f'(0)}{a} \right| \leq 4K_4$$

because $|a| \geq d_f(0) \geq |f'(0)|/4$. Since $a \in f(I)$ it follows that

$$\frac{\mathrm{dist}(0, f(I))}{\mathrm{diam}\, f(I)} \Lambda(I)^{5/2} \leq 4K_4 . \qquad \square$$

Theorem 4.20. *Let f map $\mathbb{D}$ conformally into $\mathbb{C}$. If $z_1, z_2 \in \overline{\mathbb{D}}$ then*

$$(5) \qquad \Lambda(f(S)) \leq K_5 \Lambda(f(C)),$$

$$(6) \qquad \mathrm{diam}\, f(S) \leq K_6 \,\mathrm{diam}\, f(C),$$

where S is the non-euclidean segment from z_1 to z_2 and C is any Jordan arc in $\mathbb{D}$ from z_1 to z_2.

This is the *Gehring-Hayman theorem* (GeHa62, Pom64, Jae68), see Fig. 4.8 (i). The proof is based on the following lemma, see Fig. 4.8 (ii).

Lemma 4.21. *Let $-1 \leq \xi_1 < \xi_2 \leq 1$ and assume that*

$$(7) \qquad \lambda(\xi_1, \xi_2) \geq 1 .$$

For $j = 1, 2$, let T_j be the non-euclidean line through ξ_j orthogonal to $\mathbb{R}$ if $|\xi_j| < 1$, and $T_j = \{\xi_j\}$ otherwise. Then

$$(8) \qquad \mathrm{diam}\, f(S) \leq K_7 \,\mathrm{diam}\, f(C)$$

where $S = (\xi_1, \xi_2)$ and C is any Jordan arc from T_1 to T_2.

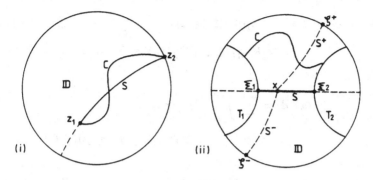

Fig. 4.8

Proof. Let $\rho = \operatorname{diam} f(C)$ and let $z_j \in T_j \cap C$. We obtain from (1.3.18) that

$$\frac{1}{4}\left|\frac{z_2 - z_1}{1 - \bar{z}_1 z_2}\right|(1 - |z_1|^2)|f'(z_1)| \leq |f(z_2) - f(z_1)| \leq \rho.$$

The quotient is $\geq K_8^{-1}$ by (7). Hence we have

(9)
$$|a - f(z_1)| = d_f(z_1) \leq K_9 \rho$$

for suitable $a \in \partial f(\mathbb{D})$.

Now let $\xi_1 \leq x \leq \xi_2$. Applying Corollary 4.18 to $g = 1/(f - a)$ we find $\zeta^{\pm}$ on the two arcs of $\mathbb{T}$ between T_1 and T_2 such that, for z on the non-euclidean segments $S^{\pm}$ from x to $\zeta^{\pm}$,

$$|g(z)| \leq |g(x)| + |g(z) - g(x)| \leq |g(x)| + K_{10}d_g(x) \leq K_{11}|g(x)|$$

because $0 \notin g(\mathbb{D})$. Since C connects T_1 and T_2 this inequality holds for some $z \in C$ and we conclude from (9) that

$$|f(x) - a| \leq K_{11}|f(z) - a| \leq K_{11}(|f(z) - f(z_1)| + |f(z_1) - a|) \leq K_{11}(\rho + K_9\rho)$$

which implies (8).   □

Proof of Theorem 4.20. We may assume that $z_j = x_j$ is real and that $x_1 < x_2$. We furthermore assume that $\lambda(x_1, x_2) \geq 1$; the case $\lambda(x_1, x_2) < 1$ can easily be handled by the Koebe distortion theorem. Then (6) follows at once from Lemma 4.21.

In order to prove (5) we divide (x_1, x_2) into a finite or infinite number of segments $S_n = (\xi_n, \xi_{n+1})$ with $1 \leq \lambda(\xi_n, \xi_{n+1}) \leq 2$. If T_n denotes the non-euclidean segment through ξ_n orthogonal to $\mathbb{R}$ then there exist arcs C_n of C that connect T_n and T_{n+1} and are disjoint except for their endpoints. Hence, by (8),

$$\operatorname{diam} f(S_n) \leq K_7 \operatorname{diam} f(C_n) \leq K_7 \Lambda(f(C_n)).$$

We see from Corollary 1.5 that, for $x \in S_n$,

$$(\xi_{n+1} - \xi_n)|f'(x)| \le \exp\left[6\lambda(\xi_n, \xi_{n+1})\right] \frac{\xi_{n+1} - \xi_n}{1 - \xi_n \xi_{n+1}} \left(1 - \xi_n^2\right)|f'(\xi_n)|$$

$$\le K_{12}|f(\xi_{n+1}) - f(\xi_n)| \le K_{12}\,\text{diam}\,f(S_n)\,.$$

It follows that

$$\int_{x_1}^{x_2} |f'(x)|\,dx = \sum_n \int_{\xi_n}^{\xi_{n+1}} |f'(x)|\,dx$$

$$\le K_7 K_{12} \sum_n \Lambda(f(C_n)) \le K_7 K_{12} \Lambda(f(C))\,. \qquad \square$$

The diameter estimate in Theorem 4.20 and related results can be generalized to quasiconformal maps in any dimension (Zin86, HeNä92).

Exercises 4.5

We assume that f maps $\mathbb{D}$ conformally onto $G \subset \mathbb{C}$. Let $K_1, K_2, \ldots$ denote suitable absolute constants.

1. Let I be an arc of $\mathbb{T}$ and let $r = 1 - \Lambda(I)/\pi > 0$. Show that there is a crosscut Q of $\mathbb{D}$ that separates 0 from I such that $\Lambda(f(Q)) \le K_1 d_f(rz)$ for $z \in I$.

2. Let C be a crosscut of G and let B be the image of a non-euclidean segment. If both endpoints of B lie in one component of $G \setminus C$, show that $\text{dist}(C, w) \le K_2\,\text{diam}\,C$ holds for all points w of B in the other component of $G \setminus C$ (Ost36).

3. Let $\zeta \in \mathbb{T}$ and suppose that the corresponding prime end p is *rectifiably accessible*, i.e. the Jordan arc of Exercise 2.5.5 has finite length. Show that the curve $\{f(r\zeta) : 0 \le r \le 1\}$ has finite length.

4. Let C be any curve in $\mathbb{D}$ from $\zeta_1 \in \mathbb{T}$ to $\zeta_2 \in \mathbb{T}$. Show that

$$d_f(0)|\zeta_1 - \zeta_2|^2 \le K_3\,\text{diam}\,f(C)\,.$$

(Consider the point nearest to 0 on the non-euclidean line from ζ_1 to ζ_2.)

4.6 The Hyperbolic Metric

Let f map $\mathbb{D}$ onto $G \subset \mathbb{C}$. The *hyperbolic metric* (or *Poincaré metric*) of G is defined by

$$(1) \qquad \lambda_G(w_1, w_2) = \lambda_{\mathbb{D}}(z_1, z_2) \quad \text{for} \quad w_j = f(z_j), \quad z_j \in \mathbb{D} \quad (j = 1, 2)$$

where $\lambda_{\mathbb{D}}$ is the non-euclidean metric in $\mathbb{D}$. By (1.2.7) this definition is independent of the choice of the function f mapping $\mathbb{D}$ onto G because all others have the form $f \circ \tau$ with $\tau \in \text{Möb}(\mathbb{D})$.

The definition can be generalized to multiply connected domains with at least three boundary points. The conformal map is then replaced by the universal covering map of $\mathbb{D}$ onto G; see e.g. Ahl73 and Hem88. The hyperbolic metric has several important monotonicity properties, see Hem79, Wei86, Min87, Hem88 and Hay89, p. 698.

It follows from (1.2.5) that

$$(2) \qquad \lambda_G(w_1, w_2) = \min_C \lambda_G(C), \quad \lambda_G(C) \equiv \int_{f^{-1}(C)} \frac{|dz|}{1 - |z|^2}$$

where the minimum is taken over all curves C in G from w_1 to w_2. It is attained for the *hyperbolic geodesic* from w_1 to w_2, that is the image of the non-euclidean segment from z_1 to z_2. It follows from Theorem 4.20 and from the definition that

$$(3) \qquad \Lambda(S) \le K\Lambda(C), \quad \lambda_G(S) \le \lambda_G(C)$$

where S is the hyperbolic geodesic and C is any curve in G from w_1 to w_2. Thus the hyperbolic geodesic is almost as short as possible in the euclidean sense; see Fig. 4.9. For non-convex domains there is not always a shortest curve in G from w_1 to w_2.

There is another geometric property (Jør56, see also Küh69, Min87), see Fig. 4.9.

Proposition 4.22. *Let $D \subset G$ be a disk and S a hyperbolic geodesic of G. If ∂D is tangential to S then $D \cap S = \emptyset$.*

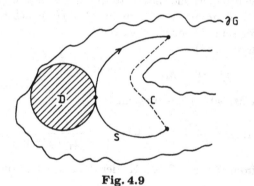

Fig. 4.9

Proof. We may assume S is the image of a real interval in $\mathbb{D}$ and that ∂D and S touch at $f(0)$, furthermore that $f(0) = 0$ and $f'(0) = 1$. Then

$$h(z) = \frac{1}{f(z)} - \frac{1}{z} - z \quad (|z| < 1)$$

is analytic in $\mathbb{D}$. Since ∂D is tangential to S and thus to $\mathbb{R}$ at 0 we see that $\{1/w : w \in D\} = \{w^* : \operatorname{Im} w^* > a\}$ (or $= \{\operatorname{Im} w^* < a\}$). Since $D \cap \partial G = \emptyset$ it follows that

$$\limsup_{|z| \to 1} \operatorname{Im} h(z) = \limsup_{|z| \to 1} \operatorname{Im} \frac{1}{f(z)} \le a.$$

The maximum principle therefore implies that $\operatorname{Im} h(z) \le a$ for $z \in \mathbb{D}$, in particular that

$$\operatorname{Im}\frac{1}{f(x)} = \operatorname{Im} h(x) \leq a \quad \text{for} \quad -1 < x < 1$$

which is equivalent to $D \cap S = \emptyset$. $\square$

The hyperbolic metric is not a geometric quantity in contrast to the *quasi-hyperbolic metric*

$$(4) \qquad \lambda_G^*(w_1, w_2) = \min_C \int_C \frac{|dw|}{\operatorname{dist}(w, \partial G)} \quad (w_1, w_2 \in G),$$

where the minimum is taken over all curves $C \subset G$ from w_1 to w_2. It follows from (4.5.2) that

$$(5) \qquad \frac{1}{1 - |z|^2} \leq \frac{|f'(z)|}{\operatorname{dist}(f(z), \partial G)} \leq \frac{4}{1 - |z|^2} \quad (z \in \mathbb{D})$$

and thus, by (2) and (4),

$$(6) \qquad \lambda_G(w_1, w_2) \leq \lambda_G^*(w_1, w_2) \leq 4\lambda_G(w_1, w_2).$$

See e.g. GeOs79 and GeHaMa89 for further results. If G is multiply connected then λ_G^* can be much larger than λ_G; see e.g. BeaPo78, Pom84.

We use the quasihyperbolic metric to give a geometric characterization (BePo82) of *Hölder domains*, i.e. of domains $G = f(\mathbb{D})$ such that f is Hölder continuous in $\overline{\mathbb{D}}$ for some exponent α with $0 < \alpha \leq 1$; see (3.3.9). This holds if and only if (Dur70, p. 74)

$$(7) \qquad |f'(z)| \leq M_1(1 - |z|)^{\alpha-1} \quad (z \in \mathbb{D})$$

for some constant M_1 which is equivalent to

$$\alpha \log \frac{1 + |z|}{1 - |z|} \leq M_2 + \log \frac{1}{(1 - |z|^2)|f'(z)|}.$$

Hence it follows from (5) that G is a Hölder domain if and only if

$$(8) \qquad 2\alpha\lambda_G(f(0), w) \leq M_3 + \log 1/\operatorname{dist}[w, \partial G] \quad (w \in G)$$

and thus, by (6), if and only if

$$(9) \qquad \lambda_G^*(f(0), w) = O(1/\operatorname{dist}[w, \partial G]) \quad \text{as} \quad w \to \partial G;$$

the exponent α is not completely determined by (9).

See e.g. NäPa83, NäPa86, SmSt87 for further results about Hölder domains. If $f(z) = \sum_{n=0}^{\infty} a_n z^n$ then (SmSt91; see also Sta89, SmSt90, JoMa91)

$$(10) \qquad \sum_{n=1}^{\infty} n^{1+\delta}|a_n|^2 < \infty \quad \text{for some } \delta > 0.$$

Hence it follows by the Schwarz inequality that the power series converges absolutely on $\mathbb{T}$. For general bounded domains, the estimate (10) holds only for $\delta = 0$ and the power series need not converge absolutely even in the case of Jordan domains; see e.g. Pir62.

Exercises 4.6

1. Let $D \subset G$ be a disk and S a hyperbolic geodesic of G. Show that $\overline{D} \cap S$ is connected (Jør56).

2. Let G be as in Fig. 4.10 and $w^{\pm} = -1 \pm i/2$. Show that the hyperbolic geodesic S intersects the positive real axis at a point u with $1/K \leq u \leq K$. Give numerical upper and lower bounds for $\lambda_G(w^+, w^-)$.

3. Show that Hölder domains have no outward pointing (zero-angle) cusps.

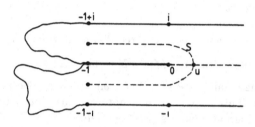

Fig. 4.10

Chapter 5. Quasidisks

5.1 An Overview

A quasicircle in $\mathbb{C}$ is a (not necessarily rectifiable) Jordan curve J such that

$$\text{diam } J(a,b) \leq M|a-b| \quad \text{for} \quad a,b \in J$$

where $J(a,b)$ is the smaller arc of J between a and b. The inner domain is called a quasidisk. This important concept was introduced by Ahlfors and appears in many different contexts (see e.g. Geh87).

There are two one-sided versions of quasidisks, the John domain and the linearly connected domain (see Fig. 5.1). We will discuss these types of domains in detail because they appear quite often. The corresponding conformal maps satisfy upper and lower Hölder conditions respectively and behave "tamely" in many respects. For example, finite angular derivatives are automatically unrestricted derivatives for John domains. A quasidisk is a linearly connected John domain.

Fig. 5.1. A John domain (i) and a linearly connected domain (ii). The definitions essentially say that the shaded domains must not be much longer than $|a-b|$

Quasicircle Theorem. *Let J be a Jordan curve in $\mathbb{C}$ and let f map $\mathbb{D}$ conformally onto the inner domain of J. Then the following conditions are equivalent:*

(a) *J is a quasicircle;*
(b) *f is quasisymmetric on $\mathbb{T}$;*
(c) *f has a quasiconformal extension to $\mathbb{C}$;*
(d) *there is a quasiconformal map of $\mathbb{C}$ onto $\mathbb{C}$ that maps $\mathbb{T}$ onto J.*

We shall now explain the concepts used here and describe the plan of proof.

A quasisymmetric map of $\mathbb{T}$ into $\mathbb{C}$ is a homeomorphism h such that, for $z_1, z_2, z_3 \in \mathbb{T}$,

$$(1) \qquad |z_1 - z_2| = |z_2 - z_3| \quad \Rightarrow \quad |h(z_1) - h(z_2)| \leq M|h(z_2) - h(z_3)|$$

(Ahlfors); we shall actually use Väisälä's formally stronger (but equivalent) definition. The implication (a) $\Rightarrow$ (b) follows from results about John domains and linearly connected domains.

There are several excellent books about quasiconformal maps, e.g. Ahl66b, LeVi73 and Leh87. We will therefore say little about the general theory and concentrate on the relation to conformal mapping.

A *quasiconformal map* of $\mathbb{C}$ onto $\mathbb{C}$ is a homeomorphism h such that small circles are approximately mapped onto small ellipses of bounded axes ratio. The analytic formulation of this condition is that $h(x + iy)$ is absolutely continuous in x for almost all y and in y for almost all x and that the partial derivatives are locally square integrable and satisfy the *Beltrami differential equation*

$$(2) \qquad \frac{\partial h}{\partial \overline{z}} = \mu(z)\frac{\partial h}{\partial z} \quad \text{for almost all } z \in \mathbb{C},$$

where μ is a complex measurable function with

$$(3) \qquad |\mu(z)| \leq \kappa < 1 \quad \text{for} \quad z \in \mathbb{C}.$$

We say then that h is κ-*quasiconformal*. A geometric definition can be given in terms of modules of curve families.

Our proof of (b) $\Rightarrow$ (c) uses the recent Douady-Earle extension method. This is more difficult than the usual Beurling-Ahlfors method but has the great advantage of giving a Möbius-invariant extension.

The implication (c) $\Rightarrow$ (d) is trivial. In order to show (d) $\Rightarrow$ (a) we use a deep theorem of Bojarski, Ahlfors and Bers without proof. It says that every quasiconformal map can be embedded into a family of quasiconformal maps that depends analytically on a complex parameter $\lambda \in \mathbb{D}$ and reduces to the identity for $\lambda = 0$. More generally we shall prove a result of Mañé, Sad and Sullivan that such analytic families of injective maps always lead to quasicircles.

Whereas there are no conformal maps in $\mathbb{R}^n (n > 2)$ except for Möbius transformations, there are many interesting quasiconformal maps in $\mathbb{R}^n$, see e.g. Väi71.

There is an extensive literature on conformal maps of $\mathbb{D}$ with quasiconformal extension to $\mathbb{C}$, see e.g. Leh87. Their Schwarzian derivatives S_f form the *universal Teichmüller space* with the norm $\sup(1 - |z|^2)^2|S_f(z)|$. Its closure is strictly less than the corresponding space for all conformal maps; see Geh78, AsGe86, Thu86, Ast88a, Kru89, Ham90.

5.2 John Domains

Let $M_1, M_2, \ldots$ denote suitable positive constants. The bounded simply connected domain G is called a *John domain* (MaSa79) if, for every rectilinear crosscut $[a, b]$ of G,

$$(1) \qquad \qquad \operatorname{diam} H \leq M_1 |a - b|$$

holds for one of the two components H of $G \setminus [a, b]$; see Proposition 2.12 and Fig. 5.1 (i). It is clear that ∂G is locally connected. If ∂G is a piecewise smooth Jordan curve then G is a John domain if and only if there are no outward-pointing cusps.

We now show that the rectilinear crosscuts of the definition can be replaced by any crosscuts.

Proposition 5.1. *Let G be a John domain. If C is a crosscut of G then*

$$(2) \qquad \qquad \operatorname{diam} H \leq M_2 \operatorname{diam} C$$

holds for one of the components H of $G \setminus C$.

Proof. Let $M_2 = 2M_1 + 2$, furthermore $\delta = \operatorname{diam} C$ and $S = [a, b]$ where a and b are the endpoints of the crosscut C. If $\operatorname{diam} G \leq M_2 \delta$ then (2) is trivial. In the other case we can find $z_0 \in G$ with $|z_0 - a| > (M_1 + 1)\delta$. Let H be the component of $G \setminus C$ that does not contain z_0 and let G_0 be the component of $G \setminus S$ that contains z_0; see Fig. 5.2.

We claim that $H \cap G_0 \subset D(a, \delta)$. Otherwise there is $z \in H \cap G_0$ with $|z - a| > \delta$. Since also $|z_0 - a| > \delta$ it follows that $S \cup C$ does not separate z_0 and z. Since $S \cup \partial G$ does not separate z_0 and z and since $(S \cup C) \cap (S \cup \partial G) = S$ is connected, it follows from Janiszewski's theorem (Section 1.1) that the union $S \cup C \cup \partial G$ does not separate z_0 and z which is false because $z \in H$.

Every component $G_k (k = 1, 2, \ldots)$ of $G \setminus \overline{G_0}$ is, by Proposition 2.13, also a component of $G \setminus S_k$ for some segment $S_k \subset S$. The other component contains z_0 and therefore has a diameter $> M_1 \delta$. Hence (1) shows that $\operatorname{diam} G_k \leq M_1 \delta$ so that $H \cap G_k \subset D(a, (M_1 + 1)\delta)$. It follows that $H \subset D(a, (M_1 + 1)\delta)$ which implies (2). $\qquad \square$

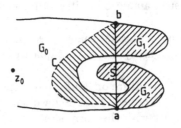

Fig. 5.2 The shaded domain is H

We now give an analytic characterization of John domains (Pom64, JeKe82a). We define (see Fig. 5.3)

$$(3) \qquad B(re^{it}) = \{\rho e^{i\vartheta} : r \le \rho \le 1, |\vartheta - t| \le \pi(1 - r)\} \quad (0 \le r < 1),$$

$$(4) \qquad I(re^{it}) = \mathbb{T} \cap \partial B(re^{it}) = \{e^{i\vartheta} : |\vartheta - t| \le \pi(1 - r)\}.$$

The factor π has been chosen so that $B(0) = \overline{\mathbb{D}}$ and $I(0) = \mathbb{T}$. Let again $d_f(z) = \text{dist}(f(z), \partial G)$.

Theorem 5.2. *Let f map $\mathbb{D}$ conformally onto G such that*

$$(5) \qquad\qquad\qquad d_f(0) \ge c \, \text{diam} \, G.$$

Then the following conditions are equivalent:

(i) *G is a John domain;*
(ii) *$\text{diam} \, f(B(z)) \le M_3 d_f(z)$ for $z \in \mathbb{D}$;*
(iii) *there exists α with $0 < \alpha \le 1$ such that*

$$|f'(\zeta)| \le M_4 |f'(z)| \left(\frac{1 - |\zeta|}{1 - |z|} \right)^{\alpha - 1} \quad \text{for} \quad \zeta \in \mathbb{D} \cap B(z), \quad z \in \mathbb{D},$$

(iv) *there exists $\beta > 0$ such that, for all arcs $A \subset I \subset \mathbb{T}$,*

$$\Lambda(A) \le \beta \Lambda(I) \quad \Rightarrow \quad \text{diam} \, f(A) \le \frac{1}{2} \, \text{diam} \, f(I).$$

The function f is continuous in $\overline{\mathbb{D}}$ by Theorem 2.1. The constants M_3, M_4, α, β depend only on each other, on M_1 and on c; we shall use the assumption (5) only in the proof of (i) $\Rightarrow$ (ii). It is possible to replace (iv) by

(iv') *for every $\varepsilon > 0$ there exists $\delta > 0$ such that*

$$\Lambda(A) \le \delta \Lambda(I) \quad \Rightarrow \quad \text{diam} \, f(A) \le \varepsilon \, \text{diam} \, f(I)$$

for all arcs $A \subset I \subset \mathbb{T}$.

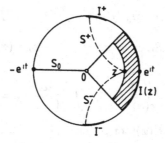

Fig. 5.3. The shaded domains is $B(z)$

Proof. Let $K_1, K_2, \dots$ denote suitable absolute constants. We write $z = re^{it}$ and $\zeta = \rho e^{i\vartheta}$.

(i) $\Rightarrow$ (ii) (see Fig. 5.3). If $r \leq 1/2$ then

$$\operatorname{diam} f(B(z)) \leq \operatorname{diam} G \leq c^{-1}d_f(0) \leq K_1 c^{-1}d_f(z)$$

by (5) and Corollary 1.6. Let now $1/2 < r < 1$. Then the two arcs $I^\pm$ of $\mathbb{T}$ with $(3\pi/2)(1-r) \leq |\vartheta - t| \leq 2\pi(1-r)$ do not intersect $S_0 = [-e^{it}, 0]$. By Corollary 4.18 there are non-euclidean segments $S^\pm$ from z to $I^\pm$ such that

$$(6) \qquad \qquad \Lambda(f(S^\pm)) \leq K_2 d_f(z).$$

Our choice of $I^\pm$ implies that $S^+ \cup S^-$ separates $B(z)$ from S_0. Hence $f(B(z)) \subset H$ and $f(S_0) \subset H'$ where H and H' are the two components of $G \setminus f(S^+ \cup S^-)$. We see therefore from (2) and (6) that

$$\operatorname{diam} f(B(z)) \leq \operatorname{diam} H \leq 2K_2 M_2 d_f(z)$$

or that

$$d_f(0) \leq \operatorname{diam} f(S_0) \leq \operatorname{diam} H' \leq 2K_2 M_2 d_f(z).$$

In the second case it follows from (5) that

$$\operatorname{diam} f(B(z)) \leq \operatorname{diam} G \leq c^{-1}d_f(0) \leq c^{-1}2K_2 M_2 d_f(z).$$

Hence (ii) holds in both cases.

(ii) $\Rightarrow$ (iii). We define

$$(7) \qquad \qquad \varphi(r) = \int_r^1 (1-x)|f'(x)|^2\, dx \qquad \text{for} \quad 0 \leq r < 1$$

and $\Delta(r) = \{z : r \leq x < 1, 0 \leq y \leq 1 - x\}$ where $z = x + iy$. Since $|f'(x)| \leq K_3 |f'(z)|$ for $z \in \Delta(r)$ by Corollary 1.6 and since f is univalent, we see that

$$\varphi(r) \leq K_3^2 \int_r^1 \int_0^{1-x} |f'(z)|^2\, dy\, dx = K_3^2 \operatorname{area} f(\Delta(r)).$$

Since $\Delta(r) \subset B(r)$ we thus obtain from (ii) that

$$(8) \qquad \qquad \varphi(r) \leq K_4 M_3^2 d_f(r)^2 \leq \frac{1}{2\alpha}(1-r)^2 |f'(r)|^2$$

where $\alpha = 1/(8K_4 M_3^2)$. Since $\varphi'(x) = -(1-x)|f'(x)|^2$ by (7), we conclude that, for $r \leq \rho < 1$,

$$\log \frac{\varphi(\rho)}{\varphi(r)} = \int_r^\rho \frac{\varphi'(x)}{\varphi(x)}\, dx \leq -\int_r^\rho \frac{2\alpha}{1-x}\, dx = 2\alpha \log \frac{1-\rho}{1-r}.$$

Using that $|f'(\rho)| \leq K_5 |f'(x)|$ for $\rho \leq x \leq (1+\rho)/2$ we deduce from (7) that

$$\frac{1}{4}(1-\rho)^2 |f'(\rho)|^2 < K_5^2 \varphi(\rho) \leq K_5^2 \varphi(r) \left(\frac{1-\rho}{1-r} \right)^{2\alpha}$$

and thus, by (8),

(9) $|f'(\rho)| \le (2/\alpha)^{1/2} K_5 |f'(r)| \left(\dfrac{1-\rho}{1-r}\right)^{\alpha-1}$ for $r \le \rho < 1$.

We now apply (9) to $f(e^{it}z)$. If $\zeta \in B(z)$ then, again by Corollary 1.6,

$$|f'(\rho e^{i\vartheta})| \le K_6 |f'(\rho e^{it})| \le K_6 K_5 \left(\frac{2}{\alpha}\right)^{1/2} |f'(re^{it})| \left(\frac{1-\rho}{1-r}\right)^{\alpha-1}$$

which is our condition (iii).

(iii) $\Rightarrow$ (iv). Let $\zeta \in B(z)$. Integrating (iii) on the circular arc from re^{it} to $re^{i\vartheta}$ and then on $[re^{i\vartheta}, \rho e^{i\vartheta}]$ we obtain

$$|f(z) - f(\zeta)| \le M_4 |f'(z)| \left(|t - \vartheta| + (1-r)^{-\alpha+1} \int_r^\rho (1-x)^{\alpha-1}\, dx\right).$$

Since $|t - \vartheta| \le \pi(1-r)$ it follows that

(10) $\operatorname{diam} f(B(z)) \le M_5 (1-r)|f'(z)|$ for $z \in \mathbb{D}$.

Let $A \subset I$ be arcs on $\mathbb{T}$ and choose ζ, z such that $A = I(\zeta), I = I(z)$. Then $\zeta \in B(z)$ and therefore, by (10) and (iii),

$$\operatorname{diam} f(A) \le M_5(1-\rho)|f'(\zeta)| \le M_6 \left(\frac{1-\rho}{1-r}\right)^\alpha (1-r)|f'(z)|$$

$$\le K_7 M_6 \left(\frac{1-\rho}{1-r}\right)^\alpha \operatorname{diam} f(I);$$

the last estimate follows from Proposition 4.19. This implies (iv') and thus (iv) because $(1-\rho)/(1-r) = \Lambda(A)/\Lambda(I)$.

(iv) $\Rightarrow$ (i). Let $[w_1, w_2]$ with $w_j = f(z_j)$, $z_j \in \mathbb{T}$ be a crosscut of G and let I be the (shorter) arc of $\mathbb{T}$ between z_1 and z_2, furthermore S_0 the non-euclidean line from z_1 to z_2 and z_0 the point of S_0 closest to 0. We set $M_7 = \exp(2\pi K_8/\beta)$ where K_8 is the constant in Corollary 4.18 and write

(11) $$\{\zeta \in I : |f(\zeta) - f(z_0)| > M_7 d_f(z_0)\} = \bigcup_k A_k$$

with open arcs A_k on I. It follows from Corollary 4.18 that $2\pi K_8/\beta \le K_8/\omega(z_0, A_k)$. Since $|\zeta - z_0| \le \sqrt{2}(1 - |z_0|)$ for $\zeta \in I$ we see from (4.4.1) that

$$\frac{\Lambda(A_k)}{\Lambda(I)} \le \frac{4\pi(1 - |z_0|)\omega(z_0, A_k)}{2(1 - |z_0|)} = 2\pi\omega(z_0, A_k) \le \beta.$$

Hence (11) and (iv) show that

$$\operatorname{diam} f(I) \le 2M_7 d_f(z_0) + \sup_k \operatorname{diam} f(A_k) \le 2M_7 d_f(z_0) + \frac{1}{2} \operatorname{diam} f(I)$$

and therefore that $\operatorname{diam} f(I) \le 4M_7 d_f(z_0)$. Finally

$$d_f(z_0) \le \Lambda(f(S_0)) \le K_9 |w_1 - w_2|$$

by Theorem 4.20. If H is the component of $G \setminus [w_1, w_2]$ bounded by $f(I)$ and $[w_1, w_2]$ it follows that

$$\operatorname{diam} H = \operatorname{diam} f(I) \leq 4K_9 M_7 |w_1 - w_2|.$$

Hence G is a John domain. This completes the proof of Theorem 5.2.   □

Corollary 5.3. *Let f map $\mathbb{D}$ conformally onto a John domain. If $z \in D$ then*

$$(12) \qquad |f(\zeta_1) - f(\zeta_2)| \leq M_8 d_f(z) \left(\frac{|\zeta_1 - \zeta_2|}{1 - |z|} \right)^{\alpha} \quad \text{for} \quad \zeta_1, \zeta_2 \in B(z).$$

Proof. Let $r = |z|$ and $\delta = |\zeta_1 - \zeta_2|$. If $1 - 2\delta < r$ then (12) follows at once from (ii). Suppose therefore that $\rho = 1 - 2\delta \geq r$ and that $\zeta_j = \rho_j e^{i\vartheta_j}$ with $\rho_1 \leq \rho_2$. If $\rho_1 < \rho = 1 - 2\delta$ then $|f'(\zeta)| \leq K_{10}|f'(\zeta_1)|$ for $|\zeta - \zeta_1| \leq \delta$ by Corollary 1.6, and (12) follows easily from (iii) by integration over $[\zeta_1, \zeta_2]$.

Let now $\rho \leq \rho_1 \leq \rho_2$. Since $(1 - r)|f'(z)| \leq 4d_f(z)$ by Corollary 1.4, it follows from (iii) that

$$|f(\rho_j e^{i\vartheta_j}) - f(\rho e^{i\vartheta_j})| \leq \frac{M_9 d_f(z)}{(1 - r)^{\alpha}} \int_{\rho}^{\rho_j} (1 - x)^{\alpha - 1} \, dx$$

$$\leq \frac{M_9 d_f(z)}{(1 - r)^{\alpha}} \frac{(1 - \rho)^{\alpha}}{\alpha} = \frac{M_9}{\alpha} d_f(z) \left(\frac{2\delta}{1 - r} \right)^{\alpha}$$

for $j = 1, 2$ and furthermore that

$$|f(\rho e^{i\vartheta_1}) - f(\rho e^{i\vartheta_2})| \leq M_9 d_f(z) \frac{(1 - \rho)^{\alpha - 1}}{(1 - r)^{\alpha}} |\vartheta_1 - \vartheta_2| \rho \leq M_{10} d_f(z) \left(\frac{\delta}{1 - r} \right)^{\alpha},$$

and these estimates together imply (12).   □

Let again $M_1, M_2, \ldots$ denote positive constants. We prove a result (SmSt90) about the average growth; compare (4.6.10).

Theorem 5.4. *If*

$$(13) \qquad f(z) = \sum_{n=0}^{\infty} a_n z^n$$

maps $\mathbb{D}$ conformally onto a John domain then

$$(14) \qquad \sum_{n=1}^{\infty} n^{1+\delta} |a_n|^2 < \infty \quad \text{for some } \delta > 0.$$

Proof. For $0 \leq r < 1$ we consider

$$(15) \qquad \varphi(r) = \frac{1}{2\pi} \int_0^{2\pi} |f'(re^{it})|^2 \, dt = \sum_{n=1}^{\infty} n^2 |a_n|^2 r^{2n-2}.$$

Since f is univalent it follows from Theorem 5.2 (ii) that

$$\int_r^1 \int_{-\pi(1-r)}^{\pi(1-r)} |f'(\rho e^{i\vartheta + it})|^2 \rho \, d\vartheta \, d\rho = \text{area } f(B(re^{it})) \le M_1 (1-r)^2 |f'(re^{it})|^2 \,.$$

If we integrate this inequality over $0 \le t \le 2\pi$ and substitute $\tau = \vartheta + t$ we see from (15) that, for $1/2 \le r < 1$,

$$\pi \int_r^1 \varphi(\rho) \, d\rho \le M_1 (1-r) \varphi(r) \,.$$

Writing $\delta = \pi/(2M_1)$ we therefore have

$$\frac{d}{dr} \left[(1-r)^{-2\delta} \int_r^1 \varphi(\rho) \, d\rho \right] \le 0 \quad \text{for} \quad 1/2 \le r < 1 \,.$$

Since φ is increasing we conclude that

$$(1-r)^{-2\delta+1} \varphi(r) \le (1-r)^{-2\delta} \int_r^1 \varphi(\rho) \, d\rho \le 2^{2\delta} \int_{1/2}^1 \varphi(\rho) \, d\rho$$

for $1/2 \le r < 1$ and thus, by (15),

$$\sum_{n=1}^{\infty} n^{1+\delta} |a_n|^2 \le M_2 \int_{1/2}^1 (1-r)^{-\delta} \varphi(r) \, dr \le M_3 \int_{1/2}^1 (1-r)^{\delta-1} \, dr < \infty \,. \quad \square$$

We now return to the angular derivative and the isogonality defined in Section 4.3. We show that for John domains the angular approach can be replaced by unrestricted approach in $\mathbb{D}$; compare War35b.

Theorem 5.5. *Let f map $\mathbb{D}$ conformally onto a John domain G and let $\zeta \in \mathbb{T}$. (i) If $f'(\zeta) \ne \infty$ exists then*

(16) $$\frac{f(z) - f(\zeta)}{z - \zeta} \to f'(\zeta) \quad \text{as} \quad z \to \zeta, \quad z \in \overline{\mathbb{D}} \,.$$

(ii) *If f is isogonal at ζ then there is γ such that*

(17) $$\arg \frac{f(z) - f(\zeta)}{z - \zeta} \to \gamma \quad \text{as} \quad z \to \zeta, \quad z \in \overline{\mathbb{D}}$$

and ∂G has a tangent at $f(\zeta)$.

Proof (see Fig. 5.4.). (i) It follows from (4.3.1) that there are ρ_n with $0 < \rho_n < 1$ such that

(18) $$\left| \frac{f(z^*) - f(\zeta)}{z^* - \zeta} - f'(\zeta) \right| < \frac{1}{n} \quad \text{for} \quad z^* \in \Delta_n \,, \quad n = 1, 2, \dots,$$

where Δ_n is the Stolz angle $\{|\arg(1 - \bar{\zeta}z)| < \pi/2 - 1/n, |z - \zeta| < \rho_n\}$. Let now $z \in \mathbb{D}$ and $|z - \zeta| < \rho_n$. By (18) it is sufficient to consider the case $z \notin \Delta_n$. We choose $z^* \in \partial\Delta_n$ with $|z^* - \zeta| = |z - \zeta|$ and write

$$\frac{f(z) - f(\zeta)}{z - \zeta} - f'(\zeta) = \left(\frac{f(z^*) - f(\zeta)}{z^* - \zeta} - f'(\zeta)\right)\frac{z^* - \zeta}{z - \zeta} + \frac{z^* - z}{z - \zeta}f'(\zeta)$$
$$- \frac{f(z^*) - f(z)}{z - \zeta}.$$

Since $z, z^* \in B(r\zeta)$ it follows from Corollary 5.3 that, with $r = 1 - |z - \zeta|$,

$$\left|\frac{f(z^*) - f(z)}{z - \zeta}\right| \leq M_4 \frac{d_f(r\zeta)}{1-r}\left|\frac{z^* - z}{1-r}\right|^{\alpha} \leq M_4 |f'(r\zeta)|\frac{2}{n^{\alpha}}$$

because $|z^* - z| \leq |z - \zeta|/n$, hence from (18) that

$$\left|\frac{f(z) - f(\zeta)}{z - \zeta} - f'(\zeta)\right| \leq (1 + |f'(\zeta)|)\frac{1}{n} + M_4 |f'(r\zeta)|\frac{2}{n^{\alpha}}$$

and (16) follows because $|f'(r\zeta)|$ is bounded (Proposition 4.7).

(ii) Let now f be isogonal at ζ. By (4.3.5) and (4.3.8), there are $\rho_n > 0$ such that

(19) $\quad \left|\arg \frac{f(z^*) - f(\zeta)}{z^* - \zeta} - \gamma\right| < \frac{1}{n}, \quad \left|\frac{f(z^*) - f(\zeta)}{(z^* - \zeta)f'(z^*)}\right| > \frac{1}{2} \quad$ for $\quad z^* \in \Delta_n$

where Δ_n is defined as in (i). Given $z \in \mathbb{D}$ with $|z - \zeta| < \rho_n$ and $z \notin \Delta_n$ we choose $z^* \in \partial\Delta_n$ with $|z^* - \zeta| = |z - \zeta|$ and write

$$\frac{f(z) - f(\zeta)}{z - \zeta} = \frac{f(z^*) - f(\zeta)}{z^* - \zeta}\frac{z^* - \zeta}{z - \zeta}\left(1 - \frac{f(z^*) - f(z)}{f(z^*) - f(\zeta)}\right).$$

It follows from Corollary 5.3 and (19) that

$$\left|\frac{f(z^*) - f(z)}{f(z^*) - f(\zeta)}\right| \leq \frac{2M_4 d_f(z^*)}{|z^* - \zeta||f'(z^*)|}\left|\frac{z - z^*}{1 - |z^*|}\right|^{\alpha} \leq M_5\frac{1 - |z^*|}{|z^* - \zeta|} \leq \frac{M_5}{n}.$$

Since $|\arg[(z^* - \zeta)/(z - \zeta)]| < 1/n$ we conclude from (19) that

$$\left|\arg \frac{f(z) - f(\zeta)}{z - \zeta} - \gamma\right| < \frac{2}{n} + \frac{M_6}{n}.$$

Hence (17) holds, and it easily follows from (17) that ∂G has a tangent at $f(\zeta)$; see Section 3.4. $\qquad\square$

It follows from Corollary 5.3 that

(20) $\qquad |f(\zeta) - f(r\zeta)| \leq M_7 d_f(r\zeta) \quad$ for $\quad 0 \leq r < 1, \quad \zeta \in \mathbb{T}.$

This means that the "curvilinear sector"

(21) $\qquad \{w \in \mathbb{C} : |w - f(r\zeta)| < M_7^{-1}|f(\zeta) - f(r\zeta)|, \ 0 \leq r < 1\}$

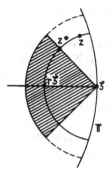

Fig. 5.4. The Stolz angle Δ_n is shaded

of "vertex" $f(\zeta)$ lies in the image domain G. If G contains however a rectilinear sector of fixed size at $f(\zeta)$ for each $\zeta \in \mathbb{T}$, then ∂G is rectifiable (and even $f' \in H^p$ for some $p > 1$; FiLe87). There exist John domains such that G contains a sector at $f(\zeta)$ for almost no $\zeta \in \mathbb{T}$ (Section 6.5). See e.g. GeHaMa89 and NäVä91 for further results on John domains.

It also follows from Corollary 5.3 that f satisfies a Hölder condition. Hence every John domain is a Hölder domain (see Section 4.6) but not conversely (BePo82).

Exercises 5.2

1. Show that every bounded convex domain is a John domain.

2. Let G_1 and G_2 be John domains. Show that $G_1 \cup G_2$ is a John domain if it is simply connected.

3. Let f map $\mathbb{D}$ conformally onto a John domain. Show that

$$\int_r^1 |f'(\rho\zeta)|\, d\rho = O(d_f(r\zeta)) = O((1-r)^\alpha) \quad \text{as} \quad r \to 1-$$

uniformly in $\zeta \in \mathbb{T}$ for some $\alpha > 0$.

4. Show that condition (ii) of Theorem 5.2 can be replaced by

$$\operatorname{diam} f(I(z)) \le M d_f(z) \quad \text{for} \quad z \in \mathbb{D}.$$

5. If the John domain is bounded by a Jordan curve without tangents, show that there is nowhere a finite nonzero angular derivative.

5.3 Linearly Connected Domains

We turn now to a concept that is dual to the concept of a John domain. We say that the simply connected domain $G \subset \mathbb{C}$ is *linearly connected* if any two points $w_1, w_2 \in G$ can be connected by a curve $A \subset G$ such that (Fig. 5.1 (ii))

(1) $$\operatorname{diam} A \leq M_1|w_1 - w_2| ;$$

by $M_1, M_2, \dots$ we denote suitable positive constants. Linearly connected domains have been studied e.g. by NäPa83 and Zin85. The inner domain of a piecewise smooth Jordan curve is linearly connected if and only if it has no inward-pointing cusps (of inner angle 2π). For John domains, we had to exclude outward-pointing cusps (of inner angle 0).

Proposition 5.6. *Let f map $\mathbb{D}$ conformally onto a linearly connected domain G. Then f is continuous in $\overline{\mathbb{D}}$ with values in $\widehat{\mathbb{C}}$ and*

(2) $$\operatorname{diam} f(S) \leq M_2|f(z_1) - f(z_2)| \quad for \quad z_1, z_2 \in \overline{\mathbb{D}}$$

where S is the non-euclidean segment from z_1 to z_2.

Hence every finite value can be assumed only once. (The value ∞ can be assumed finitely often, see Pom64, Satz 3.9.) In particular a bounded linearly connected domain is a Jordan domain.

Proof. Let $\zeta \in \mathbb{T}$ and let p be the prime end corresponding to ζ. Suppose that its impression is not a single point, see Section 2.5. Let ω_0 be a principal point of p (possibly $= \infty$) and $\omega \neq \infty$ some other point in the impression. We choose ρ with $0 < 4\rho < |\omega - \omega_0|$. By Theorem 2.16 there exists $z_1 \in \mathbb{D}$ such that $|f(z_1) - \omega| < \rho/M_1$. Since ω_0 is a principal point, there is a crosscut Q of $\mathbb{D}$ with $\operatorname{dist}(f(Q), \omega) > 3\rho$ that separates z_1 from ζ and therefore also from some $z_2 \in \mathbb{D}$ with $|f(z_2) - \omega| < \rho/M_1$; see Fig. 5.5. By (1) we can find a curve $A \subset G$ from $f(z_1)$ to $f(z_2)$ such that $\operatorname{diam} A \leq M_1|f(z_1) - f(z_2)| < 2\rho$. Hence $A \subset D(\omega, 3\rho)$ so that $f(Q) \cap A = \emptyset$ which is false because $f^{-1}(A)$ has to intersect Q. Hence the impression of p reduces to a point so that f has an unrestricted limit at ζ by Corollary 2.17. It follows that f has a continuous extension to $\overline{\mathbb{D}}$. The estimate (2) with $M_2 = K_6 M_1$ is therefore an immediate consequence of (1) and Theorem 4.20 with $C = f^{-1}(A)$.   □

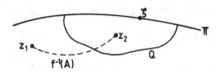

Fig. 5.5

The next result is a dual to Theorem 5.2 (iii) and Corollary 5.3 on John domains. Let $I(z)$ again be defined by (5.2.4).

Theorem 5.7. *Let f map $\mathbb{D}$ conformally onto a linearly connected domain. Then there exists $\beta < 2$ such that*

(3) $\quad |f'(\rho\zeta)| \geq \frac{1}{8}|f'(r\zeta)| \left(\frac{1-\rho}{1-r}\right)^{\beta-1} \quad$ for $\quad \zeta \in \mathbb{T}, \quad 0 \leq r \leq \rho < 1,$

(4) $\quad |f(\zeta_1) - f(\zeta_2)| \geq M_3^{-1} d_f(z) \left(\frac{|\zeta_1 - \zeta_2|}{1-|z|}\right)^{\beta} \quad$ for $\quad \zeta_1, \zeta_2 \in I(z), \quad z \in \mathbb{D}.$

The last inequality (for $z = 0$) implies the lower Hölder condition

(5) $\quad\quad\quad |f(\zeta_1) - f(\zeta_2)| \geq M_4^{-1}|\zeta_1 - \zeta_2|^{\beta} \quad$ for $\quad \zeta_1, \zeta_2 \in \mathbb{T}$

which is equivalent to the Hölder condition

(6) $\quad |f^{-1}(w_1) - f^{-1}(w_2)| \leq M_5|w_1 - w_2|^{1/\beta} \quad$ for $\quad w_1, w_2 \in \partial G$

for the inverse function. See Gai72 and NäPa83 for explicit bounds in terms of the constant M_1 in (1).

Proof. (a) For fixed b we consider the family $\mathfrak{M}$ of all functions $f(z) = z + a_2z^2 + \dots$ that are analytic and univalent in $\mathbb{D}$ and satisfy

(7) $\quad\quad\quad \text{diam } f(S) \leq b|f(z_1) - f(z_2)| \quad$ for $\quad z_1, z_2 \in \mathbb{D}$

where S is the non-euclidean segment from z_1 to z_2. It is clear that the family $\mathfrak{M}$ is compact with respect to locally uniform convergence in $\mathbb{D}$. Hence

(8) $\quad\quad\quad\quad\quad\quad \beta = \max_{f \in \mathfrak{M}} |a_2|$

is assumed. If $|a_2| = 2$ then $f(z) = z(1 - az)^{-2}$ with $|a| = 1$ so that $f(\mathbb{D})$ is a slit plane and (7) does not hold. It follows that $\beta < 2$.

The Koebe transform (1.3.12) of $f \in \mathfrak{M}$ also satisfies (7) and thus belongs to $\mathfrak{M}$. Hence we see from (8) and (1.3.12) that

(9) $\quad \left|\frac{1}{2}(1 - |z|^2)\frac{f''(z)}{f'(z)} - \bar{z}\right| \leq \beta \quad$ for $\quad z \in \mathbb{D}, \quad f \in \mathfrak{M}$

and we conclude that

(10) $\quad \left|\frac{\partial}{\partial r}\log[(1 - r^2)f'(r\zeta)]\right| = \left|\frac{f''(r\zeta)}{f'(r\zeta)} - \frac{2r\bar{\zeta}}{1 - r^2}\right| \leq \frac{2\beta}{1 - r^2}.$

If we integrate and then exponentiate we obtain

$\quad\quad \frac{(1 - \rho^2)|f'(\rho\zeta)|}{(1 - r^2)|f'(r\zeta)|} \geq \left(\frac{1-\rho}{1+\rho}\frac{1+r}{1-r}\right)^{\beta} \quad$ for $\quad 0 \leq r < \rho < 1$

which implies (3) for $f \in \mathfrak{M}$. Furthermore Corollary 1.4 shows that

(11) $\quad d_f(\rho\zeta) \geq \frac{1}{16}d_f(r\zeta)\left(\frac{1-\rho}{1-r}\right)^{\beta} \quad$ for $\quad 0 \leq r \leq \rho < 1, \quad \zeta \in \mathbb{T}.$

(b) Let now $f(\mathbb{D})$ be linearly connected. To show (3) and (4) we may assume that $f(0) = 0$ and $f'(0) = 1$. Then $f \in \mathfrak{M}$ with $b = M_2$ by (2), and (3) follows from part (a). Let $\rho\zeta$ be the point on the non-euclidean segment from ζ_1 to ζ_2 that is nearest to 0. Let first $|z| = r < \rho < 1$. We see from (2) that

$$M_2|f(\zeta_1) - f(\zeta_2)| \geq \operatorname{diam} f(S) \geq d_f(\rho\zeta)$$

and (4) follows from (11) because $1 - \rho \geq |\zeta_1 - \zeta_2|/2$ and $d_f(r\zeta) \geq M_4^{-1}d_f(z)$ by Corollary 1.6. The case $\rho < r$ follows easily from Corollary 1.6 and (2). $\square$

Remark. A family $\mathfrak{F}$ of analytic functions

(12) $$f(z) = z + a_2 z^2 + \ldots, \quad f'(z) \neq 0 \quad (z \in \mathbb{D})$$

is called *linearly invariant* (Pom64) if

(13) $$f \in \mathfrak{F}, \quad \tau \in \operatorname{M\ddot{o}b}(\mathbb{D}) \quad \Rightarrow \quad \frac{f \circ \tau - f(\tau(0))}{\tau'(0)f'(\tau(0))} \in \mathfrak{F}.$$

The *order* of $\mathfrak{F}$ is defined by $\sup\{|a_2| : f \in \mathfrak{F}\}$. The family S has order 2.

In part (a) of the above proof, we have shown first that $\mathfrak{M}$ is linearly invariant of order $\beta < 2$ and next that (3) holds (with a different constant) for any linearly invariant family of order $\beta < \infty$.

Let $0 \leq \alpha \leq 1$. The function f is called α-*close-to-convex* if it is analytic and univalent in $\mathbb{D}$ and if

(14) $$|\arg[f'(z)/h'(z)]| \leq \alpha\pi/2 \quad (z \in \mathbb{D})$$

for some convex univalent function h. The case $\alpha = 1$ has been considered in Section 3.6.

Proposition 5.8. *Let $0 \leq \alpha < 1$. If f is α-close-to-convex then $f(\mathbb{D})$ is linearly connected.*

Proof. The function $\varphi = f \circ h^{-1}$ is analytic in the convex domain $H = h(\mathbb{D})$ and satisfies $|\arg \varphi'(v)| \leq \alpha\pi/2$ for $v \in H$ by (14). If $z_j \in \mathbb{D}$ $(j = 1, 2)$ and $v_j = h(z_j)$ then

$$\frac{\varphi(v_2) - \varphi(v_1)}{v_2 - v_1} = \int_0^1 \varphi'(v_1 + t(v_2 - v_1)) \, dt = \int_0^1 e^{i \arg \varphi'}|\varphi'| \, dt$$

and thus, because $\varphi(v_j) = f(z_j)$,

$$\operatorname{Re} \frac{f(z_2) - f(z_1)}{v_2 - v_1} \geq \cos \frac{\alpha\pi}{2} \int_0^1 |\varphi'| \, dt \geq \frac{\cos(\alpha\pi/2)}{|v_1 - v_2|} \Lambda(C)$$

where $C = f \circ h^{-1}([v_1, v_2])$ is a curve in $f(\mathbb{D})$ from $f(z_1)$ to $f(z_2)$. Hence (1) holds with $M_1 = 1/\cos(\alpha\pi/2) < \infty$. $\square$

Exercises 5.3

1. Let J be a Jordan curve in $\mathbb{C}$. Show that the inner domain of J is a John domain if and only if the outer domain is linearly connected (allowing now domains that contain ∞).

2. Let $0 \leq \alpha \leq 1$. Show that the α-close-to-convex functions form a linearly invariant family of order $1 + \alpha$ (Rea56).

3. Let $\mathfrak{F}$ be a linearly invariant family of order β. Show that (Pom64, Satz 2.4)

$$2\beta^2 - 2 \leq \sup_{f \in \mathfrak{F}} \sup_{z \in \mathbb{D}} \left(1 - |z|^2\right)^2 |S_f(z)| < 2\beta^2 + K\beta$$

where S_f is the Schwarzian derivative and K is an absolute constant. (For the lower bound, consider a function where $|a_2|$ is close to β.)

5.4 Quasidisks and Quasisymmetric Functions

Let J be a Jordan curve in $\mathbb{C}$ and let $M_1, M_2, \ldots$ denote suitable positive constants that depend only on J and the function f.

We say that J is a *quasicircle* (or *quasiconformal curve*) if

$$(1) \qquad \operatorname{diam} J(a,b) \leq M_1 |a - b| \qquad \text{for} \quad a, b \in J$$

where $J(a,b)$ is the arc (of smaller diameter) of J between a and b. The inner domain G of a quasicircle J is called a *quasidisk*. Note that J need not be rectifiable. Domains where (1) holds with diameter replaced by length are called Lavrentiev domains and will be studied in Section 7.4.

A piecewise smooth Jordan curve is a quasicircle if and only if it has no cusps (of angle 0 or 2π). As a further example consider the domain

$$G = \{u + iv : 0 < u < 1, \quad u^\alpha \sin(\pi/u) < v < 1\} \quad (\alpha > 0).$$

If $\alpha > 2$ then $J = \partial G$ is piecewise smooth and has no cusps. If $\alpha = 2$ then J is still a quasicircle. If however $0 < \alpha < 2$ then $a_n = 1/(2n) \in J$, $b_n = 1/(2n+1) \in J$ and

$$\frac{\operatorname{diam} J(a_n, b_n)}{|a_n - b_n|} > \frac{2n(2n+1)}{(2n+1)^\alpha} \to \infty \qquad \text{as} \quad n \to \infty$$

so that J is not a quasidisk.

For $z \in \mathbb{D}$ we define again $d_f(z) = \operatorname{dist}(f(z), \partial G)$ and the arc $I(z) = \{e^{it} : |t - \arg z| \leq \pi(1 - |z|)\}$.

Theorem 5.9. *Let f map $\mathbb{D}$ conformally onto G. Then the following conditions are equivalent:*

(i) *G is a quasidisk;*
(ii) *G is a linearly connected John domain;*

(iii) f *is continuous in* $\overline{\mathbb{D}}$, *and if* $I(z)(1/2 \leq |z| < 1)$ *has the endpoints* e^{it_1}, e^{it_2}
$(t_1 \leq t_2 \leq t_1 + \pi)$ *then*

$$\operatorname{diam} f(I(z)) \leq M_2 d_f(z) \leq M_3 |f(e^{it_1}) - f(e^{it_2})|.$$

If G is a quasidisk then $J = \partial G$ is a Jordan curve so that f is continuous and injective in $\overline{\mathbb{D}}$. By Theorem 5.5, a finite angular derivative is an unrestricted derivative, and f is isogonal at $\zeta \in \mathbb{T}$ if and only if J has a tangent at $f(\zeta)$. Because of (ii) we can use all the estimates obtained in Sections 5.2 and 5.3. We shall establish further estimates in Section 5.6.

Proof. (i) $\Rightarrow$ (ii). Let (i) be satisfied. If $[a, b]$ is a crosscut of G then $J(a, b) \cup [a, b]$ bounds a component of $G \setminus [a, b]$ that has diameter $\leq M_1 |a - b|$. Hence G is a John domain, see (5.2.1).

In order to show that G is linearly connected (see (5.3.1)), consider $w_1, w_2 \in G$; see Fig. 5.6. We may assume that $[w_1, w_2] \not\subset G$; otherwise we choose $A = [w_1, w_2]$. Let $f(\zeta_j)$ $(\zeta_j \in \mathbb{T})$ be the boundary point on $[w_1, w_2]$ nearest to w_j. By (1) one of the arcs I of $\mathbb{T}$ from ζ_1 to ζ_2 satisfies $\operatorname{diam} f(I) \leq M_1 |f(\zeta_1) - f(\zeta_2)|$. If r is close enough to 1 then $f(rI)$ is a curve in G of diameter $2M_1 |f(\zeta_1) - f(\zeta_2)| < 2M_1 |w_1 - w_2|$ which can be connected within G by curves of length $< |w_1 - w_2|$ to w_j. Together this gives a curve $A \subset G$ from w_1 to w_2 with $\operatorname{diam} A \leq (2M_1 + 2)|w_1 - w_2|$.

(ii) $\Rightarrow$ (iii). Since G is a John domain we see from Theorem 5.2 that (see (5.2.3))

$$(2) \qquad \operatorname{diam}\{f(e^{it}) : t_1 \leq t \leq t_2\} \leq \operatorname{diam} f(B(z)) \leq M_2 d_f(z).$$

Since G is linearly connected it follows from Theorem 5.7 with $\zeta_j = e^{it_j}$ that

$$(3) \qquad d_f(z) \leq M_4 |f(e^{it_1}) - f(e^{it_2})|.$$

This gives (iii) with $M_3 = M_2 M_4$.

(iii) $\Rightarrow$ (i). To see this we choose $f(e^{it_1}) = a$ and $f(e^{it_2}) = b$. $\square$

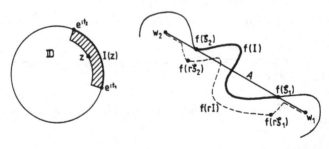

Fig. 5.6. Condition (iii) and the proof of (i) $\Rightarrow$ (ii)

The map h of $\mathbb{T}$ into $\mathbb{C}$ is called *quasisymmetric* if it is injective (one-to-one) and if there exists a strictly increasing continuous function $\lambda(x)$ ($0 \leq x < +\infty$) with $\lambda(0) = 0$ such that

$$(4) \qquad \left| \frac{h(z_1) - h(z_2)}{h(z_2) - h(z_3)} \right| \leq \lambda \left(\left| \frac{z_1 - z_2}{z_2 - z_3} \right| \right) \qquad \text{for} \quad z_1, z_2, z_3 \in \mathbb{T};$$

see BeuAh56, TuVä80, Väi81, Väi84. Reversing the roles of z_1 and z_3 we deduce that

$$(5) \qquad 1/\lambda \left(\left| \frac{z_3 - z_2}{z_2 - z_1} \right| \right) \leq \left| \frac{h(z_1) - h(z_2)}{h(z_2) - h(z_3)} \right|.$$

Since $\lambda(x) \to 0$ as $x \to 0$ it follows from (4) that h is continuous. Another consequence is (5.1.1) which conversely implies (TuVä80) that h is quasisymmetric.

Proposition 5.10. *If h is a quasisymmetric map of $\mathbb{T}$ into $\mathbb{C}$ then $J = h(\mathbb{T})$ is a quasicircle.*

Proof. We write $w_j = h(z_j)$, $z_j \in \mathbb{T}$ ($j = 1, \ldots, 4$). If w_1 and w_3 are given let z_2 lie on the (smaller) arc of $\mathbb{T}$ between z_1 and z_3. We choose z_4 on the opposite side of z_3 such that $|z_4 - z_3| = |z_3 - z_1|$. Then $|z_2 - z_3| \leq |z_3 - z_4|$ and thus, by (4),

$$\left| \frac{w_2 - w_3}{w_1 - w_3} \right| = \left| \frac{w_4 - w_3}{w_3 - w_1} \right| \left| \frac{w_2 - w_3}{w_3 - w_4} \right| \leq \lambda(1)^2. \qquad \square$$

We have thus proved (b) $\Rightarrow$ (a) of Section 5.1, and we now prove (a) $\Rightarrow$ (b).

Theorem 5.11. *If f maps $\mathbb{D}$ conformally onto a quasidisk then f restricted to $\mathbb{T}$ is quasisymmetric.*

We prove (4) with

$$(6) \qquad \lambda(x) \leq \begin{cases} M_5 x^\alpha & \text{for } 0 \leq x \leq 1, \\ M_6 x^\beta & \text{for } 1 < x < \infty, \end{cases}$$

where $\alpha > 0$ and $\beta < 2$ are the constants of Theorem 5.2 (iii) and Theorem 5.7. It is easy to modify the right-hand side such that it becomes continuous.

Proof. Let $w_j = f(z_j)$, $z_j \in \mathbb{T}$ for $j = 1, 2, 3$. We choose $z_0 \in \mathbb{D}$ such that $I(z_0)$ is the arc of $\mathbb{T}$ between z_1 and z_3 that contains z_2.

We first consider the case that $|z_1 - z_2| \leq |z_2 - z_3|$. It follows from Corollary 5.3 that

$$(7) \qquad |w_1 - w_2| \leq M_7 d_f(z_0) \left(\frac{|z_1 - z_2|}{1 - |z_0|} \right)^\alpha \leq M_8 d_f(z_0) \left| \frac{z_1 - z_2}{z_2 - z_3} \right|^\alpha$$

because $2\pi(1 - |z_0|) \geq |z_2 - z_3|$. Since $|w_2 - w_3| \geq M_9^{-1}d_f(z_0)$ by (5.3.4), we conclude that

$$\left|\frac{w_1 - w_2}{w_2 - w_3}\right| \leq M_8 M_9 \left|\frac{z_1 - z_2}{z_2 - z_3}\right|^\alpha \quad \text{if} \quad \left|\frac{z_1 - z_2}{z_2 - z_3}\right| \leq 1.$$

Now let $|z_1 - z_2| > |z_2 - z_3|$ and thus $|z_1 - z_2| \geq 1 - |z_0|$. Then

$$|w_2 - w_3| \geq M_{10}^{-1}d_f(z_0)\left(\frac{|z_2 - z_3|}{1 - |z_0|}\right)^\beta \geq M_{10}^{-1}d_f(z_0)\left|\frac{z_2 - z_3}{z_1 - z_2}\right|^\beta$$

by (5.3.4) while $|w_1 - w_2| \leq M_{11}d_f(z_0)$ by the first inequality in (7). Hence

$$\left|\frac{w_1 - w_2}{w_2 - w_3}\right| \leq M_{10}M_{11}\left|\frac{z_1 - z_2}{z_2 - z_3}\right|^\beta \quad \text{if} \quad \left|\frac{z_1 - z_2}{z_2 - z_3}\right| \geq 1. \qquad \Box$$

We need some further results about general quasisymmetric maps. The first result will be used for the proof of (b) $\Rightarrow$ (c).

Proposition 5.12. *If h and h^* are quasisymmetric maps of $\mathbb{T}$ onto the same Jordan curve then $\varphi = h^{*-1} \circ h$ is a quasisymmetric map of $\mathbb{T}$ onto $\mathbb{T}$.*

Proof. Since $h = h^* \circ \varphi$ it follows from (5) and (4) that, with $z_j^* = \varphi(z_j)$,

$$1/\lambda^*\left(\left|\frac{z_3^* - z_2^*}{z_2^* - z_1^*}\right|\right) \leq \left|\frac{h^*(z_1^*) - h^*(z_2^*)}{h^*(z_2^*) - h^*(z_3^*)}\right|$$
$$= \left|\frac{h(z_1) - h(z_2)}{h(z_2) - h(z_3)}\right| \leq \lambda\left(\left|\frac{z_1 - z_2}{z_2 - z_3}\right|\right)$$

so that (4) for φ holds with λ replaced by $1/\lambda^{*-1}(1/\lambda)$. $\qquad \Box$

Corollary 5.13. *If f and g map $\mathbb{D}$ and $\mathbb{D}^*$ conformally onto the inner and outer domains of a quasicircle, then $\varphi = g^{-1} \circ f$ is a quasisymmetric map of $\mathbb{T}$ onto $\mathbb{T}$.*

Proof. Let w_0 lie in the inner domain. It is easy to see that the function $h(z) = 1/(g(z^{-1}) - w_0)$ maps $\mathbb{D}$ conformally onto the inner domain of a quasicircle. Hence its restriction to $\mathbb{T}$ is quasisymmetric by Theorem 5.11, and since $M_{12}^{-1} \leq |h(z)| \leq M_{12}$ for $z \in \mathbb{T}$, it follows that g is quasisymmetric on $\mathbb{T}$. Hence $g^{-1} \circ f$ is quasisymmetric by Theorem 5.11 and Proposition 5.12. $\Box$

Now we give a formulation (Väi84) in terms of crossratios; these are invariant under all Möbius transformations.

Proposition 5.14. *If h is a quasisymmetric map of $\mathbb{T}$ into $\mathbb{C}$ then, for $z_j \in \mathbb{T}$ $(j = 1, \ldots, 4)$,*

(8)
$$\frac{|h(z_1) - h(z_3)|\, |h(z_2) - h(z_4)|}{|h(z_1) - h(z_4)|\, |h(z_2) - h(z_3)|} \leq \lambda^* \left(\frac{|z_1 - z_3|\, |z_2 - z_4|}{|z_1 - z_4|\, |z_2 - z_3|} \right)$$

where the function λ^* depends only on λ.

Proof. We write

(9)
$$a = \left| \frac{z_1 - z_3}{z_1 - z_4} \right|, \qquad b = \frac{|z_1 - z_3|\, |z_2 - z_4|}{|z_1 - z_4|\, |z_2 - z_3|}.$$

By symmetry we may assume that

(10)
$$|z_1 - z_3| \leq |z_2 - z_4|, \qquad |z_2 - z_3| \leq |z_1 - z_4|$$

hence that $a^2 \leq b$. The quasisymmetry and (9) show that

$$b^* \equiv \left| \frac{h(z_1) - h(z_3)}{h(z_1) - h(z_4)} \right| \left| \frac{h(z_2) - h(z_4)}{h(z_2) - h(z_3)} \right| \leq \lambda(a)\lambda\left(\frac{b}{a} \right).$$

If $a \geq 1$ it follows that $b^* \leq \lambda(\sqrt{b})\lambda(b)$; if $a \geq b$ it follows that $b^* \leq \lambda(\sqrt{b})\lambda(1)$. This leaves the case that $a < 1$ and $a < b$. Then

$$|z_2 - z_4| \leq |z_2 - z_3| + |z_3 - z_1| + |z_1 - z_4| \leq (2 + a)|z_1 - z_4| < 3|z_1 - z_4|$$

by (9) and (10), and since furthermore

$$|z_1 - z_4| \leq |z_1 - z_3| + |z_3 - z_2| + |z_2 - z_4| \leq (2 + a/b)|z_2 - z_4| < 3|z_2 - z_4|$$

we conclude from (9) and the quasisymmetry that

$$b^* = \left| \frac{h(z_2) - h(z_4)}{h(z_1) - h(z_4)} \right| \left| \frac{h(z_1) - h(z_3)}{h(z_2) - h(z_3)} \right| \leq \lambda(3)\lambda(3b). \qquad \square$$

Exercises 5.4

1. Write $w = u + iv$ and consider the domain

$$G = \{0 < u < 1, \, 0 < v < 1\} \setminus \bigcup_{n=1}^{\infty} \left\{ \frac{1}{2^n} \leq u \leq \frac{a_n}{2^n}, \, 0 \leq v \leq \frac{1}{2^n} \right\}$$

where $1 < a_n < 2$. When is G a John domain, linearly connected or a quasidisk?

2. The *snowflake* curve is defined by the following infinite construction: Start with an equilateral triangle and replace the middle third of each side by the other two sides of an outward-pointing equilateral triangle. In the resulting polygon, replace the middle third of each of the 12 sides by the other two sides of an equilateral triangle, and so on. Show that the snowflake curve is a non-rectifiable quasicircle.

3. Let f map $\mathbb{D}$ conformally onto a quasidisk. Show that there are constants $0 < \alpha < \beta < 2$ and $c > 0$, $M < \infty$ such that

$$c \left| \frac{\zeta_1 - \zeta_2}{z_1 - z_2} \right|^\beta \leq \left| \frac{f(\zeta_1) - f(\zeta_2)}{f(z_1) - f(z_2)} \right| \leq M \left| \frac{\zeta_1 - \zeta_2}{z_1 - z_2} \right|^\alpha$$

for $z_1, z_2 \in \mathbb{T}$ and for ζ_1, ζ_2 on the smaller arc of $\mathbb{T}$ between z_1 and z_2.

4. If G is a quasidisk show that $\Lambda(C) \leq M|w_1 - w_2|$ where C is the hyperbolic geodesic from w_1 to w_2 in $\overline{G}$. (Use e.g. Theorem 5.2 (iii) and Proposition 5.6.)

5. Let h be a quasisymmetric map of $\mathbb{T}$ onto J and φ of $\mathbb{T}$ onto $\mathbb{T}$. If g is univalent in a domain $G \supset J$, show that $g \circ h \circ \varphi$ is quasisymmetric.

6. Let h_n be a sequence of quasisymmetric maps of $\mathbb{T}$ into a fixed disk D with a fixed function λ in (4). Show that (h_n) is equicontinuous and deduce that some subsequence converges uniformly to a quasisymmetric map or a constant.

5.5 Quasiconformal Extension

1. We first consider the extension of quasisymmetric maps of $\mathbb{T}$ to quasiconformal maps of $\mathbb{D} \cup \mathbb{T}$.

Theorem 5.15. *The sense-preserving homeomorphism φ of $\mathbb{T}$ onto $\mathbb{T}$ can be extended to a homeomorphism $\widetilde{\varphi}$ of $\overline{\mathbb{D}}$ onto $\overline{\mathbb{D}}$ that is real-analytic in $\mathbb{D}$ and has the following properties:*
   (i) *If $\sigma, \tau \in \mathrm{M\ddot{o}b}(\mathbb{D})$ then the extension of $\sigma \circ \varphi \circ \tau$ is given by $\sigma \circ \widetilde{\varphi} \circ \tau$.*
   (ii) *If φ is quasisymmetric then $\widetilde{\varphi}$ is quasiconformal in $\mathbb{D}$.*

A complex function is real-analytic if its real and imaginary parts are real-analytic functions of x and y. We shall show that

$$(1) \qquad \left| \frac{\partial \widetilde{\varphi}}{\partial \bar{z}} \middle/ \frac{\partial \widetilde{\varphi}}{\partial z} \right| \leq \kappa < 1 \quad \text{for} \quad z \in \mathbb{D}$$

where the constant κ depends only on the function λ in the definition (5.4.4) of quasisymmetry.

The proof of this important result (BeuAh56, DoEa86) will be postponed to later in this section. The Beurling-Ahlfors extension does not satisfy (i) and is not difficult to construct (see e.g. LeVi73, p. 83); it would suffice to prove (b) $\Rightarrow$ (c) in Section 5.1.

We shall construct the Douady-Earle extension. It is uniquely determined by the *conformal naturalness* property (i) together with

$$(2) \qquad \int_{\mathbb{T}} \varphi(\zeta)|d\zeta| = 0 \quad \Rightarrow \quad \widetilde{\varphi}(0) = 0 \,.$$

We now deduce (BeuAh56, Tuk81, DoEa86):

Theorem 5.16. *Every quasisymmetric map h of $\mathbb{T}$ into $\mathbb{C}$ can be extended to a quasiconformal map $\tilde{h}$ of $\mathbb{C}$ onto $\widehat{\mathbb{C}}$ such that the extension of*

(3) $$\sigma \circ h \circ \tau, \quad \sigma \in \text{Möb}, \quad \tau \in \text{Möb}(\mathbb{D})$$

is given by $\sigma \circ \tilde{h} \circ \tau$.

Our extension $\tilde{h}$ is a homeomorphism of $\mathbb{C}$ onto $\mathbb{C}$ that is real-analytic in $\mathbb{C} \setminus \mathbb{T}$ and satisfies

(4) $$\left| \frac{\partial \tilde{h}}{\partial \bar{z}} \Big/ \frac{\partial \tilde{h}}{\partial z} \right| \leq \kappa < 1 \quad \text{for} \quad |z| \neq 1.$$

We will not need the general theory of quasiconformal maps at this stage; we remark only that a homeomorphism $\tilde{h}$ of $\mathbb{C}$ onto $\mathbb{C}$ satisfying (4) is quasiconformal because $\mathbb{T}$ is a removable set for quasiconformality.

Proof. Since h is quasisymmetric it follows from Proposition 5.10 that $J = h(\mathbb{T})$ is a quasicircle which we may assume to be positively oriented. Let f and g map $\mathbb{D}$ and $\mathbb{D}^*$ onto the inner and outer domain of J such that $g(\infty) = \infty$. The restrictions of f and g to $\mathbb{T}$ are quasisymmetric by Theorem 5.11. Hence

(5) $$\varphi = f^{-1} \circ h, \quad \psi = g^{-1} \circ h$$

are quasisymmetric (Proposition 5.12).

Let $\tilde{\varphi}$ denote the Douady-Earle extension of φ to $\overline{\mathbb{D}}$ (see Theorem 5.15). Similarly there is an extension $\tilde{\psi}$ of ψ to $\overline{\mathbb{D}}^*$. We define

(6) $$\tilde{h}(z) = \begin{cases} f(\tilde{\varphi}(z)) & \text{for } |z| \leq 1, \\ g(\tilde{\psi}(z)) & \text{for } |z| \geq 1; \end{cases}$$

if $|z| = 1$ then both definitions agree and $\tilde{h}(z) = h(z)$ by (5). It follows that $\tilde{h}$ is a homeomorphism of $\widehat{\mathbb{C}}$ onto $\widehat{\mathbb{C}}$. This function is real-analytic in $\mathbb{C} \setminus \mathbb{T}$ by Theorem 5.15. If $|z| < 1$ then

$$\frac{\partial \tilde{h}}{\partial \bar{z}} = f'(\tilde{\varphi}) \frac{\partial \tilde{\varphi}}{\partial \bar{z}}, \quad \frac{\partial \tilde{h}}{\partial z} = f'(\tilde{\varphi}) \frac{\partial \tilde{\varphi}}{\partial z},$$

so that (4) follows from (1); the case $|z| > 1$ is handled in a similar manner.

Let $\sigma \in \text{Möb}$ and $\tau \in \text{Möb}(\mathbb{D})$, thus also $\tau \in \text{Möb}(\mathbb{D}^*)$. Consider $h^* = \sigma \circ h \circ \tau$ and the conformal maps $f^* = \sigma \circ f \circ \tau$ and $g^* = \sigma \circ g \circ \tau$. The functions corresponding to (5) are $\varphi^* = \tau^{-1} \circ \varphi \circ \tau$ and $\psi^* = \tau^{-1} \circ \psi \circ \tau$. Hence it follows from the conformal naturalness property (i) that

$$\tilde{h}^*(z) = \begin{cases} f^* \circ \tilde{\varphi}^*(z) = \sigma \circ f \circ \tilde{\varphi} \circ \tau(z) & \text{for } |z| \leq 1, \\ g^* \circ \tilde{\psi}^*(z) = \sigma \circ g \circ \tilde{\psi} \circ \tau(z) & \text{for } |z| \geq 1. \end{cases}$$

Hence $\tilde{h}^* = \sigma \circ \tilde{h} \circ \tau$ by (6). $\qquad \square$

Theorem 5.17. *Every conformal map f of $\mathbb{D}$ onto a quasidisk can be extended to a quasiconformal map $\tilde{f}$ of $\widehat{\mathbb{C}}$ onto $\widehat{\mathbb{C}}$. The extension of $\sigma \circ f \circ \tau$ with $\sigma \in \mathrm{M\ddot{o}b}$ and $\tau \in \mathrm{M\ddot{o}b}(\mathbb{D})$ is given by $\sigma \circ \tilde{f} \circ \tau$.*

This result (Ahl63, DoEa86) establishes (a) $\Rightarrow$ (c) and thus (b) $\Rightarrow$ (c) in Section 5.1 because (a) $\Leftrightarrow$ (b) by Section 5.4.

Proof. We choose h in (5) as the restriction of f to $\mathbb{T}$ which is quasisymmetric by Theorem 5.11. Then φ is the identity and it follows from (i) and (2) that $\tilde{\varphi}$ is also the identity, and (6) therefore defines the quasiconformal extension of f to $\widehat{\mathbb{C}}$. $\square$

2. Let now φ be a sense-preserving homeomorphism of $\mathbb{T}$ onto $\mathbb{T}$. The Douady-Earle extension of φ is based on the function

$$(7) \qquad F(z,w) \equiv F_\varphi(z,w) = \int_{\mathbb{T}} \frac{w - \varphi(\zeta)}{1 - \overline{w}\varphi(\zeta)} p(z,\zeta)|d\zeta| \quad (z,w \in \mathbb{D})$$

where $p(z,\zeta)$ is the Poisson kernel (3.1.4). Since $\int_{\mathbb{T}} p(z,\zeta)|d\zeta| = 1$ we can rewrite (7) as

$$(8) \qquad F(z,w) = w - \frac{1 - |w|^2}{2\pi} \int_{\mathbb{T}} \frac{\varphi(\zeta)}{1 - \overline{w}\varphi(\zeta)} \frac{1 - |z|^2}{|\zeta - z|^2} |d\zeta|.$$

Let $\sigma, \tau \in \mathrm{M\ddot{o}b}(\mathbb{D})$. We have

$$(9) \qquad \frac{\sigma(w) - \sigma(\omega)}{1 - \overline{\sigma(w)}\sigma(\omega)} q(w) = \frac{w - \omega}{1 - \overline{w}\omega}, \quad |q(w)| = 1$$

where q is independent of ω. Hence it follows from (7) with the substitution $\zeta = \tau(s)$ that

$$F(\tau(z),w) = \int_{\mathbb{T}} \frac{w - \varphi \circ \tau(s)}{1 - \overline{w}\varphi \circ \tau(s)} p(\tau(z),\tau(s))|\tau'(s)| \, |ds|$$

$$= q(w) \int_{\mathbb{T}} \frac{\sigma(w) - \sigma \circ \varphi \circ \tau(s)}{1 - \overline{\sigma(w)}\sigma \circ \varphi \circ \tau(s)} p(z,s)|ds|$$

and thus, by (7),

$$(10) \qquad F(\tau(z),\sigma^{-1}(w)) = \tilde{q}(w) F_{\sigma \circ \varphi \circ \tau}(z,w), \quad |\tilde{q}(w)| = 1.$$

This property will make it possible to transfer information from $(0,0)$ to any point of $\mathbb{D} \times \mathbb{D}$.

We now study (DoEa86) the local behaviour of F at $(0,0)$. The Fourier coefficients of φ are

$$(11) \qquad a_n = \frac{1}{2\pi} \int_{\mathbb{T}} \overline{\zeta}^n \varphi(\zeta)|d\zeta| \quad (n \in \mathbb{Z}).$$

We assume that

(12)
$$a_0 = \frac{1}{2\pi} \int_{\mathbb{T}} \varphi(\zeta)|d\zeta| = 0.$$

It follows from (8) that the partial derivatives satisfy

(13)
$$F_z(0,0) = -\frac{1}{2\pi} \int_{\mathbb{T}} \overline{\zeta}\varphi(\zeta)|d\zeta| = -a_1,$$

(14)
$$F_{\overline{z}}(0,0) = -\frac{1}{2\pi} \int_{\mathbb{T}} \zeta\varphi(\zeta)|d\zeta| = -a_{-1},$$

(15)
$$F_w(0,0) = 1,$$

(16)
$$F_{\overline{w}}(0,0) = -\frac{1}{2\pi} \int_{\mathbb{T}} \varphi(\zeta)^2|d\zeta| = -2\sum_{n=1}^{\infty} a_n a_{-n} \equiv b.$$

We deduce from Lemma 5.18 below that

(17)
$$\delta \equiv |a_1|^2 - |a_{-1}|^2 > 0.$$

We see from (16) that

(18)
$$|b|^2 \le |b| \le 2|a_1 a_{-1}| + \sum_{n=2}^{\infty}\left(|a_n|^2 + |a_{-n}|^2\right)$$
$$= 1 - \left(|a_1| - |a_{-1}|\right)^2 \le 1 - \delta^2/4$$

where we have used Parseval's formula and $|\varphi(\zeta)| = 1$. Hence it follows from (14), (15) and (16) that the Jacobian satisfies

(19)
$$1 \ge J_w(0,0) \equiv |F_w(0,0)|^2 - |F_{\overline{w}}(0,0)|^2 = 1 - |b|^2 \ge \frac{\delta^2}{4}.$$

Since $F(0,0) = 0$ by (12), it follows from the implicit function theorem that $F(z,w) = 0$ has a unique solution $w = \widetilde{\varphi}(z)$ near 0 with $\widetilde{\varphi}(0) = 0$. The partial derivatives are

(20)
$$\widetilde{\varphi}_z = (\overline{F}_{\overline{z}}F_{\overline{w}} - F_z\overline{F}_w)/J_w, \quad \widetilde{\varphi}_{\overline{z}} = (\overline{F}_z F_{\overline{w}} - F_{\overline{z}}\overline{F}_w)/J_w$$

so that, by (13)–(16),

(21)
$$\widetilde{\varphi}_z(0) = \frac{a_1 - \overline{a}_{-1}b}{1 - |b|^2}, \quad \widetilde{\varphi}_{\overline{z}}(0) = \frac{a_{-1} - \overline{a}_1 b}{1 - |b|^2}.$$

Hence we deduce from (17) and (19) that

(22)
$$|\widetilde{\varphi}_z(0)|^2 - |\widetilde{\varphi}_{\overline{z}}(0)|^2 = \frac{|a_1|^2 - |a_{-1}|^2}{1 - |b|^2} \ge \delta$$

so that $\widetilde{\varphi}$ is a local diffeomorphism near 0.

Lemma 5.18. *Write* $\varphi(e^{it}) = e^{iu(t)}$ *where* u *is real and continuous with* $u(t + 2\pi) = u(t) + 2\pi$ *and suppose that*

(23) $u(t + s) - u(t) \geq \alpha$ *for* $\dfrac{\pi}{3} \leq s \leq \dfrac{2\pi}{3}$, $t \in \mathbb{R}$

with $0 < \alpha \leq \pi/3$. *Then*

(24) $\delta = |a_1|^2 - |a_{-1}|^2 \geq 2\pi^{-1} \sin^3 \alpha > 0$

(25) $|\widetilde{\varphi}_{\bar{z}}(0)/\widetilde{\varphi}_z(0)| \leq 1 - \dfrac{1}{125} \sin^9 \alpha$.

Since φ is a sense preserving homeomorphism of $\mathbb{T}$ onto $\mathbb{T}$ it is clear that (23) holds for some $\alpha > 0$ and (17) thus follows from (24).

Proof. We see from (11) that

$$\delta = \frac{1}{4\pi^2} \int_0^{2\pi} \int_0^{2\pi} \left(e^{i(t - \tau - u(t) + u(\tau))} - e^{i(-t + \tau - u(t) + u(\tau))} \right) d\tau \, dt \,.$$

Writing $\tau = t + s$ and taking the real part we obtain

(26)
$$\delta = \frac{1}{2\pi^2} \int_0^{2\pi} \int_0^{2\pi} \sin s \, \sin[u(t + s) - u(t)] \, dt \, ds$$
$$= \frac{1}{2\pi^2} \int_0^{\pi} \sin s \left(\int_0^{\pi} v(t, s) \, dt \right) ds$$

where, because $u(t + 2\pi) = u(t) + 2\pi$,

$$v(t, s) = \sin[u(t + s) - u(t)] + \sin[u(t + 2\pi) - u(t + s + \pi)]$$
$$+ \sin[u(t + s + \pi) - u(t + \pi)] + \sin[u(t + \pi) - u(t + s)] \,.$$

The terms in the square brackets sum up to 2π. Each is nonnegative for $0 \leq s \leq \pi$ and $\geq \alpha$ for $\pi/3 \leq s \leq 2\pi/3$ by (23). Hence Lemma 5.19 below shows that

$$v(t, s) \geq \begin{cases} 0 & \text{for } 0 \leq s \leq \pi, \\ 4\sin^3 \alpha & \text{for } \pi/3 \leq s \leq 2\pi/3 \end{cases}$$

and (24) follows from (26). Furthermore, by (21) and (18),

$$1 - \left| \frac{\widetilde{\varphi}_{\bar{z}}(0)}{\widetilde{\varphi}_z(0)} \right|^2 = 1 - \left| \frac{a_{-1} - \bar{a}_1 b}{a_1 - \bar{a}_{-1} b} \right|^2 = \frac{\left(|a_1|^2 - |a_{-1}|^2 \right) \left(1 - |b|^2 \right)}{|a_1 - \bar{a}_{-1} b|^2} \geq \frac{\delta^3}{16}$$

which implies (25).   □

Lemma 5.19. *If $x_j \geq \alpha > 0$ $(j = 1, 2, 3, 4)$ and $x_1 + x_2 + x_3 + x_4 = 2\pi$ then*

$$(27) \qquad \sigma \equiv \sin x_1 + \sin x_2 + \sin x_3 + \sin x_4 \geq \begin{cases} 4\sin^3 \alpha & \text{if } \alpha \leq \pi/3, \\ 0 & \text{otherwise}. \end{cases}$$

Proof. We consider only the case $\alpha \leq \pi/3$ and may assume that $\alpha \leq x_1 \leq x_2 \leq x_3 \leq x_4$. We consider first the case that $x_4 \leq \pi$. Then all terms in (27) are non negative, furthermore $2x_1 + 2x_3 \leq 2\pi$ and thus $x_3 \leq \pi - \alpha$ and $\sin x_j \geq \sin \alpha$ for $j = 1, 2, 3$. It follows that

$$\sigma \geq 3\sin \alpha \geq 4\sin^3 \alpha.$$

Consider next the case that $x_4 > \pi$. Then $x_4 = 2\pi - y$ with $0 \leq y < \pi$. Hence $y = x_1 + x_2 + x_3$ and thus $\alpha \leq x_1 \leq x_2 \leq \pi/2$, $\alpha \leq x_3 \leq \pi - 2\alpha$. It follows that

$$\begin{aligned} \sin y &= \cos(x_1 + x_2)\sin x_3 + \cos x_3 \sin(x_1 + x_2) \\ &\leq \cos 2\alpha \sin x_3 + (\cos x_1 \sin x_2 + \cos x_2 \sin x_1)\cos \alpha \\ &\leq (\sin x_3 + \sin x_2 + \sin x_1)\cos^2 \alpha - \sin^3 \alpha \end{aligned}$$

therefore

$$\begin{aligned} \sigma &= \sin x_1 + \sin x_2 + \sin x_3 - \sin y \\ &\geq (\sin x_1 + \sin x_2 + \sin x_3)\sin^2 \alpha + \sin^3 \alpha \geq 4\sin^3 \alpha. \qquad \square \end{aligned}$$

Lemma 5.20. *For each $z \in \mathbb{D}$ there exists a unique $w \in \mathbb{D}$ with $F_\varphi(z, w) = 0$.*

Proof. Let $z \in \mathbb{D}$ be fixed. We see from (8) that

$$(28) \qquad |F(z, w) - w| \leq \frac{1}{2\pi} \int_{\mathbb{T}} \frac{1 - |w|^2}{|\varphi(\zeta) - w|} \frac{1 - |z|^2}{|\zeta - z|^2} |d\zeta|.$$

Let now $\omega_0 \in \mathbb{T}$ and $w_n \in \mathbb{D}$ such that $w_n \to w_0$ as $n \to \infty$. For $\zeta \notin \varphi^{-1}(w_0)$ the integrand tends $\to 0$, and since it is bounded by $4/(1 - |z|)$, it follows from Lebesgue's bounded convergence theorem that $|F(z, w_n) - w_n| \to 0$ as $n \to \infty$. Hence $F(z, w)$ becomes continuous in $\overline{\mathbb{D}}$ if we define $F(z, w) = w$ for $w \in \mathbb{T}$.

We conclude from (10) and (19) that $w \mapsto F(z, w)$ has a positive Jacobian for all w with $F(z, w) = 0$. Hence our assertion is a consequence of the following special case of the *Poincaré-Hopf theorem* (Mil65, p. 35):

Consider a vector field that is continuous in $\overline{\mathbb{D}}$ and continuously differentiable in $\mathbb{D}$. If it points outward at all points of $\mathbb{T}$ and if its Jacobian is positive at all its zeros, then it has exactly one zero in $\mathbb{D}$. $\qquad \square$

3. Let again φ be a sense-preserving homeomorphism of $\mathbb{T}$ onto $\mathbb{T}$. We define

$$(29) \qquad \widetilde{\varphi}(z) = \begin{cases} \varphi(z) & \text{for } z \in \mathbb{T}. \\ w & \text{where } F_\varphi(z, w) = 0 \text{ for } z \in \mathbb{D}; \end{cases}$$

there is a unique $w \in \mathbb{D}$ with $F_\varphi(z, w) = 0$ by Lemma 5.20. We call $\widetilde{\varphi}$ the *Douady-Earle extension* (or *barycentric extension*) of φ (DoEa86). It is thus uniquely determined by

$$(30) \qquad F_\varphi(z, \widetilde{\varphi}(z)) \equiv \frac{1}{2\pi} \int_\mathbb{T} \frac{\widetilde{\varphi}(z) - \varphi(\zeta)}{1 - \overline{\widetilde{\varphi}(z)}\varphi(\zeta)} \frac{1 - |z|^2}{|\zeta - z|^2} |d\zeta| = 0 \,.$$

The complex harmonic extension

$$\varphi^*(z) = \int_\mathbb{T} \varphi(\zeta) p(z, \zeta) |d\zeta| \ (z \in \mathbb{D}), \quad \varphi^*(z) = \varphi(z) \ (z \in \mathbb{T})$$

is continuous in $\overline{\mathbb{D}}$. We remark that φ^* is a homeomorphism of $\overline{\mathbb{D}}$ onto $\overline{\mathbb{D}}$; we shall not use this result (Cho45). See e.g. ClShS84 for the theory of injective complex harmonic functions.

Proof of Theorem 5.15. (a) We deduce from (8) and (30) that, for $\zeta_0 \in \mathbb{T}$,

$$\widetilde{\varphi}(z) - \varphi^*(z) = \int_\mathbb{T} \frac{\varphi(\zeta) - \widetilde{\varphi}(z)}{1 - \overline{\widetilde{\varphi}(z)}\varphi(\zeta)} \overline{\widetilde{\varphi}}(z)(\varphi(\zeta) - \varphi(\zeta_0)) p(z, \zeta) |d\zeta|$$

and thus

$$|\widetilde{\varphi}(z) - \varphi^*(z)| \leq \int_\mathbb{T} |\varphi(\zeta) - \varphi(\zeta_0)| p(z, \zeta) |d\zeta| \,.$$

If now $z \to \zeta_0$, $z \in \mathbb{D}$ then the right-hand side tends $\to 0$ because φ is continuous. Since φ^* is continuous in $\overline{\mathbb{D}}$ it follows that $\widetilde{\varphi}$ is continuous in $\overline{\mathbb{D}}$. The conformal naturalness condition (i) is a consequence of (30) and (10). We denote the extension $\widetilde{\varphi}$ also by φ.

Given $z_0 \in \mathbb{D}$ we choose $\sigma, \tau \in \text{Möb}(\mathbb{D})$ such that $\tau(0) = z_0$, $\sigma(0) = \varphi(z_0)$ and define $h = \sigma^{-1} \circ \varphi \circ \tau$. Then $h(0) = 0$ and

$$(31) \qquad \begin{aligned} h_z(0) &= \varphi_z(z_0)\tau'(0)/\sigma'(0) \,, \\ h_{\overline{z}}(0) &= \varphi_{\overline{z}}(z_0)\overline{\tau'(0)}/\sigma'(0) \end{aligned}$$

and thus $|\varphi_z(z_0)|^2 - |\varphi_{\overline{z}}(z_0)|^2 > 0$ by (22). Since $\varphi(\mathbb{D}) \subset \mathbb{D}$ and $\varphi(\mathbb{T}) = \mathbb{T}$ it follows that φ is an unbranched proper covering and thus a homeomorphism of $\overline{\mathbb{D}}$ onto $\overline{\mathbb{D}}$.

(b) Now let φ be quasisymmetric on $\mathbb{T}$. We see from (31) that

$$(32) \qquad |\varphi_{\overline{z}}(z_0)/\varphi_z(z_0)| = |h_{\overline{z}}(0)/h_z(0)| \,.$$

We write $h = e^{iu}$. Suppose that (23) does not hold so that $u(t + s) - u(t) < \alpha$ for some t and s with $\pi/3 \leq s \leq 2\pi/3$. Then $|h(e^{i(t+s)}) - h(e^{it})| < 2 \sin \frac{\alpha}{2} < \alpha$. If

$$z_1 = e^{it+is} \,, \quad z_2 = e^{it+10\pi i/9} \,, \quad z_3 = e^{it+11\pi i/9} \,, \quad z_4 = e^{it}$$

then

$$\frac{|z_1 - z_3| \, |z_2 - z_4|}{|z_1 - z_4| \, |z_2 - z_3|} \le \frac{2}{\sin(\pi/18)} = M_1 \,.$$

We apply Proposition 5.14 to the quasisymmetric function φ. Since (5.4.8) is Möbius invariant and $h = \sigma^{-1} \circ \varphi \circ \tau$ we therefore see that

$$|h(z_1) - h(z_3)| \, |h(z_2) - h(z_4)| \le 2\lambda^*(M_1)|h(z_1) - h(z_4)| < M_2\alpha$$

where λ^* depends only on λ. We conclude that

$$\min(|h(z_1) - h(z_3)|, \, |h(z_2) - h(z_4)|) < (M_2\alpha)^{1/2} \,.$$

Hence one of the two arcs of $\mathbb{T}$ either between z_1 and z_3 or between z_2 and z_4 is mapped onto an arc of $\mathbb{T}$ of length $< M_3\alpha^{1/2}$. Adding the arc between z_1 and z_4 we get, in all four possible cases, an arc $I \subset \mathbb{T}$ such that

(33) $$\Lambda(I) \ge \frac{10\pi}{9} \,, \quad \Lambda(h(I)) < \alpha + M_3\alpha^{1/2} < M_4\alpha^{1/2} \,.$$

Since $|h(z_4)| = 1$ we conclude that

$$\left| \int_{\mathbb{T}} h|dz| \right| = \left| \int_{I} h|dz| + \int_{\mathbb{T} \setminus I} h|dz| \right| \ge \frac{10\pi}{9}(1 - M_4\alpha^{1/2}) - \frac{8\pi}{9} \,.$$

The first integral is zero by (30) because $h(0) = 0$. It follows that (23) holds if $\alpha < M_5^{-1}$. Note that M_5 depends only on λ^* and thus λ and not on z_0. Consequently (32) and (25) show that

$$|\varphi_{\bar{z}}(z_0)/\varphi_z(z_0)| \le \kappa < 1 \quad \text{for} \quad z_0 \in \mathbb{D}$$

for some constant κ. Thus φ is quasiconformal in $\mathbb{D}$. $\qquad\qquad \square$

For a later application we need a further property; again we use the Möbius invariant condition (5.4.8).

Proposition 5.21. *Let φ be a sense-preserving quasisymmetric map of $\mathbb{T}$ onto $\mathbb{T}$ and suppose that*

(34) $$\omega_\nu = e^{i\vartheta + 2\pi i\nu/3} \,, \quad \varphi(\omega_\nu) = \omega_\nu \quad \text{for} \quad \nu = 1, 2, 3 \,.$$

Then the Douady-Earle extension $\widetilde{\varphi}$ of φ satisfies

(35) $$|\widetilde{\varphi}(0)| \le r_0 < 1$$

where the constant r_0 depends only on λ.

Proof. We may assume that $\widetilde{\varphi}(0) = r$ with $0 \le r < 1$. It follows from (30) that

(36) $$2\pi = \int_{\mathbb{T}} \left(1 - \frac{r - \varphi(\zeta)}{1 - r\varphi(\zeta)} \right) |d\zeta| = \int_{\mathbb{T}} \frac{(1 - r)(\varphi(\zeta) + 1)}{1 - r\varphi(\zeta)} |d\zeta| \,.$$

We consider the interval

(37) $$I = \{\zeta \in \mathbb{T} : |\varphi(\zeta) - r| \leq 6(1-r)\}$$

and obtain from (36) that

$$2\pi \leq \int_I 2|d\zeta| + \int_{\mathbb{T}\setminus I} \frac{2(1-r)}{6(1-r)}|d\zeta| \leq 2\Lambda(I) + \frac{2\pi}{3}$$

and thus $\Lambda(I) \geq 2\pi/3$. Hence $\omega_\nu \in I$ for some ν, say $\nu = 3$, and there exists $z_1 \in I$ with $|z_1 - \omega_3| = 1$. We obtain from (34) and from Proposition 5.14 with $z_2 = \omega_1$, $z_3 = \omega_2$ and $z_4 = \omega_3$ that

(38) $$\left|\frac{\varphi(z_1) - \omega_2}{\varphi(z_1) - \omega_3}\right| \left|\frac{\omega_1 - \omega_3}{\omega_1 - \omega_2}\right| \leq \lambda^* \left(\left|\frac{z_1 - \omega_2}{z_1 - \omega_3}\right| \left|\frac{\omega_1 - \omega_3}{\omega_3 - \omega_2}\right| \right) ;$$

the second factors are $= 1$. Since $|\varphi(z_1) - \omega_3| \leq |\varphi(z_1) - r| + |\varphi(\omega_3) - r| \leq 12(1-r)$ by (37) it follows from (38) that

$$1 < |\omega_2 - \omega_3| \leq |\varphi(z_1) - \omega_2| + |\varphi(z_1) - \omega_3| \leq 12(1-r)(1 + \lambda^*(2))$$

and thus $\varphi(0) = r < 1 - [12(1 + \lambda^*(2))]^{-1}$. □

Exercises 5.5.

1. Let f map $\mathbb{D}$ conformally onto a quasidisk in $\mathbb{C}$ and let $\widetilde{f}$ be the quasiconformal extension of Theorem 5.17. Show that the extension of the Koebe transform of f is formally the Koebe transform of $\widetilde{f}$. If f is odd, show that $\widetilde{f}$ is odd.

2. Let φ_n be quasisymmetric maps of $\mathbb{T}$ onto $\mathbb{T}$ with the same function λ in (5.4.4). If $\varphi_n \to \varphi$ as $n \to \infty$ uniformly on $\mathbb{T}$, show that $\widetilde{\varphi}_n \to \widetilde{\varphi}$ uniformly in $\mathbb{D}$.

3. Let φ be a sense-preserving homeomorphism of $\mathbb{T}$ onto $\mathbb{T}$ such that $|\varphi(\zeta) - \varphi(\zeta')| \leq M|\zeta - \zeta'|$. Prove that

$$\widetilde{\varphi}(r\zeta) - \varphi(\zeta) = O\left((1-r)\log\frac{1}{1-r}\right) \qquad \text{as} \quad r \to 1-.$$

See Ear89 for more precise results.

5.6 Analytic Families of Injections

We now prove a surprising result (MaSaSu83) of a different nature.

Theorem 5.22. *Suppose that*

(i) $f_\lambda : \mathbb{D} \to \mathbb{C}$ *is injective for each* $\lambda \in \mathbb{D}$,
(ii) $f_\lambda(z)$ *is analytic in* $\lambda \in \mathbb{D}$ *for each fixed* $z \in \mathbb{D}$,
(iii) $f_0(z) = z$ *for* $z \in \mathbb{D}$.

Then each $f_\lambda(\lambda \in \mathbb{D})$ can be extended to a homeomorphism of $\overline{\mathbb{D}}$ onto a closed quasidisk in $\mathbb{C}$.

Thus f_λ, $\lambda \in \mathbb{D}$, is an *analytic family* of injections of $\mathbb{D}$ into $\mathbb{C}$ that reduce to the identity for $\lambda = 0$. Note that the continuity of f_λ is a conclusion, not an assumption.

Even more is true though we do not need this (BeRo86): Under the above assumptions, f_λ is the restriction to $\mathbb{D}$ of a $|\lambda|$-quasiconformal map of $\widehat{\mathbb{C}}$. Furthermore (SuThu86, BeRo86) the extension can be made such that it is analytic in $\{|\lambda| < 1/3\}$.

The proof of these additional facts is quite deep whereas Theorem 5.22 is an easy consequence of *Schottky's theorem*: If h is analytic and $\neq 0, 1$ in $\mathbb{D}$ then

(1) $$|h(\lambda)| \le \Phi(|h(0)|, (1 + |\lambda|)/(1 - |\lambda|)) \quad \text{for} \quad |\lambda| < 1$$

where $\Phi(x, t)$ $(0 \le x < +\infty)$ is a (universal) strictly increasing continuous function with $\Phi(0, t) = 0$. Very good bounds (Hem80, Hem88) are

(2) $$\Phi(x, t) < \begin{cases} \exp[t\pi^2/\log(16/x)] & \text{for } 0 < x \le 1 \\ (e^\pi x)^t & \text{for } 1 < x < \infty \end{cases}$$

which we will however not need.

Proof of Theorem 5.22. Let z_1, z_2, z_3 be distinct points in $\mathbb{D}$. The function

$$h(\lambda) = (f_\lambda(z_1) - f_\lambda(z_2))/(f_\lambda(z_3) - f_\lambda(z_2)) \quad (\lambda \in \mathbb{D})$$

is analytic and $\neq 0, 1$ with $h(0) = (z_1 - z_2)/(z_3 - z_2)$. Hence we obtain from (1) that

(3) $$\left| \frac{f_\lambda(z_1) - f_\lambda(z_2)}{f_\lambda(z_2) - f_\lambda(z_3)} \right| \le \Phi\left(\left| \frac{z_1 - z_2}{z_2 - z_3} \right|, \frac{1 + |\lambda|}{1 - |\lambda|} \right).$$

First let $\lambda \in \mathbb{D}$ be fixed. If also z_2, z_3 are fixed we deduce from (3) that f_λ is bounded in $\mathbb{D}$. If we let $z_1 \to z_2$ and use that $\Phi(0, t) = 0$ we see that f_λ is uniformly continuous in $\mathbb{D}$ and thus has a continuous extension to $\overline{\mathbb{D}}$. Hence (3) holds for $z_1, z_2, z_3 \in \overline{\mathbb{D}}$. It follows that f_λ is injective in $\overline{\mathbb{D}}$ and thus a homeomorphism. Furthermore (3) shows that f_λ is quasisymmetric on $\mathbb{T}$; see (5.4.4). Therefore Proposition 5.10 shows that $f_\lambda(\mathbb{T})$ is a quasicircle and thus $f_\lambda(\overline{\mathbb{D}})$ a closed quasidisk in $\mathbb{C}$. $\square$

The first application is to the Becker univalence criterion (Bec72):

Corollary 5.23. *Let f be analytic and locally univalent in $\mathbb{D}$. If*

(4) $$(1 - |z|^2) \left| z \frac{f''(z)}{f'(z)} \right| \le \kappa < 1 \quad \text{for} \quad z \in \mathbb{D}$$

then f maps $\mathbb{D}$ conformally onto a quasidisk.

Proof. We may assume that $f(0) = 0$ and $f'(0) = 1$. For $\lambda \in \mathbb{D}$ let f_λ be defined by

$$(5) \qquad\qquad \log f'_\lambda = \frac{\lambda}{\kappa} \log f', \quad f_\lambda(0) = 0.$$

Then $(1 - |z|^2)|z f''_\lambda(z)/f'_\lambda(z)| \le |\lambda| < 1$ by (4) so that f_λ is univalent in $\mathbb{D}$ by Theorem 1.11. It is clear that f_λ is analytic in λ and (5) implies $f_0(z) = z$. Hence our assertion follows from Theorem 5.22 because $f_\kappa = f$. $\qquad\square$

We turn now to the Nehari univalence criterion (AhWe62) on the Schwarzian derivative S_f defined in (1.2.15).

Corollary 5.24. *Let f be meromorphic and locally univalent in $\mathbb{D}$ and let*

$$(6) \qquad\qquad \left(1 - |z|^2\right)^2 |S_f(z)| \le 2\kappa < 2 \quad \text{for} \quad z \in \mathbb{D}.$$

If $f(0) \in \mathbb{C}$ and $f''(0) = 0$ then f maps $\mathbb{D}$ conformally onto a (bounded) quasidisk.

If $f(0) = \infty$ or $f''(0) \ne 0$ we can find a Möbius transformation σ such that $g = \sigma \circ f$ satisfies $g(0) = g''(0) = 0$. Since $S_g = S_f$ we conclude from the corollary that $f(\mathbb{D})$ is a quasidisk in $\widehat{\mathbb{C}}$, i.e. the image of a bounded quasidisk under a Möbius transformation.

Proof. We may additionally assume that $f(0) = 0$ and $f'(0) = 1$. Let u_λ and v_λ be the solutions of the differential equation

$$(7) \qquad\qquad w'' + \frac{\lambda}{2\kappa} S_f(z) w = 0 \quad (z \in \mathbb{D})$$

that satisfy the initial conditions $u_\lambda(0) = 0$, $u'_\lambda(0) = 1$ and $v_\lambda(0) = 1$, $v'_\lambda(0) = 0$. Since also $u''_\lambda(0) = 0$ by (7) we see that

$$(8) \qquad f_\lambda(z) \equiv \frac{u_\lambda(z)}{v_\lambda(z)} = \frac{z + O(z^3)}{1 + O(z^2)} = z + O(z^3) \quad \text{as} \quad z \to 0.$$

Since $w'' = 0$ for $\lambda = 0$ we have $u_0(z) \equiv z$, $v_0(z) \equiv 1$ and thus $f_0(z) \equiv z$. Using (7) it is easy to see that $S_{f_\lambda} = \frac{\lambda}{\kappa} S_f$. Thus $S_{f_\kappa} = S_f$ and since $f(z) = z + O(z^3)$ we conclude from (8) that $f_\kappa = f$. If $\lambda \in \mathbb{D}$ then

$$\left(1 - |z|^2\right)^2 |S_{f_\lambda}(z)| = \left(1 - |z|^2\right)^2 \frac{|\lambda|}{\kappa} |S_f(z)| \le 2|\lambda| < 2 \quad (z \in \mathbb{D})$$

by (6). Hence f_λ is univalent in $\mathbb{D}$ by the Nehari univalence criterion (Theorem 1.12) and has no pole because $f''_\lambda(0) = 0$. Finally u_λ and v_λ depend analytically on λ and therefore also f_λ. Hence we conclude from Theorem 5.22 that $f_\lambda(\mathbb{D})$ is a (bounded) quasidisk, hence also $f(\mathbb{D}) = f_\kappa(\mathbb{D})$. $\qquad\square$

Related univalence criteria hold for any quasidisk instead of $\mathbb{D}$ (and only for quasidisks); see e.g. Ahl63, MaSa78 and Geh87.

We now need a deep *embedding theorem* for general quasiconformal maps (AhBer60; Leh87, p. 70):

Let h be a quasiconformal map of $\widehat{\mathbb{C}}$ onto $\widehat{\mathbb{C}}$ that keeps 0, 1 and ∞ fixed. For $\lambda \in \mathbb{D}$ let h_λ be the quasiconformal map that satisfies

(9) $$\frac{\partial h_\lambda}{\partial \overline{z}} \bigg/ \frac{\partial h_\lambda}{\partial z} = \frac{\lambda}{\kappa}\mu(z) \quad \text{for almost all } z \in \mathbb{C}, \quad |\mu(z)| \leq \kappa < 1$$

(see (5.1.2) and (5.1.3)) and keeps $0, 1, \infty$ fixed. Then $h_\kappa = h$ and $h_\lambda(z)$ depends analytically on $\lambda \in \mathbb{D}$ for each $z \in \mathbb{D}$.

Using this result we can now prove (d) $\Rightarrow$ (a) and thus complete the proof of the quasicircle theorem stated in Section 5.1.

Theorem 5.25. *If h is a quasiconformal map of $\widehat{\mathbb{C}}$ onto $\widehat{\mathbb{C}}$ that keeps ∞ fixed, then $h(\mathbb{T})$ is a quasicircle.*

Proof. We may assume that $h(0) = 0$ and $h(1) = 1$. We consider the quasiconformal maps h_λ ($\lambda \in \mathbb{D}$) defined above. It follows from (9) and the normalization that $h_0(z) = z$. Hence $h(\mathbb{T}) = h_\kappa(\mathbb{T})$ is a quasicircle by Theorem 5.22. $\square$

Another consequence of the embedding theorem is the *Lehto majorant principle* (see e.g. Leh87, p. 77) that we formulate here for the class S of normalized analytic univalent functions.

Theorem 5.26. *Let $\Phi : S \to \mathbb{C}$ be an analytic functional that vanishes for the identity and satisfies*

(10) $$|\Phi(f)| \leq M \quad \text{for all} \quad f \in S.$$

If $f \in S$ and if f has a κ-quasiconformal extension to $\widehat{\mathbb{C}}$ keeping ∞ fixed then

$$|\Phi(f)| \leq \kappa M.$$

Here we call Φ an *analytic functional* if $\Phi(f_\lambda)$ is an analytic function whenever f_λ depends analytically on λ. If $f_\lambda \in S$ then $|f_\lambda(z)| \leq (1 - |z|)^{-2}$ for $z \in \mathbb{D}$ by the Koebe distortion theorem. An application of Vitali's theorem therefore shows that $f_\lambda^{(k)}(z)$ is analytic in λ for $k \in \mathbb{N}$, $z \in \mathbb{D}$. It follows that any rational combination of $f_\lambda^{(k)}(z_j)$ with $z_j \in \mathbb{D}$ is an analytic functional provided that it is finite for all $f \in S$.

Proof. We apply the embedding theorem to the extension h of $f/f(1)$. It follows from (9) that $\partial h_\lambda(z)/\partial \overline{z} = 0$ for almost all $z \in \mathbb{D}$ so that h_λ is conformal in $\mathbb{D}$ (LeVi57, p. 183). The functions

$$f_\lambda(z) = h_\lambda(z)/h_\lambda'(0) \quad (z \in \mathbb{D}), \quad \lambda \in \mathbb{D}$$

belong therefore to S and depend analytically on λ; as above it can be shown that $h'_\lambda(0)$ is analytic. Hence $\varphi(\lambda) = \Phi(f_\lambda)$ is analytic in $\mathbb{D}$ and satisfies $\varphi(0) = 0$ and $|\varphi(\lambda)| \leq M$ by (10). Hence $|\varphi(\lambda)| \leq |\lambda|M$ by Schwarz's lemma and thus $|\Phi(f)| = |\varphi(\kappa)| \leq \kappa M$ because $f_\kappa = f$. $\square$

Applications. We assume that $f \in S$ has a κ-quasiconformal extension to $\widehat{\mathbb{C}}$ that keeps ∞ fixed. Then the Schwarzian derivative satisfies (Küh69b)

$$(11) \qquad \left(1 - |z|^2\right)^2 |S_f(z)| \leq 6\kappa \quad \text{for} \quad z \in \mathbb{D};$$

this follows by considering the analytic functional $S_f(z)$ (z fixed) and using that the left-hand side of (11) is bounded by 6 for $f \in S$. Furthermore

$$(12) \qquad \left| \log z \frac{f'(z)}{f(z)} \right| \leq \kappa \log \frac{1 + |z|}{1 - |z|} \quad \text{for} \quad z \in \mathbb{D}$$

because $\log[z f'(z)/f(z)]$ is an analytic functional vanishing for the identity which is bounded by $\log[(1+|z|)/(1-|z|)]$ for all $f \in S$ (Gru32; Pom75, p. 66; Dur83, p. 126). If $z, \zeta \in \mathbb{D}$ then (Küh71)

$$(13) \qquad \left| \log \frac{(z - \zeta)^2 f'(z) f'(\zeta)}{(f(z) - f(\zeta))^2} \right| \leq \kappa \log \frac{|1 - \bar{z}\zeta|^2}{(1 - |z|^2)(1 - |\zeta|^2)};$$

the corresponding estimate for the full class S is a consequence of the Golusin inequality (see e.g. Pom75, p. 64; Dur83, p. 126). See e.g. Scho75 for further estimates.

Exercises 5.6.

1. Show that Theorem 5.22 remains true if (iii) is replaced by the assumption that f_0 maps $\mathbb{D}$ conformally onto a quasidisk.

2. Let f be analytic and $f'(z) \in A$ for $z \in \mathbb{D}$ where A is a compact convex set with $0 \notin A$. Show that f maps $\mathbb{D}$ conformally onto a quasidisk. (Let $\psi(\lambda)$ map $\mathbb{D}$ conformally onto a flat rectangle containing 0 and 1 with $\psi(0) = 0$ and apply Proposition 1.10 to $(1 - \psi(\lambda))z + \psi(\lambda)f(z)$.)

3. Let $b_n(\lambda)$ ($n = 0, 1, \dots$) be analytic in $\mathbb{D}$ and $b_n(0) = 0$. If

$$g_\lambda(z) = z + \sum_{n=0}^{\infty} b_n(\lambda) z^{-n} \quad (|z| > 1)$$

is univalent for $\lambda \in \mathbb{D}$ show that $g_\lambda(\mathbb{T})$ is a quasicircle and (Küh71) that $\Sigma n |b_n(\lambda)|^2 \leq |\lambda|^2$.

4. In the Lehto majorant principle, replace $\Phi(\text{id}) = 0$ by $\Phi(\text{id}) = a$ and show that now (Leh76)

$$|\Phi(f)| \leq (|a| + \kappa M)/(1 + |a|\kappa/M).$$

5. Let $\zeta \in \mathbb{T}$ and $0 \leq r < \rho < 1$. If f maps $\mathbb{D}$ conformally onto a John domain, show that

$$\frac{1}{4} \leq |f(\rho\zeta) - f(r\zeta)|(\rho - r)^{-1}|f'(r\zeta)|^{-1} \leq M_1$$

for some constant M_1. If f has a κ-quasiconformal extension to $\widehat{\mathbb{C}}$ keeping ∞ fixed, deduce that (compare Theorems 5.2 and 5.7)

$$\frac{1}{M_2}\left(\frac{1-\rho}{1-r}\right)^\kappa \leq \left|\frac{f'(\rho\zeta)}{f'(r\zeta)}\right| \leq M_3\left(\frac{1-r}{1-\rho}\right)^\kappa.$$

Chapter 6. Linear Measure

6.1 An Overview

Linear measure Λ is a generalization of length and was studied in detail by Besicovitch (Bes38). We follow the excellent presentation in Fal85. The linear measure is an important special case of a Hausdorff measure to be discussed in Section 10.2; see Rog70. We shall write $E \overset{\circ}{=} E'$ if the plane sets E and E' differ only by a set of zero linear measure.

We assume that f maps $\mathbb{D}$ conformally onto $G \subset \mathbb{C}$. Then the angular limit $f(\zeta)$ exists and is finite for almost all $\zeta \in \mathbb{T}$.

Riesz-Privalov Theorem. *If G is bounded by a rectifiable Jordan curve then f' belongs to the Hardy space H^1 and, for $E \subset \mathbb{T}$,*

$$(1) \qquad \Lambda(E) = 0 \quad \Leftrightarrow \quad \Lambda(f(E)) = 0 \,.$$

We shall study domains with rectifiable boundary in more detail in Chapter 7. In the present chapter we are more concerned with domains that have a "bad" boundary.

The Riesz-Privalov (or *F. and M. Riesz*) theorem together with Fatou's theorem leads to a fundamental result in the general theory of boundary behaviour.

Plessner's Theorem. *If g is meromorphic in $\mathbb{D}$ then, for almost all $\zeta \in \mathbb{T}$, either g has a finite angular limit at ζ, or $g(\Delta)$ is dense in $\widehat{\mathbb{C}}$ for every Stolz angle Δ at ζ.*

Thus the behaviour is, at almost all points, either very good or very bad. A consequence is:

Privalov Uniqueness Theorem. *If a meromorphic function in $\mathbb{D}$ has the angular limit 0 on a set of positive measure on $\mathbb{T}$, then it vanishes identically.*

We defined that f is *conformal* at $\zeta \in \mathbb{T}$ if

$$(2) \qquad f'(\zeta) = \lim_{z \to \zeta, z \in \Delta} \frac{f(z) - f(\zeta)}{z - \zeta} = \lim_{z \to \zeta, z \in \Delta} f'(z) \neq 0, \infty$$

for every Stolz angle Δ at ζ. We say that f is *twisting* at ζ if $f(z)$ winds around the boundary point $f(\zeta)$ infinitely often in both directions as $z \to \zeta$, $\zeta \in \Gamma$ for every curve Γ ending at ζ. This is the very opposite of conformal.

McMillan Twist Theorem. *At almost all points of* $\mathbb{T}$, *the map* f *is either conformal or twisting.*

Let $\text{Sect}(f)$ denote the set of all $\zeta \in \mathbb{T}$ such that $f(\zeta)$ is *sectorially accessible* from G; this means that G contains an open triangle of vertex $f(\zeta)$. Then $\text{Sect}(f) \overset{\circ}{=} \{\zeta \in \mathbb{T} : f \text{ is conformal at } \zeta\}$ by the twist theorem. The next result generalizes the absolute continuity property (1).

McMillan Sector Theorem. *If* $E \subset \text{Sect}(f)$ *then*

$$\Lambda(E) = 0 \quad \Leftrightarrow \quad \Lambda(f(E)) = 0 \,.$$

The image of $\text{Sect}(f)$ has σ-finite linear measure, i.e. is the union of countably many sets of finite linear measure.

Makarov Compression Theorem. *There is a partition*

$$\mathbb{T} = \text{Sect}(f) \cup A_0 \cup A_1 \quad \text{with} \quad \Lambda(f(A_0)) = 0, \quad \Lambda(A_1) = 0 \,.$$

If almost no point of ∂G is sectorially accessible then it follows that almost all of $\mathbb{T}$ is compressed by f into a set of zero linear measure. This is an extreme case of the "crowding effect" in numerical conformal mapping. Another result of Makarov (Theorem 10.6) will show that a set of positive measure on $\mathbb{T}$ cannot be compressed much further.

These results can be expressed as relations between the linear measure Λ and the harmonic measure ω of sets on ∂G. The Makarov compression theorem shows that ω is concentrated on a set of σ-finite linear measure which proves a conjecture of Øxendal (Øxe81).

The case that G is bounded by a Jordan curve J is of particular interest. Then we can also consider the harmonic measure ω^* with respect to the outer domain of J. If T is the set of points on J where J has a tangent then there is a partition (BiCaGaJo89)

(3) $$J = T \cup B \cup B^* \quad \text{with} \quad \omega(B) = 0, \quad \omega^*(B^*) = 0 \,.$$

Finally we mention a theorem about tangent directions of arbitrary plane sets E. Its proof is similar to that of Proposition 6.23; see Sak64, Chapt. IX for details. For $w \in E$, let $C(w)$ denote the union of all halflines $L : w + e^{i\vartheta}s$, $0 \le s < +\infty$ such that there exist $w_n \in E$, $w_n \ne w$ with $\arg(w_n - w) \to \vartheta$, $w_n \to w$ as $n \to \infty$.

Kolmogoroff-Verčenko Theorem. *For every set* $E \subset \mathbb{C}$ *there is a canonical partition* $E = E_0 \cup E_1 \cup E_2 \cup E_3$ *with*

(0) $\Lambda(E_0) = 0$;
(i) $C(w)$ *is a (complete) straight line for* $w \in E_1$;
(ii) $C(w)$ *is a halfplane for* $w \in E_2$;
(iii) $C(w) = \mathbb{C}$ *for* $w \in E_3$.

Furthermore $E_1 \cup E_2$ *has* σ-*finite linear measure. If* $B \subset E_1 \cup E_2$ *and if* $\partial C(w)$ *is orthogonal to* $\mathbb{R}$ *for all* $w \in B$ *then the projection of* B *onto* $\mathbb{R}$ *has zero measure.*

As an example, consider

$$E = [0,1] \cup \bigcup_{n \in \mathbb{Z}, n \neq 0} [0, e^{\imath/n}].$$

Then $E_0 = \{0\} \cup \{e^{i/n} : n \in \mathbb{Z}, n \neq 0\}$, $E_1 = \bigcup(0, e^{i/n})$, $E_2 = \{1\}$ and $E_3 = (0,1)$.

6.2 Linear Measure and Measurability

1. We begin with some concepts from general measure theory (Fal85). An *outer measure* μ in $\mathbb{C}$ assigns to every set $E \subset \mathbb{C}$ a number $\mu(E) \geq 0$ which may be infinite, such that $\mu(\emptyset) = 0$,

(1) $\qquad\qquad E_1 \subset E_2 \quad \Rightarrow \quad \mu(E_1) \leq \mu(E_2) \quad \text{(monotonicity)},$

(2) $\qquad\qquad \mu\left(\bigcup_{n=1}^{\infty} E_n\right) \leq \sum_{n=1}^{\infty} \mu(E_n) \quad \text{(subaddivity)}.$

The outer measure μ is called *metric* if

(3) $\qquad\qquad \text{dist}(E_1, E_2) > 0 \quad \Rightarrow \quad \mu(E_1 \cup E_2) = \mu(E_1) + \mu(E_2).$

The set E is called μ-*measurable* if

(4) $\qquad\qquad \mu(X) = \mu(X \cap E) + \mu(X \setminus E) \quad \text{for all sets } X \subset \mathbb{C};$

note that $\leq$ always holds by (2). Every set E with $\mu(E) = 0$ is μ-measurable because $\mu(X \cap E) + \mu(X \setminus E) \leq \mu(E) + \mu(X)$ by (1).

The σ-algebra of *Borel* sets is, by definition, the smallest σ-algebra containing the open sets, i.e. the intersection of all such σ-algebras.

We now state a basic fact (Fal85, Th. 1.2, Th. 1.5): Let μ be an outer measure. The collection $\mathfrak{M}$ of all μ-measurable sets is a σ-algebra and the restriction of μ to $\mathfrak{M}$ is a (possibly infinite) *measure*, i.e.

(5) $\qquad\qquad \mu\left(\bigcup_{n=1}^{\infty} E_n\right) = \sum_{n=1}^{\infty} \mu(E_n) \quad \text{for disjoint } E_n \in \mathfrak{M}.$

If μ is metric then every Borel set belongs to $\mathfrak{M}$.

2. Next we introduce the concept of linear measure. We say that (B_k) is an ε-*cover* of E if

(6) $$E \subset \bigcup_k B_k, \quad \text{diam } B_k \leq \varepsilon \quad \text{for all } k;$$

the countably many sets B_k are arbitrary but may be assumed to be convex. The *outer linear measure* $\Lambda(E)$ of $E \subset \mathbb{C}$ is defined by

(7) $$\Lambda(E) = \lim_{\varepsilon \to 0+} \inf_{(B_k)} \sum_k \text{diam } B_k$$

where the infimum is taken over all ε-covers of E. This infimum increases as ε decreases so that the limit exists but may be infinite.

Proposition 6.1. *The outer linear measure Λ is a metric outer measure. The collection $\mathfrak{M}$ of all Λ-measurable sets is a σ-algebra that contains all Borel sets, and Λ is a measure on $\mathfrak{M}$.*

This is the *linear measure*. We shall say *linearly measurable* instead of Λ-measurable.

Proof. In view of the fact stated above it suffices to show that Λ is a metric outer measure.

It is trivial that $\Lambda(\emptyset) = 0$ and that (1) holds. To prove (2) let $(B_{nk})_k$ be ε-covers of E_n. Their totality is an ε-cover of $E = \bigcup_n E_n$. Hence

$$\inf_{(B_k)} \sum_k \text{diam } B_k \leq \sum_n \inf_{(B_{nk})} \sum_k \text{diam } B_{nk}$$

and (2) follows for $\varepsilon \to 0$. Hence Λ is an outer measure.

Now let $\rho = \text{dist}(E_1, E_2) > 0$ and let (B_k) be an ε-cover of $E = E_1 \cup E_2$. If $\varepsilon < \rho$ then no B_k intersects both E_1 and E_2 so that we have a disjoint partition of (B_k) into ε-covers (B_{jk}) of E_j $(j = 1, 2)$. It follows that

$$\inf \sum_k \text{diam } B_k \geq \inf \sum_k \text{diam } B_{1k} + \inf \sum_k \text{diam } B_{2k}$$

and therefore that $\Lambda(E) \geq \Lambda(E_1) + \Lambda(E_2)$. Together with (2) this implies that (3) holds. Hence Λ is metric. $\qquad\qquad\square$

For sets on $\mathbb{R}$ the definitions of outer linear measure and outer Lebesgue measures coincide. Hence the linear measure on $\mathbb{R}$ is identical with the Lebesgue measure. The same is true for sets on the unit circle $\mathbb{T}$.

Every countable set in $\mathbb{C}$ has linear measure 0 by (2) because a point has zero measure. Every set E with $\Lambda(E) = 0$ is linearly measurable.

Every non-empty open set in $\mathbb{C}$ has infinite linear measure because it contains uncountably many disjoint line segments each of which has positive measure. Furthermore linear measure is not σ-finite, i.e. it is not possible to write $\mathbb{C}$ as the union of countably many sets of finite linear measure.

We write $E \overset{\circ}{=} E'$ if the sets E and E' in $\mathbb{C}$ differ only by a set of measure zero, i.e.

$$(8) \qquad E \overset{\circ}{=} E' \quad \Leftrightarrow \quad \Lambda(E \setminus E') = \Lambda(E' \setminus E) = 0 \, .$$

This is clearly an equivalence relation.

It is obvious that Λ is linear in the sense that

$$\Lambda(h(E)) = |a| \Lambda(E) \quad \text{for} \quad h(z) = az + b \, .$$

If h is a *contraction*, i.e. a map from E into $\mathbb{C}$ with

$$(9) \qquad |h(z_1) - h(z_2)| \leq |z_1 - z_2| \quad \text{for} \quad z_1, z_2 \in E,$$

then $\operatorname{diam} h(B) \leq \operatorname{diam} B$ and therefore

$$(10) \qquad \Lambda(h(E)) \leq \Lambda(E) \, .$$

In particular it follows that $\Lambda(P) \leq \Lambda(E)$ if P is the projection of E onto any line in $\mathbb{C}$.

Proposition 6.2. *Let J be a Jordan arc. Then $\Lambda(J)$ is equal to the length of J. If $\Lambda(J) < \infty$ and if $J : \psi(s)$, $0 \leq s \leq \Lambda(J)$ is the arc length parametrization of J then*

$$(11) \qquad \Lambda(\psi(A)) = \Lambda(A) \quad \textit{for measurable} \quad A \subset [0, \Lambda(J)] \, .$$

Thus our present notation agrees with the notation introduced in Section 1.1. In spite of this result, "irregular sets" of finite linear measure have nothing to do with length; see e.g. Bes38 and Fal85.

Proof. Let $b \leq +\infty$ be the length of J and let $J = J_1 \cup \cdots \cup J_n$ be any partition into subarcs J_ν from $w_{\nu-1}$ to w_ν ($\nu = 1, \ldots, n$). The projection P_ν of J_ν onto the line through $w_{\nu-1}$ and w_ν satisfies $[w_{\nu-1}, w_\nu] \subset P_\nu$ and $\Lambda(P_\nu) \leq \Lambda(J_\nu)$ so that

$$\sum_{\nu=1}^n |w_\nu - w_{\nu-1}| \leq \sum_{\nu=1}^n \Lambda(P_\nu) \leq \sum_{\nu=1}^n \Lambda(J_\nu) = \Lambda(J)$$

because the sets J_ν are linearly measurable and disjoint except for the endpoints. Hence $b \leq \Lambda(J)$ by (1.1.7). In particular $b = \infty$ implies $\Lambda(J) = \infty$.

Now let $b < \infty$. The arc length parametrization satisfies

$$|\psi(s_2) - \psi(s_1)| \leq s_2 - s_1 \quad \text{for} \quad 0 \leq s_1 \leq s_2 \leq b$$

and is thus a contraction of $I = [0, b]$ onto J. Hence (10) shows that $\Lambda(J) \leq \Lambda(I) = b$ and thus $\Lambda(J) = b$. If $A \subset I$ then $\Lambda(\psi(A)) \leq \Lambda(A)$ and $\Lambda(J \setminus \psi(A)) \leq \Lambda(I \setminus A)$ by (10). Hence the subadditivity (2) shows that

$$\Lambda(J) \leq \Lambda(\psi(A)) + \Lambda(J \setminus \psi(A)) \leq \Lambda(I) = b = \Lambda(J)$$

if A is measurable; see (4) with $X = I$. Hence equality holds in all these inequalities, in particular (11) holds. $\qquad\square$

Proposition 6.3. (i) *If E is any set in $\mathbb{C}$ then there are open sets $H_n \supset E$ such that*

$$(12) \qquad\qquad \Lambda(E) = \Lambda\left(\bigcap_{n=1}^{\infty} H_n\right).$$

   (ii) *If $E \subset \mathbb{C}$ is linearly measurable and $\Lambda(E) < \infty$ then there are closed sets $A_n \subset E$ such that*

$$(13) \qquad\qquad \Lambda(E) = \Lambda\left(\bigcup_{n=1}^{\infty} A_n\right).$$

See Fal85, p. 8 for the proof of these results (Bes38). Part (i) shows that Λ is a regular outer measure. Intersections of countably many open sets are called G_δ-sets. Part (ii) says that every linearly measurable set E of finite measure can be written as

$$(14) \qquad\qquad E = B \cup E_0, \quad B \text{ Borel set}, \quad \Lambda(E_0) = 0;$$

more specifically B is an F_σ-set, i.e. the union of countably many closed sets.

Proposition 6.4. *Let A be any set on $\mathbb{T}$ with $\Lambda(A) > 0$. Then there exists $\zeta \in A$ such that*

$$(15) \qquad\qquad \Lambda(A \cap I)/\Lambda(I) \to 1 \quad \text{as} \quad \Lambda(I) \to 0,$$

where I are the arcs of $\mathbb{T}$ with $\zeta \in I$.

Here Λ is the outer measure. If (15) holds we say that ζ is a *point of density* of A. See e.g HeSt69, p. 274 for the proof. The concept of density plays an important role also for sets in $\mathbb{C}$; see e.g. Fal85.

3. Finally we consider the problem of measurability. Most questions dealing with preimages can be handled by the following result together with Proposition 6.1 and Proposition 6.3.

Proposition 6.5. *Let f be a continuous map of $\mathbb{D}$ into $\widehat{\mathbb{C}}$. Then the points $\zeta \in \mathbb{T}$ where the radial limit $f(\zeta)$ exists form a Borel set, and if $B \subset \widehat{\mathbb{C}}$ is a Borel set then*

$$(16) \qquad\qquad f^{-1}(B) = \{\zeta \in \mathbb{T} : f(\zeta) \text{ exists}, \ f(\zeta) \in B\}$$

is also a Borel set.

Proof. Let $d^{\#}$ denote the spherical distance on $\widehat{\mathbb{C}}$ and let $r_n = 1 - 1/n$. For $m \le n$ and $k = 1, 2, \ldots$, the set

$$V_{mnk} = \left\{ \zeta \in \mathbb{T} : d^{\#}(f(r\zeta), f(r_m\zeta)) < \frac{1}{k} \quad \text{for} \quad r_m \le r \le r_n \right\}$$

is open because f is continuous. Hence

$$(17) \qquad A = \bigcap_{k=1}^{\infty} \bigcup_{m=1}^{\infty} \bigcap_{n=m}^{\infty} V_{mnk}$$

is a Borel set, and it is easy to see that $f(\zeta)$ exists if and only if $\zeta \in A$.

Now let B be an open set and (B_k) an exhaustion of B by open sets with $\overline{B}_k \subset B$. Then

$$V'_{mnk} = \{ \zeta \in A : f(r\zeta) \in B_k \quad \text{for} \quad r_m \le r \le r_n \}$$

is a Borel set by the above result and because f is continuous. Hence

$$A' = \bigcap_{k=1}^{\infty} \bigcup_{m=1}^{\infty} \bigcap_{n=m}^{\infty} V'_{mnk}$$

is a Borel set and $\zeta \in A'$ if and only if $\zeta \in A$ and $f(\zeta) \in B$.

Consider finally the class $\mathfrak{M}$ of all sets $B \subset \widehat{\mathbb{C}}$ such that $f^{-1}(B)$ is a Borel set. Then $\mathfrak{M}$ is a σ-algebra and we have shown above that $\mathfrak{M}$ contains all open sets. Hence $\mathfrak{M}$ contains the σ-algebra of Borel sets. $\qquad \square$

The situation is more complicated with respect to images.

Proposition 6.6. *Let f be continuous in $\mathbb{D}$ and let the radial limit $f(\zeta)$ exist for all ζ in the measurable set $A \subset \mathbb{T}$. Then there are closed sets $A_n \subset A$ such that $f(A_n)$ is closed for $n = 1, 2, \ldots$ and*

$$(18) \qquad A = E \cup \bigcup_{n=1}^{\infty} A_n , \quad \Lambda(E) = 0 .$$

Proof. By Egorov's theorem there are closed sets A_n with $A_n \subset A_{n+1} \subset A$ such that $\Lambda(A \setminus A_n) < 1/n$ and

$$f((1 - 1/k)\zeta) \to f(\zeta) \quad \text{as} \quad k \to \infty \quad \text{uniformly for} \quad \zeta \in A_n .$$

Thus $f(A_n)$ is closed because f is continuous in $\mathbb{D}$ and (18) holds. $\qquad \square$

The continuous image of a Borel set need not be a Borel set. A *Souslin set* (or *analytic set*) has the form

$$(19) \qquad E = \bigcup [A(n_1) \cap A(n_1, n_2) \cap A(n_1, n_2, n_3) \cap \ldots]$$

where the sets $A(n_1, \ldots, n_k)$ are compact and where the union is taken over $\mathbb{N}^{\mathbb{N}}$, that is over the uncountably many sequences $(n_k)_{k \in \mathbb{N}}$ of natural numbers. Every Borel set is a Souslin set, moreover the continuous image of a Borel set is a Souslin set (Kur52, p. 360, p. 390). Every Souslin set is linearly measurable (Rog70, p. 48).

If f is continuous in $\mathbb{D}$ and if $A \subset \mathbb{T}$ is a Borel or Souslin set then (see e.g. McM66, Theorem 7 (i))

$$(20) \qquad f(A) = \{f(\zeta) : \zeta \in A, \, f(\zeta) \text{ exists}\}$$

is a Souslin set and therefore linearly measurable, possibly with $\Lambda(f(A)) = \infty$. Conversely (BerNi92), given any nowhere dense perfect set $A \subset \mathbb{T}$ and any Souslin set $E \subset \widehat{\mathbb{C}}$, there is a function f analytic in $\mathbb{D}$ such that the set of radial limits on A is equal to E.

The image of a measurable set under a continuous mapping need not be measurable (assuming the axiom of choice):

Proposition 6.7. *Let f be continuous in $\mathbb{D}$ and let $B \subset \mathbb{T}$ be measurable. If the radial limit $f(\zeta)$ exists for $\zeta \in B$ and if*

$$(21) \qquad E \subset B, \quad \Lambda(E) = 0 \quad \Rightarrow \quad \Lambda(f(E)) = 0$$

then

$$(22) \qquad A \subset B, \quad A \text{ measurable} \quad \Rightarrow \quad f(A) \text{ linearly measurable}.$$

If $f(B) \subset \mathbb{T}$ and if (21) is false then (22) is false.

Condition (21) is a weak version of absolute continuity and will play a great role in the following sections.

Proof. By Proposition 6.6 we can write $A = E \cup A^*$ where $\Lambda(E) = 0$ and $f(A^*)$ is a Borel set. Hence $f(A)$ is linearly measurable by (21).

Assume now that $f(B) \subset \mathbb{T}$ and that $E \subset B$ has zero measure but $f(E)$ has positive outer measure. Since every set on $\mathbb{R}$ with positive outer Lebesgue measure contains a non-measurable set (Oxt80, p. 24), we can find $A \subset E$ such that $f(A)$ is non-measurable; the set A is measurable because it has outer measure 0. $\qquad \square$

Exercises 6.2

1. Let E be a continuum. Use projection to show that $\Lambda(E) \geq \operatorname{diam} E$.

2. Let J and J' be Jordan curves such that J lies in the inner domain of J'. If J is convex show that $\Lambda(J) < \Lambda(J')$.

3. Let f map $\mathbb{D}$ conformally onto G and let $F(\zeta)$ denote the impression of the prime end of G corresponding to $\zeta \in \mathbb{T}$. Show that the sets $A = \{\zeta \in \mathbb{T} : F(\zeta) \text{ is one point}\}$ and $\bigcup_{\zeta \in A} F(\zeta)$ are linearly measurable.

4. Let $\Lambda'(E) = \inf \sum_k \operatorname{diam} B_k$ where the infimum is taken over all covers of E (of any diameter). Show that Λ' is a non-metric outer measure. Show that $\Lambda'(E) = \operatorname{diam} E$ if E is a continuum and deduce that compact sets are in general not Λ'-measurable.

5. Let g be an increasing function on the interval I and let A_0, A_1, A_∞ be the subsets where g has a zero, finite nonzero or infinite derivative respectively. Use the Kolmogoroff-Verčenko theorem (Section 6.1) to show that

$$\Lambda(A_0) + \Lambda(A_1) = \Lambda(I), \quad \Lambda(A_\infty) = 0,$$

$$\Lambda(g(A_0)) = 0, \quad \Lambda(g(A_1)) + \Lambda(g(A_\infty)) = \Lambda(g(I)).$$

6.3 Rectifiable Curves

Let J be a Jordan curve in $\mathbb{C}$ and G its inner domain. If f maps $\mathbb{D}$ conformally onto G then f has a continuous extension to $\overline{\mathbb{D}}$ and f maps $\mathbb{T}$ one-to-one onto J; see Theorem 2.6. Let Λ denote the linear measure; it is the same as the length for rectifiable curves by Proposition 6.2.

The following theorem (Rie16, Pri19) is the basis of many proofs in the following sections.

Theorem 6.8. *Let f map $\mathbb{D}$ conformally onto the inner domain of the Jordan curve J. Then*

$$(1) \qquad\qquad \Lambda(J) < \infty \quad \Leftrightarrow \quad f' \in H^1 .$$

If $\Lambda(J) < \infty$ and if $A \subset \mathbb{T}$ is measurable then

$$(2) \qquad\qquad \Lambda(f(A)) = \int_A |f'(\zeta)|\,|d\zeta|$$

and $f(A)$ is linearly measurable, furthermore

$$(3) \qquad\qquad \Lambda(E) = 0 \quad \Leftrightarrow \quad \Lambda(f(E)) = 0$$

for $E \subset \mathbb{T}$ and

$$(4) \qquad \lim_{z \to \zeta, z \in \overline{\mathbb{D}}} \frac{f(z) - f(\zeta)}{z - \zeta} \neq 0, \infty \quad \text{exists} \ = f'(\zeta) \ \text{for almost all } \zeta \in \mathbb{T}.$$

The Hardy space H^1 consists of all functions g analytic in $\mathbb{D}$ with

$$(5) \qquad\qquad \sup_{0 < r < 1} \int_{\mathbb{T}} |g(r\zeta)|\,|d\zeta| < \infty\,;$$

see e.g. Dur70. If $g \in H^1$ then $g(\zeta)$ exists and is $\neq 0$ for almost all $\zeta \in \mathbb{T}$ and is integrable, furthermore

$$(6) \quad \int_{\mathbb{T}} |g(r\zeta) - g(\zeta)|\,|d\zeta| \to 0, \quad \int_{\mathbb{T}} |g(r\zeta)|\,|d\zeta| \to \int_{\mathbb{T}} |g(\zeta)|\,|d\zeta| \quad \text{as } r \to 1.$$

Proof. (a) First let $\Lambda(J) < \infty$. Since

$$u_n(z) = \sum_{\nu=1}^{n} |f(ze^{2\pi i\nu/n}) - f(ze^{2\pi i(\nu-1)/n})|$$

is subharmonic in $\mathbb{D}$ and continuous in $\overline{\mathbb{D}}$, it follows from the maximum principle and from the definition (1.1.7) of the length that $u_n(z) \leq \Lambda(J)$ for $z \in \mathbb{D}$. We conclude that

$$(7) \qquad r\int_0^{2\pi} |f'(re^{it})|\, dt = \lim_{n\to\infty} u_n(r) \leq \Lambda(J)$$

for $0 \leq r < 1$. Hence $f' \in H^1$.

(b) Conversely let $f' \in H^1$. Let A be an arc of $\mathbb{T}$ and let $t_0 < t_1 < \cdots < t_n$ where e^{it_0} and e^{it_n} are the endpoints of A. Then

$$\sum_{\nu=1}^{n} |f(e^{it_\nu}) - f(e^{it_{\nu-1}})| = \lim_{r\to 1} \sum_{\nu=1}^{n} |f(re^{it_\nu}) - f(re^{it_{\nu-1}})|$$

$$\leq \lim_{r\to 1} \int_A |f'(r\zeta)|\, |d\zeta|\,.$$

Hence the definition of length and (6) show that

$$(8) \qquad \Lambda(f(A)) \leq \int_A |f'(\zeta)|\, |d\zeta|\,.$$

We next show that this inequality holds for all measurable $A \subset \mathbb{T}$. If the arcs I_k cover A then, by (6.2.1) and (6.2.2),

$$\Lambda(f(A)) \leq \sum_k \Lambda(f(I_k)) \leq \sum_k \int_{I_k} |f'|\, |d\zeta| < \int_A |f'|\, |d\zeta| + \varepsilon$$

if the cover is suitably chosen. This implies (8).

We see from the subadditivity and from (8) that, for measurable A,

$$\Lambda(J) = \Lambda(f(\mathbb{T})) \leq \Lambda(f(A)) + \Lambda(f(\mathbb{T} \setminus A))$$

$$\leq \int_A |f'|\, |d\zeta| + \int_{\mathbb{T}\setminus A} |f'|\, |d\zeta| = \int_{\mathbb{T}} |f'|\, |d\zeta| \leq \Lambda(J)$$

because of (7). It follows that $\Lambda(J) < \infty$ and furthermore that equality holds in (8), i.e. that (2) holds.

If $\Lambda(E) = 0$ then $\Lambda(f(E)) = 0$ by (2). Conversely suppose that $\Lambda(f(E)) = 0$. Since f is a homeomorphism of $\mathbb{T}$ onto J it follows from Proposition 6.3 (i) that there is a Borel set $B \supset E$ such that $\Lambda(f(B)) = 0$. We conclude from (2) that $f'(\zeta) = 0$ for almost all $\zeta \in B$ so that $\Lambda(E) \leq \Lambda(B) = 0$ because $f' \in H^1$. Thus (3) holds, and $f(A)$ is measurable by Proposition 6.7.

Since $f' \in H^1$ it easily follows from (6) that

$$f(e^{it}) - f(1) = i \int_0^t e^{i\tau} f'(e^{i\tau}) \, d\tau$$

so that $f(e^{it})$ is absolutely continuous and has a finite nonzero derivative (as a function of t) almost everywhere. Hence (4) is a consequence of the following proposition.   □

Proposition 6.9. *Let f be analytic and bounded in $\mathbb{D}$. If $\zeta \in \mathbb{T}$ and if*

$$(9) \qquad \frac{f(z) - \omega}{z - \zeta} \to a \in \mathbb{C} \qquad as \quad z \to \zeta$$

restricted to $z \in \mathbb{T}$, then (9) holds for all $z \in \mathbb{D}$.

Proof. We write $g(z) = (f(z) - \omega)/(z - \zeta)$ and $g(\zeta) = a$. For $n = 1, 2, \ldots$ the functions $(z - \zeta)^{1/n} g(z)$ $(z \in \mathbb{D})$ belong to H^1 because $(z - \zeta)^{1/n-1}$ does and f is bounded. They can therefore be represented as Poisson integrals

$$(z - \zeta)^{1/n} g(z) = \int_{\mathbb{T}} p(z, z')(z' - \zeta)^{1/n} g(z') |dz'| \qquad \text{for} \quad z \in \mathbb{D}.$$

Since g is, by assumption, bounded on $\mathbb{T}$, we obtain for fixed $z \in \mathbb{D}$ and $n \to \infty$ that

$$g(z) = \int_{\mathbb{T}} p(z, z') g(z') |dz'| \qquad \text{for} \quad z \in \mathbb{D}.$$

Since g restricted to $\mathbb{T}$ is continuous at ζ by (9), it follows (Ahl66a, p. 168) that $g(z) \to g(\zeta) = a$ as $z \to \zeta$, $z \in \mathbb{D}$.   □

The assumption of Theorem 6.8 that $G = f(\mathbb{D})$ is a Jordan domain can be relaxed. Applying Theorem 6.8 to suitable Jordan domains obtained by covering ∂G by finitely many small disks it is not difficult to show (Pom75, p. 320) that

$$(10) \qquad \Lambda(\partial G) < \infty \quad \Leftrightarrow \quad f' \in H^1.$$

The following theorem (Ale89) is deeper; we shall not give the proof.

Theorem 6.10. *Let B be a continuum in $\mathbb{C}$ with $\Lambda(B) < \infty$ and let G_k be the components of $\widehat{\mathbb{C}} \setminus B$. If f_k maps $\mathbb{D}$ conformally onto G_k then $f_k' \in H^1$ (with modifications for the unbounded component) and*

$$(11) \qquad 2\Lambda(B) = \sum_k \int_{\mathbb{T}} |f_k'(z)| \, |dz|.$$

In particular, if f maps $\mathbb{D}$ conformally onto G and if $\Lambda(\partial G) < \infty$ then

$$(12) \qquad \Lambda(\partial G) \leq \int_{\mathbb{T}} |f'(z)| \, |dz| \leq 2\Lambda(\partial G);$$

the left-hand estimate can be established as in the proof of Theorem 6.8.

We now prove an estimate (Lav36) that sharpens the implication $\Lambda(E) > 0 \Rightarrow \Lambda(f(E)) > 0$.

Proposition 6.11. *Let f map $\mathbb{D}$ conformally onto the inner domain of a Jordan curve of finite length L. If $E \subset \mathbb{T}$ is measurable then*

$$
(13) \qquad \Lambda(f(E)) \geq L \exp\left(-\frac{2\pi}{\Lambda(E)} \log \frac{L}{\pi|f'(0)|}\right).
$$

Proof. We may assume that $L = 1$. It follows from the inequality

$$
(14) \qquad \exp\left(\frac{1}{\Lambda(A)} \int_A \log \varphi(t)\, dt\right) \leq \frac{1}{\Lambda(A)} \int_A \varphi(t)\, dt
$$

between the geometric and the arithmetic mean (HaLiPo67, p. 137) that, with $\lambda = \Lambda(E)$ and $0 < r < 1$,

$$
\begin{aligned}
2\pi \log |f'(0)| &= \int_E \log |f'(rz)|\, |dz| + \int_{\mathbb{T}\setminus E} \log |f'(rz)|\, |dz| \\
&\leq \lambda \log\left(\frac{1}{\lambda} \int_E |f'(rz)|\, |dz|\right) \\
&\quad + (2\pi - \lambda) \log\left(\frac{1}{2\pi - \lambda} \int_{\mathbb{T}\setminus E} |f'(rz)|\, |dz|\right).
\end{aligned}
$$

If we let $r \to 1$ and then use (2) we obtain from $L = 1$ that

$$
2\pi \log |f'(0)| \leq \lambda \log \Lambda(f(E)) - \lambda \log \lambda - (2\pi - \lambda) \log(2\pi - \lambda)
$$

and (13) follows because the last two terms together are bounded by $-2\pi \log \pi$. $\qquad \square$

Finally we consider sequences of conformal maps (War30).

Theorem 6.12. *Let $f_n (n = 1, 2, \dots)$ and f map $\mathbb{D}$ conformally onto the inner domains of the rectifiable Jordan curves J_n and J. If*

$$
(15) \qquad f_n \to f \quad as \quad n \to \infty \quad locally\ uniformly\ in\ \mathbb{D},
$$

$$
(16) \qquad \Lambda(J_n) \to \Lambda(J) \quad as \quad n \to \infty,
$$

then

$$
(17) \qquad \int_{\mathbb{T}} |f_n'(z) - f'(z)|\, |dz| \to 0 \quad as \quad n \to \infty.
$$

In Theorem 2.11 we proved that (15) holds (even uniformly) if there are parametrizations $J_n : \varphi_n(\zeta),\ \zeta \in \mathbb{T}$ and $J : \varphi(\zeta),\ \zeta \in \mathbb{T}$ such that $\varphi_n \to \varphi$ as $n \to \infty$ uniformly in $\mathbb{T}$ and if $f_n(0) \to f(0)$, $f_n'(0) > 0$, $f'(0) > 0$.

Proof. Let $n = 1, 2, \ldots$. By Theorem 6.8 the analytic functions

$$(18) \qquad g_n(z) = \sqrt{f_n'(z)} = \sum_{k=0}^{\infty} b_{nk} z^k, \quad g(z) = \sqrt{f'(z)} = \sum_{k=0}^{\infty} b_k z^k$$

belong to the Hardy space H^2 and satisfy, by Parseval's formula,

$$(19) \qquad \frac{\Lambda(J_n)}{2\pi} = \sum_{k=0}^{\infty} |b_{nk}|^2, \quad \frac{\Lambda(J)}{2\pi} = \sum_{k=0}^{\infty} |b_k|^2.$$

Since $|f_n' - f'| \leq (2|g_n|^2 + 2|g|^2)^{1/2} |g_n - g|$ it follows from the Schwarz inequality that

$$(20) \qquad \begin{aligned} \left(\int_{\mathbb{T}} |f_n' - f'| \, |dz| \right)^2 &\leq 2 \int_{\mathbb{T}} (|g_n|^2 + |g|^2) |dz| \int_{\mathbb{T}} |g_n - g|^2 |dz| \\ &= 4\pi \left(\Lambda(J_n) + \Lambda(J) \right) \sum_{k=0}^{\infty} |b_{nk} - b_k|^2 \end{aligned}$$

by (18) and Parseval's formula.

Given $\varepsilon > 0$ we choose m so large that $\sum_{k=m}^{\infty} |b_k|^2 < \varepsilon$ and therefore

$$\sum_{k=m}^{\infty} |b_{nk}|^2 < \left(\sum_{k=0}^{\infty} |b_{nk}|^2 - \sum_{k=0}^{\infty} |b_k|^2 \right) + \sum_{k=0}^{m-1} \left(|b_k|^2 - |b_{nk}|^2 \right) + \varepsilon.$$

For large n, the first term on the right-hand side is $< \varepsilon$ by (19) and (16) while the second term is $< \varepsilon$ by (15). Using again (15) we conclude that

$$\begin{aligned} \sum_{k=0}^{\infty} |b_{nk} - b_k|^2 &\leq \sum_{k=0}^{m-1} |b_{nk} - b_k|^2 + 2 \sum_{k=m}^{\infty} |b_{nk}|^2 + 2 \sum_{k=m}^{\infty} |b_k|^2 \\ &< \varepsilon + 6\varepsilon + 2\varepsilon = 9\varepsilon \end{aligned}$$

for large n. Hence (17) follows from (20) and (16). $\square$

Exercises 6.3

1. Show that a rectifiable Jordan curve has a tangent except possibly on a set of zero linear measure.

2. Let f map $\mathbb{D}$ conformally onto a Jordan domain. Show that f is absolutely continuous on $\mathbb{T}$ if and only if f has bounded variation on $\mathbb{T}$.

3. Show that every real function φ of bounded variation can be written as $\varphi = \psi \circ \chi$ where ψ is absolutely continuous and χ is continuous and strictly increasing. (Map $\mathbb{D}$ onto a domain bounded in part by the graph of φ.)

4. Prove the converse of Theorem 6.12 assuming that $f_n(0) \to f(0)$, i.e. show that (17) implies (15) and (16), even $f_n \to f$ as $n \to \infty$ uniformly in $\overline{\mathbb{D}}$.

6.4 Plessner's Theorem and Twisting

1. One of the central results (Ple27) on the boundary behaviour of general meromorphic functions is *Plessner's theorem*.

Theorem 6.13. *If g is meromorphic in $\mathbb{D}$ then, for almost all $\zeta \in \mathbb{T}$, either*

(i) *g has a finite angular limit at ζ, or*
(ii) *$g(\Delta)$ is dense in $\widehat{\mathbb{C}}$ for every Stolz angle Δ at ζ.*

Hence $\{|\zeta| = 1\}$ can be partitioned into a nullset, a set where $g(z)$ tends to a finite limit $g(\zeta)$ as $z \to \zeta$ in every Stolz angle however broad it may be, and a set where $g(z)$ comes close to every $c \in \widehat{\mathbb{C}}$ as $z \to \zeta$ in every Stolz angle however thin it may be. In fact even more is correct (CoLo66, p. 147): The Stolz angle in (ii) can be replaced by any triangle in $\mathbb{D}$ with vertex ζ. We say that ζ is a *Plessner point* if (ii) holds.

If g is bounded and analytic in $\mathbb{D}$ then (ii) is impossible so that g has an angular limit almost everywhere. This is *Fatou's theorem*; we will however need it in our proof below. Another key ingredient will be the *Lusin-Privalov construction* of a suitable subdomain of $\mathbb{D}$ with rectifiable boundary.

Proof. We consider all disks D with rational center $c \in \mathbb{Q} \times i\mathbb{Q}$ and rational radius $\delta > 0$, furthermore all rational numbers α and r with $0 < \alpha < \pi/6$ and $1/2 < r < 1$. Let

$$(D_n, \alpha_n, r_n), \quad n = 1, 2, \ldots$$

be an enumeration of these countably many objects. Let E_n be the set of all $\zeta \in \mathbb{T}$ where g has no finite angular limit and where also

$$(1) \qquad \{g(z) : |\arg(1 - \bar{\zeta}z)| < \alpha_n,\ r_n \le |z| < 1\} \cap D_n = \emptyset.$$

The union of these sets consists of the points where neither (i) nor (ii) hold. Hence it suffices to show that $\Lambda(E_n) = 0$ for all n.

We now drop the index n and consider the starlike Jordan domain

$$(2) \qquad H = D(r, 0) \cup \bigcup_{\zeta \in E} \{|\arg(1 - \bar{\zeta}z)| < \alpha,\ r \le |z| < 1\};$$

see Fig. 6.1. It is easy to see that

$$(3) \qquad \Lambda(\partial H) \le 2\pi/\sin \alpha < \infty.$$

Let φ map $\mathbb{D}$ conformally onto H.

If $z_1, \ldots, z_m$ are the zeros of $g - c$ in $\overline{D}(0, r)$ then

$$(4) \qquad h(z) = (z - z_1) \cdots (z - z_m)/(g(z) - c) \not\equiv 0$$

is bounded in H, by (1) and (2); note that c is the center of D. Hence $h \circ \varphi$ is bounded in $\mathbb{D}$ and therefore has a nonzero angular limit almost everywhere on $\mathbb{T}$ by Fatou's theorem and the Riesz uniqueness theorem (CoLo66, p. 22).

Now let

(5) $$A = \{\zeta \in \mathbb{T} : \varphi \text{ conformal at } \zeta, \varphi(\zeta) \in E\}.$$

If $\zeta \in A$ then h has no nonzero angular limit at $\varphi(\zeta)$ by (4) and the definition of E. Since φ is conformal and thus isogonal at ζ it follows from Proposition 4.10 that $h \circ \varphi$ has no nonzero angular limit at ζ; the domain $H = \varphi(\mathbb{D})$ is tangential to $\mathbb{T}$ at $\varphi(\zeta)$. Hence $\Lambda(A) = 0$ by the last paragraph. Since φ is conformal almost everywhere on $\mathbb{T}$ by (3) and Theorem 6.8, we conclude from (5) and (6.3.3) that $\Lambda(E) = 0$.   □

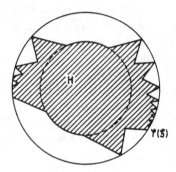

Fig. 6.1. The starlike domain $H = \varphi(\mathbb{D})$ obtained by the Lusin-Privalov construction

An easy consequence (Pri19, Pri56) is the very general *Privalov uniqueness theorem*.

Corollary 6.14. *Let* $g(z) \not\equiv 0$ *be meromorphic in* $\mathbb{D}$*. Then* g *has the angular limit* 0 *almost nowhere on* $\mathbb{T}$*.*

Proof. The function $1/g$ is meromorphic in $\mathbb{D}$. If g has the angular limit 0 at $\zeta \in \mathbb{T}$ then $1/g$ has the angular limit ∞ so that neither (i) nor (ii) hold. The set of these ζ has therefore measure zero by Plessner's theorem.   □

In the case of Bloch functions the Stolz angle in (ii) can be replaced by the radius.

Corollary 6.15. *If* $g \in \mathcal{B}$ *then, for almost all* $\zeta \in \mathbb{T}$*, either* g *has a finite angular limit, or the radial cluster set* $C_{[0,\zeta)}(g, \zeta)$ *is* $\widehat{\mathbb{C}}$*.*

Proof. Suppose that $C_{[0,\zeta)}(g, \zeta) \neq \widehat{\mathbb{C}}$. Then there exist $c \in \mathbb{C}$, $\delta > 0$ and $r_0 < 1$ such that $|g(r\zeta) - c| > \delta$ for $r_0 < r < 1$. Since $g \in \mathcal{B}$ we can, by (4.2.5), choose the Stolz angle Δ so thin that $|g(z) - g(|z|\zeta)| < \delta/2$ for $z \in \Delta$. It follows that $|g(z) - c| > \delta/2$ for $z \in \Delta$, $r_0 < |z| < 1$ so that (ii) is not satisfied. Hence our assertion follows from Plessner's theorem.   □

Applying this result to the Bloch function $g = \log f'$ (see Proposition 4.1) we obtain:

Corollary 6.16. *If f maps $\mathbb{D}$ conformally into $\mathbb{C}$ then, for almost all $\zeta \in \mathbb{T}$, either f has a finite nonzero angular derivative $f'(\zeta)$, or*

$$(6) \qquad \liminf_{r \to 1} |f'(r\zeta)| = 0, \quad \limsup_{r \to 1} |f'(r\zeta)| = +\infty.$$

In the general case it is not possible to replace the Stolz angle in (ii) by the radius. There exist (non-normal) functions g analytic in $\mathbb{D}$ such that

$$(7) \qquad |g(r\zeta)| \to +\infty \quad \text{as} \quad r \to 1 \quad \text{for almost all } \zeta \in \mathbb{T}.$$

An example comes from iteration theory (Fat20, p. 272): If $0 < |a| < 1$ then there is an analytic function g such that

$$(8) \qquad g\left(z\frac{a+z}{1+\bar{a}z} \right) = ag(z) \quad \text{for} \quad z \in \mathbb{D}$$

and this function satisfies (7). It is clear by Plessner's theorem that almost none of the radial limits can be an angular limit.

Another surprising example (CoLo66, p. 166) states that for every set $E \subset \mathbb{T}$ of first category there is a non-constant analytic function g in $\mathbb{D}$ such that $g(r\zeta) \to 0$ $(r \to 1)$ for $\zeta \in E$. There are sets of first category with measure 2π; see Section 2.6. Thus it is not possible to replace angular by radial limit in the Privalov uniqueness theorem.

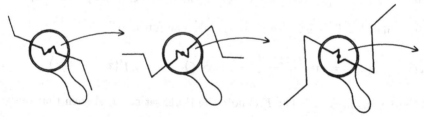

Fig. 6.2. An example of twisting: Local pictorial representation of the boundary at each stage blowing up by some factor

2. Now let f map $\mathbb{D}$ conformally onto $G \subset \mathbb{C}$. We say that f is *twisting* at $\zeta \in \mathbb{T}$ if the angular limit $f(\zeta) \neq \infty$ exists and if

$$(9) \qquad \liminf_{z \to \zeta, z \in \Gamma} \arg[f(z) - f(\zeta)] = -\infty, \quad \limsup_{z \to \zeta, z \in \Gamma} \arg[f(z) - f(\zeta)] = +\infty$$

for every curve $\Gamma \subset \mathbb{D}$ ending at ζ. Thus G twists around the boundary point $f(\zeta)$ infinitely often in both directions.

Proposition 6.17. *Let f map $\mathbb{D}$ conformally into $\mathbb{C}$ and let $f(\zeta) \neq \infty$ exist at $\zeta \in \mathbb{T}$. Then f is twisting at ζ if and only if (9) holds with Γ replaced by a Stolz angle Δ.*

Proof. Let f not be twisting at ζ so that (9) does not hold for some curve Γ. Since $\log(f - f(\zeta))$ is a Bloch function by Proposition 4.1, it follows from Proposition 4.4 that $\arg(f(z) - f(\zeta))$ is bounded below or above in Δ. The converse is trivial because $[0, \zeta) \subset \Delta$. $\square$

Theorem 6.18. *Every conformal map of $\mathbb{D}$ into $\mathbb{C}$ is either conformal or twisting at almost all points of $\mathbb{T}$.*

This is the *McMillan twist theorem* (McM69b). The points of $\mathbb{T}$ where the behaviour is neither good nor exceedingly bad thus form a set of zero measure.

For the proof we need the following stronger version of Corollary 4.18; see Theorem 9.19 below or Pom75, p. 311, p. 314.

Lemma 6.19. *Let f map $\mathbb{D}$ conformally into $\mathbb{C}$ and let $0 < \delta < 1$. If $z \in \mathbb{D}$ and I is an arc of $\mathbb{T}$ with $\omega(z, I) \geq \alpha > 0$ then there exists a Borel set $B \subset I$ with $\omega(z, B) > (1 - \delta)\omega(z, I)$ such that*

$$|f(\zeta) - f(z)| \leq \Lambda(f(S)) < K(\delta, \alpha)d_f(\zeta) \quad \text{for} \quad \zeta \in B$$

where S is the non-euclidean segment from z to ζ and where $K(\delta, \alpha)$ depends only on δ and α.

Proof of Theorem 6.18. We consider the Stolz angle

$$(10) \qquad \Delta(\zeta) = \left\{ z \in \mathbb{D} : |\arg(1 - \bar{\zeta}z)| < \pi/4, |z - \zeta| < 1/2 \right\} \quad (\zeta \in \mathbb{T})$$

and define $A = \{\zeta \in \mathbb{T} : f(\zeta) \neq \infty \text{ exists}\}$. The function

$$(11) \qquad \varphi(z, \zeta) = \arg \frac{f(z) - f(\zeta)}{z - \zeta} \quad (z \neq \zeta), \quad = \arg f'(z) \quad (z = \zeta)$$

is continuous in $\mathbb{D} \times \mathbb{D}$. Let E denote the Borel set of $\zeta \in A$ such that both

$$(12) \qquad \arg f'(r\zeta) \quad \text{is unbounded above and below for } 0 < r < 1,$$

$$(13) \qquad \arg[f(z) - f(\zeta)] \text{ . is bounded above or below in } \Delta(\zeta).$$

Suppose that $\Lambda(E) > 0$. It follows from (13) that there exist $M < \infty$ and $E' \subset E$ with $\Lambda(E') > 0$ such that, say,

$$(14) \qquad \varphi(z, \zeta) < M \quad \text{for} \quad z \in \Delta(\zeta), \quad \zeta \in E'.$$

The set E' has a point of density by Proposition 6.4 which we may assume to be $\zeta = 1$. By (11) and (12) there exist $r_n \to 1-$ such that

$$(15) \qquad \varphi(r_n, r_n) = \arg f'(r_n) \to +\infty \quad \text{as} \quad n \to \infty.$$

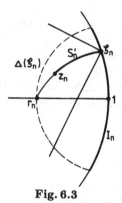

Fig. 6.3

Let $K_1, \ldots$ denote suitable absolute constants. Let I_n be the arc of $\mathbb{T}$ indicated in Fig. 6.3. By Lemma 6.19, there exists $\zeta_n \in I_n \cap E'$ such that

$$\int_{S_n} |f'(z)| \, |dz| \leq K_1 d_f(r_n)$$

where S_n is the non-euclidean segment from r_n to ζ_n. If S'_n is the subsegment (say from z_n to ζ_n) that has non-euclidean distance ≥ 1 from r_n, then $|f(z) - f(r_n)| \geq K_2^{-1} d_f(r_n)$ for $z \in S'_n$ by (1.3.18). Hence

$$\int_{S'_n} |d \arg[f(z) - f(r_n)]| \leq \int_{S'_n} \left| \frac{f'(z)}{f(z) - f(r_n)} \right| |dz| \leq K_1 K_2$$

and it easily follows from (11) and Exercise 1.3.4 that

$$|\varphi(\zeta_n, r_n) - \varphi(r_n, r_n)| \leq |\varphi(\zeta_n, r_n) - \varphi(z_n, r_n)| + |\varphi(z_n, r_n) - \varphi(r_n, r_n)| \leq K_3 \, .$$

Since $r_n \in \Delta(\zeta_n)$ for large n, we therefore deduce from (14) that $\varphi(r_n, r_n) < K_3 + M$ which contradicts (15). Hence we have shown that $\Lambda(E) = 0$.

Now let $\zeta \in A \setminus E$ so that (12) or (13) are false. If (12) does not hold then the radial cluster set $C_{[0,\zeta)}(\log f', \zeta)$ is different from $\widehat{\mathbb{C}}$ so that, by Proposition 4.1 and Corollary 6.15, the Bloch function $\log f'$ has a finite angular limit at ζ except possibly for a set of zero measure. On the other hand, if (13) does not hold then f is twisting at ζ by Proposition 6.17. This proves Theorem 6.18 because $\Lambda(A) = 2\pi$. $\qquad\qquad \square$

Corollary 6.20. *Let f map $\mathbb{D}$ conformally into $\mathbb{D}$ and let $A \subset \mathbb{T}$ be measurable. If $f(\zeta)$ exists for $\zeta \in A$ and if $f(A) \subset \mathbb{T}$ then f is conformal at almost all points of A.*

This is an immediate consequence (McM69b) of Theorem 6.18: If $\zeta \in A$ then $f(\zeta) \in \mathbb{T}$ but $f(\mathbb{D}) \subset \mathbb{D}$ so that f cannot be twisting at ζ.

Corollary 6.21 *A conformal map of $\mathbb{D}$ into $\mathbb{C}$ is conformal at almost all points where it is isogonal.*

The map f was defined (Section 4.3) to be isogonal at $\zeta \in \mathbb{T}$ if $f(\zeta) \neq \infty$ exists and if $\arg(f(z) - f(\zeta)) - \arg(z - \zeta)$ has a finite angular limit. Then f cannot be twisting at ζ because (9) does not hold for $\Gamma = [0, \zeta)$. Hence the assertion follows from Theorem 6.18.

Note that conformal always implies isogonal. Hence it follows from Proposition 4.12 that there exists a conformal map onto a quasidisk that is twisting almost everywhere.

Exercises 6.4

1. Let g be non-constant and meromorphic in $\mathbb{D}$ and suppose that all angular limits $g(\zeta)$ lie in the same countable set. Show that g has angular limits almost nowhere.

2. Consider a meromorphic function in $\mathbb{D}$ that has angular limits almost nowhere. Use the Collingwood maximality theorem to show that all points of $\mathbb{T} \setminus E$ are Plessner points where the exceptional set E is of first category.

In the following problems let f map $\mathbb{D}$ conformally onto $G \subset \mathbb{C}$.

3. Show that f cannot be twisting everywhere.

4. Show that G contains a spiral

$$f(\zeta) + r_\zeta(t)e^{it}, \quad 0 \le t < +\infty \quad \text{with} \quad r_\zeta(t) \to 0 \quad (t \to \infty)$$

for almost no $\zeta \in \mathbb{T}$.

5. Let G lie in the inner domain of the rectifiable Jordan curve J. Show that f is conformal at almost all points $\zeta \in \mathbb{T}$ for which $f(\zeta) \in J$.

6. Let J be a rectifiable Jordan curve. If $\Lambda(J \cap \partial G) = 0$ show that $\Lambda(f^{-1}(J \cap \partial G)) = 0$ provided that G lies in the inner domain of J. (It is a deep result that the final assumption is redundant; see BiJo90.)

6.5 Sectorial Accessibility

1. Let f map $\mathbb{D}$ conformally onto G. We define $\text{Sect}(f)$ as the set of all $\zeta \in \mathbb{T}$ such that $f(\zeta)$ is *sectorially accessible*, i.e. the angular limit $f(\zeta) \neq \infty$ exists and f maps a domain $V \subset \mathbb{D}$ with $\zeta \in \overline{V}$ onto an open triangle of vertex $f(\zeta)$; see Fig. 6.4.

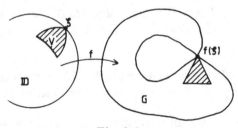

Fig. 6.4

Proposition 6.22. *All points where f is conformal belong to $\mathrm{Sect}(f)$, and f is conformal at almost all points of $\mathrm{Sect}(f)$.*

This follows at once from Proposition 4.10 and from Theorem 6.18 because f is not twisting at any point of $\mathrm{Sect}(f)$.

We now use a variant of the Lusin-Privalov construction (Sak64, Chapt. IX; McM69b).

Proposition 6.23. *There exist countably many domains $G_n \subset G$ and closed sets $A_n \subset \mathbb{T} \cap \partial f^{-1}(G_n)$ such that ∂G_n is a rectifiable Jordan curve, $f(A_n)$ is closed and*

$$(1) \qquad \mathrm{Sect}(f) = \bigcup_n A_n .$$

It follows that $\mathrm{Sect}(f)$ is an F_σ-set, furthermore that

$$(2) \qquad f(\mathrm{Sect}(f)) = \bigcup_n f(A_n) ,$$

where $f(A_n)$, the set of angular limits, is a closed set of finite linear measure. Hence

$$(3) \qquad f(\mathrm{Sect}(f)) \text{ is an } F_\sigma\text{-set of } \sigma\text{-finite linear measure}.$$

Proof. We consider all rational numbers $\alpha \in (0, \pi)$ and all lines $L : at + b$, $-\infty < t < \infty$ with $a, b \in \mathbb{Q} \times i\mathbb{Q}$, $a \neq 0$. These countably many pairs can be enumerated as (L_n, α_n). Let A'_n denote the set of all $\zeta \in \mathrm{Sect}(f)$ with $f(\zeta) \subset \overline{D}(0, n)$ for which there is an open isosceles triangle $T \subset G$ of base line L_n, height $\geq 1/n$ and vertex angle α_n at $f(\zeta)$; see Fig. 6.5. It follows from the definition that

$$(4) \qquad \mathrm{Sect}(f) = \bigcup_{n=1}^{\infty} A'_n .$$

We now show that $B'_n = f(A'_n)$ is closed. Let $w_k \in B'_n$ and $w_k \to w_0$ as $k \to \infty$. The corresponding triangles T_k converge to a triangle T_0 of vertex w_0 because the heights are $\geq 1/n$. We have $T_0 \subset G$ because T_0 is open, and $V_0 = f^{-1}(T_0)$ has a boundary point $\zeta_0 \in \mathbb{T}$ with $f(\zeta_0) = w_0$; see Exercise 2.5.5. Since $\zeta_0 \in A'_n$ by the definition of A'_n it follows that $w_0 \in B'_n$.

Let G'_n be the union of all the open triangles corresponding to any $\zeta \in A'_n$. This is an open set bounded by part of L_n together with a rotated Lipschitz graph because all triangles are isosceles and have the same vertex angle. The open set G'_n has only finitely many components because the vertices lie in $\overline{D}(0, n)$ and the heights are $\geq 1/n$. Each component is bounded by a rectifiable Jordan curve. These components partition B'_n into finitely many closed sets.

We now renumber these components and corresponding sets as G_n and B_n. Since any two points $w_1, w_2 \in B_n$ can be connected by a curve of diameter

$< M_n|w_1 - w_2|$ for some constant M_n we see from Exercise 4.5.4 that $A_n = f^{-1}(B_n) \cap \text{Sect}(f)$ is also closed. Finally (1) follows from (4).   □

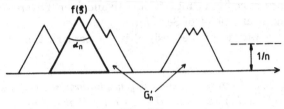

Fig. 6.5

We deduce a generalization (McM69b) of (6.3.3), the *McMillan sector theorem*.

Theorem 6.24. *Let f map $\mathbb{D}$ conformally into $\mathbb{C}$. If $E \subset \text{Sect}(f)$ then*

(5)
$$\Lambda(E) = 0 \quad \Leftrightarrow \quad \Lambda(f(E)) = 0.$$

Proof. (a) Let $E \subset \text{Sect}(f)$. We suppose first that $\Lambda(E) = 0$ and determine domains G_n and closed sets A_n according to Proposition 6.23. Let h_n map $\mathbb{D}$ conformally onto $H_n = f^{-1}(G_n) \subset \mathbb{D}$; see Fig. 6.6. Then $g_n = f \circ h_n$ maps $\mathbb{D}$ conformally onto G_n. We define

$$E_n = A_n \cap E, \quad Z_n = h_n^{-1}(E_n) \equiv \{\zeta : h_n(\zeta) \text{ exists}, h_n(\zeta) \in E_n\}$$

so that $E = \bigcup E_n$ by (1). Since $E_n = h_n(Z_n) \subset \mathbb{T}$ and $h_n(\mathbb{D}) \subset \mathbb{D}$ it follows from Löwner's lemma (Proposition 4.15) and $\Lambda(E_n) = 0$ that $\Lambda(Z_n) = 0$. Since ∂G_n is rectifiable we conclude from Theorem 6.8 that

$$\Lambda(f(E_n)) = \Lambda(g_n(Z_n)) = 0 \quad \text{for all} \quad n$$

which implies $\Lambda(f(E)) = 0$.

(b) Conversely let $\Lambda(f(E)) = 0$. Using the notation of part (a) we define

(6)
$$E_n^* = \{\zeta \in E_n : f \text{ is conformal at } \zeta\}, \quad Z_n^* = h_n^{-1}(E_n^*).$$

Let $\zeta \in E_n^*$. Since G_n contains a triangle of vertex $f(\zeta)$ and since f is conformal and thus isogonal at ζ, it follows from Proposition 4.10 that H_n contains a triangle at ζ. Hence

(7)
$$Z_n^* \subset \text{Sect}(h_n).$$

Furthermore $\Lambda(g_n(Z_n^*)) \leq \Lambda(g_n(Z_n)) = \Lambda(f(E_n)) \leq \Lambda(f(E)) = 0$ and thus $\Lambda(Z_n^*) = 0$ by Theorem 6.8. Hence we conclude from (7) and part (a) that $\Lambda(E_n^*) = \Lambda(h_n(Z_n^*)) = 0$ and thus from (6) and Proposition 6.22 that $\Lambda(E_n) = 0$ for all n. Since $E = \bigcup E_n$ it follows that $\Lambda(E) = 0$.   □

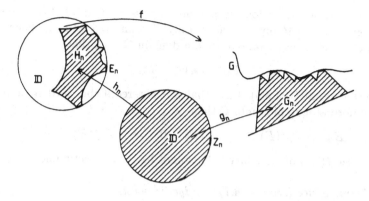

Fig. 6.6

2. Now we come to the *Makarov compression theorem* (Mak84, Mak85).

Theorem 6.25. *If f maps $\mathbb{D}$ conformally onto $G \subset \mathbb{C}$ then*

$$(8) \qquad\qquad \mathbb{T} = \mathrm{Sect}(f) \cup A_0 \cup A_1 \,,$$

where $\Lambda(f(A_0)) = 0$ and $\Lambda(A_1) = 0$.

Together with Theorem 6.24 this shows that

$$(9) \qquad E \subset \mathrm{Sect}(f) \cup A_0 \,, \quad \Lambda(E) = 0 \quad \Rightarrow \quad \Lambda(f(E)) = 0 \,,$$

$$(10) \qquad E \subset \mathrm{Sect}(f) \cup A_1 \,, \quad \Lambda(f(E)) = 0 \quad \Rightarrow \quad \Lambda(E) = 0 \,.$$

Furthermore it follows from (3) that

$$(11) \qquad f(\mathrm{Sect}(f) \cup A_0) \text{ has } \sigma\text{-finite linear measure} \,;$$

see JoWo88 for a generalization to multiply connected domains.

If the total boundary ∂G does not have σ-finite linear measure we conclude from (8) and (11) that $f(A_1)$ is much larger than $\partial G \setminus f(A_1)$ in spite of the fact that $\Lambda(A_1) = 0$. Another consequence of (5) and (8) is:

Corollary 6.26. *If the points on ∂G that are vertex of an open triangle in G form a set of zero linear measure then there is a set $A_0 \subset \mathbb{T}$ such that*

$$(12) \qquad\qquad \Lambda(A_0) = 2\pi, \quad \Lambda(f(A_0)) = 0 \,.$$

This surprising fact (Mak84) says that, for very bad boundaries, the conformal map f compresses almost the entire unit circle into a set of zero linear measure. The first example for this was given in McMPi73.

In the opposite direction, even more is possible (Pir72): Let A be any perfect set on $\mathbb{T}$ and B any compact totally disconnected set in $\mathbb{C}$. Then there exists a conformal map f onto a Jordan domain G such that

$$B \subset f(A) \subset \partial G.$$

A recent result (BiJo90) states that if J is a rectifiable Jordan curve and f any conformal map of $\mathbb{D}$ into $\mathbb{C}$ then

(13) $E \subset \mathbb{T}, \quad f(E) \subset J, \quad \Lambda(E) > 0 \quad \Rightarrow \quad \Lambda(f(E)) > 0.$

Thus the set $f(A_0)$ of Corollary 6.26 cannot lie on any rectifiable curve.

3. We introduce (see Fig. 6.7) the *dyadic points*

(14) $z_{n\nu} = \left(1 - \dfrac{1}{2^n}\right) \exp \dfrac{i\pi(2\nu - 1)}{2^n} \quad (\nu = 1, \ldots, 2^n \,; \ n = 1, 2, \ldots)$

and the corresponding *dyadic arcs*

(15) $I_{n\nu} = I(z_{n\nu}) = \left\{ e^{it} : \dfrac{2\pi(\nu - 1)}{2^n} \leq t \leq \dfrac{2\pi\nu}{2^n} \right\}.$

Dyadic arcs have the technical advantage that any two can intersect at inner points only if one is contained in the other. Every point of $\mathbb{D}$ is within a bounded non-euclidean distance of some dyadic point and the harmonic measure satisfies $\omega(z_{n\nu}, I_{n\nu}) > 1/2$.

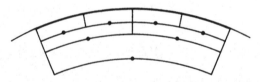

Fig. 6.7. Some dyadic points; the heavily drawn four arcs correspond to the outer four points (lengths are distorted)

Proof of Theorem 6.25. Let $K_1, K_2, \ldots$ denote suitable absolute constants. It follows from Proposition 6.22 and Corollary 6.16 that

(16) $\mathbb{T} \overset{\circ}{=} \mathrm{Sect}(f) \cup A, \quad A = \left\{ \zeta \in \mathbb{T} : \liminf_{r \to 1} |f'(r\zeta)| = 0 \right\}.$

Let $k = 1, 2, \ldots$. For $\zeta \in A$ we can find $r < 1$ such that $|f'(r\zeta)| < 2^{-k}$. We choose n such that $1 - 2^{-n} \leq r < 1 - 2^{-n-1}$ and ν such that ζ lies in the dyadic interval $I_{n\nu}$. Let $z_{n\nu}$ be the corresponding dyadic point. It follows from (1.3.22) that $|f'(z_{n\nu})| < K_1 2^{-k}$. For fixed k we consider all dyadic arcs obtained this way and we delete those that are contained in a larger one. This gives us countably many dyadic points that we now write as ζ_{kj} and corresponding dyadic arcs J_{kj} that cover A and are disjoint except for their endpoints. Note that $|f'(\zeta_{kj})| < K_1 2^{-k}$.

By Lemma 6.19 we can choose K_2 so large that, if

(17) $$D_{kj} = D(f(\zeta_{kj}), K_2(1 - |\zeta_{kj}|)|f'(\zeta_{kj})|),$$

then $f(B_{kj}) \subset D_{kj}$ with suitable Borel sets $B_{kj} \subset J_{kj}$ such that

(18) $$\Lambda(B_{kj}) > \Lambda(J_{kj})/2 > 1 - |\zeta_{kj}|.$$

We define the Borel sets

(19) $$Z_m = A \cap \bigcup_{k \geq m} \bigcup_j B_{kj}, \quad A_0 = \bigcap_{m=1}^{\infty} Z_m.$$

It follows that, for all m, the set $f(A_0)$ is covered by the disks D_{kj} with $k \geq m$ where, by (17),

$$\sum_{k \geq m} \sum_j \operatorname{diam} D_{kj} \leq 2K_1 K_2 \sum_{k=m}^{\infty} 2^{-k} \sum_j (1 - |\zeta_{kj}|) \leq \frac{K_3}{2^m};$$

indeed $1 - |\zeta_{kj}| < \Lambda(J_{kj})$ and the arcs $J_{kj} \subset \mathbb{T}$ (k fixed) are essentially disjoint so that the inner sum is $< 2\pi$. It follows that $\Lambda(f(A_0)) = 0$.

Let $\zeta \in A \setminus Z_m$ and $k \geq m$. Then $\zeta \in J_{kj}$ for some $j = j(\zeta)$. Since $J_{kj} \cap (A \setminus Z_m) \subset J_{kj} \setminus B_{kj}$ by (19) and since $B_{kj} \subset J_{kj}$, we see that

$$\frac{\Lambda(J_{kj} \cap (A \setminus Z_m))}{\Lambda(J_{kj})} \leq 1 - \frac{\Lambda(B_{kj})}{\Lambda(J_{kj})} < \frac{1}{2}$$

because of (18). Since $\zeta \in J_{kj}$ and $\Lambda(J_{kj}) \to 0$ as $k \to \infty$ we conclude that $A \setminus Z_m$ has no point of density so that $\Lambda(A \setminus Z_m) = 0$ by Proposition 6.4. Since $Z_m \subset A$ we therefore see from (19) and (16) that

$$A_0 \overset{\circ}{=} A \overset{\circ}{=} \mathbb{T} \setminus \operatorname{Sect}(f). \qquad \square$$

4. Now let f map $\mathbb{D}$ conformally onto a domain G that is starlike with respect to 0 such that $f(0) = 0$. Then, by Theorem 3.18,

(20) $$f(z) = zf'(0) \exp\left(-\frac{1}{\pi} \int_0^{2\pi} \log(1 - e^{-it}z)d\beta(t) \right) \quad \text{for} \quad z \in \mathbb{D}$$

where the increasing function β is given by

(21) $$\beta(t) = \lim_{r \to 1} \arg f(re^{it}) = \arg f(e^{it}).$$

Theorem 6.27. *Let f be starlike in $\mathbb{D}$. Then f is conformal and satisfies*

(22) $$\operatorname{Re}[e^{it}f'(e^{it})/f(e^{it})] = \beta'(t) \neq \infty$$

for almost all e^{it}. Furthermore β is absolutely continuous if and only if, for almost all ϑ, there exist $w \in \partial G$ with $\arg w = \vartheta$ such that G contains some triangle of vertex w symmetric to $[0, w]$.

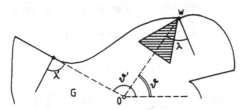

Fig. 6.8. The condition of Theorem 6.27 holds for ϑ but not for ϑ'

Proof. Let A denote the set of all t such that f is conformal at e^{it} and (22) holds. The starlike function f is nowhere twisting and is thus conformal almost everywhere on $\mathbb{T}$ by Theorem 6.18; this also follows from the fact that $\operatorname{Re}[zf'(z)/f(z)] > 0$. Taking the logarithmic derivative in (20) we see that

$$(23) \qquad \operatorname{Re}\left[z\frac{f'(z)}{f(z)}\right] = \frac{1}{2\pi}\int_0^{2\pi} \frac{1 - |z|^2}{|e^{it} - z|^2}d\beta(t) \qquad \text{for} \quad z \in \mathbb{D}.$$

This Poisson-Stieltjes integral tends to $\beta'(t)$ for $z = re^{it}$, $r \to 1-$ if β has a finite derivative (Dur70, p. 4), hence almost everywhere by Lebesgue's theorem. It follows that $\Lambda(A) = 2\pi$.

Let B denote the set of t such that, with $\zeta = e^{it}$, the domain G contains a triangle of vertex $f(\zeta)$ symmetric to $[0, f(\zeta)]$. If $t \in A$ then G contains a sector of vertex $f(\zeta)$ and of any angle $< \pi$ but not of angle $> \pi$. Its midline and $[0, f(\zeta)]$ form the angle $\lambda = \arg[\zeta f'(\zeta)/f(\zeta)]$; see Fig. 6.8. It follows by (22) that $A \cap B = \{t \in A : \beta'(t) > 0\}$.

We write $\beta = \alpha + \gamma$ where α is absolutely continuous and γ is increasing with $\gamma'(t) = 0$ for almost all $t \in I = [0, 2\pi]$. We can interpret α, β, γ also as measures. Since $\gamma'(t) \le \beta'(t) < \infty$ for $t \in A$ it follows (Sak64, Chapt. IX) that $\gamma(A) = 0$ and thus $\beta(A \cap B) = \alpha(A \cap B)$. Since α is absolutely continuous and $\alpha'(t) \le \beta'(t) = 0$ for $t \in A \setminus B$, we see that $\alpha(A \cap B) = \alpha(A)$. Since $A \stackrel{\circ}{=} I$ we conclude that $\beta(A \cap B) = \alpha(A) = \alpha(I)$. Hence it follows from (21) that the set of ϑ satisfying the last condition of our theorem has measure $\alpha(I)$; note that $\Lambda(B \setminus A) = 0$ by Theorem 6.18 so that $f(B \setminus A)$ has linear measure 0 by Theorem 6.24 and therefore makes no contribution to the angular measure. Finally β is absolutely continuous if and only if $\alpha(I) = \beta(I) = 2\pi$. □

Exercises 6.5

1. Let f map $\mathbb{D}$ conformally into $\mathbb{C}$. Show that, for $E \subset \operatorname{Sect}(f)$,

$$E \text{ measurable} \quad \Leftrightarrow \quad f(E) \text{ is linearly measurable}.$$

Construct a function f and a measurable set $E \subset \mathbb{T} \setminus \operatorname{Sect}(f)$ such that $f(E)$ is not linearly measurable.

2. Let $B \subset \mathbb{C}$ be compact. Deduce from Proposition 6.23 that the points of B that are vertex of an open triangle in $\mathbb{C} \setminus B$ form a set of σ-finite linear measure. (This also follows from the Kolmogoroff-Verčenko theorem in Section 6.1.)

3. Let $J : w(t)$, $0 \leq t \leq 2\pi$ be a Jordan curve in $\mathbb{C}$ such that the right-hand derivative $w'_+(t)$ exists and is finite for almost all t. Show that the derivative $w'(t)$ exists for almost all t. (Apply the Kolmogoroff-Verčenko theorem to the graphs of $\operatorname{Re} w$ and $\operatorname{Im} w$.)

4. Show that the set A_0 in the Makarov compression theorem can be chosen such that f is twisting at every point of A_0.

5. Let f be starlike with $f(\mathbb{D}) = \{\rho e^{i\vartheta} : 0 \leq \rho < r(\vartheta)\}$. Show that β is singular (i.e. $\beta'(t) = 0$ for almost all t) if and only if the last condition of Theorem 6.27 is satisfied for almost no ϑ.

6. Under the same assumption, suppose that $r(\vartheta) \leq 1$ for all ϑ and that $E = \{\vartheta : r(\vartheta) < 1\}$ is countable. If E is closed show that β is absolutely continuous. If E is dense show that β may be singular.

6.6 Jordan Curves and Harmonic Measure

We assume throughout this section that J is a Jordan curve in $\mathbb{C}$ with inner domain G and outer domain G^*. Let f map $\mathbb{D}$ conformally onto G while f^* maps $\mathbb{D}^* = \{|z| > 1\} \cup \{\infty\}$ conformally onto G^*; see Fig. 6.9. Then f and f^* have continuous injective extensions to $\mathbb{T}$ by Theorem 2.6. We consider the harmonic measures

(1)
$$\omega(E) \equiv \omega(f(0), E, G) = \Lambda(f^{-1}(E))/(2\pi),$$
$$\omega^*(E) =. \omega(f^*(\infty), E, G^*) = \Lambda(f^{*-1}(E))/(2\pi),$$

and define

(2) $\quad S = \{w \in J : \text{ there is an open triangle of vertex } w \text{ in } G\} = f(\operatorname{Sect}(f)),$

similarly S^* with G^* instead of G, and

(3) $$T = \{w \in J : J \text{ has a tangent at } w\}.$$

It is clear that $T \subset S \cap S^*$.

The McMillan sector theorem states that

(4) $\quad \Lambda(E) = 0 \quad \Leftrightarrow \quad \omega(E) = 0, \quad \Lambda(E^*) = 0 \quad \Leftrightarrow \quad \omega^*(E^*) = 0$

for $E \subset S$ and $E^* \subset S^*$. It follows that, for $E \subset T$,

(5) $$\Lambda(E) = 0 \quad \Leftrightarrow \quad \omega(E) = 0 \quad \Leftrightarrow \quad \omega^*(E) = 0.$$

Let again $\overset{\circ}{=}$ mean "equal up to a set of zero linear measure". We see from (4) and Proposition 6.22 that f is conformal at $f^{-1}(w)$ for almost all $w \in S$ and that f^* is conformal at $f^{*-1}(w)$ for almost all $w \in S^*$. It follows that $S \cap S^* \overset{\circ}{=} T$.

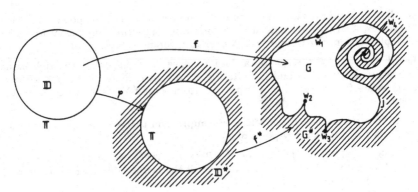

Fig. 6.9. Here $w_1 \in T$, $w_2 \in S \setminus S^*$, $w_3 \in S^* \setminus S$, $w_4 \notin S \cup S^*$

The Makarov compression theorem now reads

$$(6) \qquad J = S \cup B_0 \cup B_1 = S^* \cup B_0^* \cup B_1^*$$

where $\Lambda(B_0) = 0$, $\omega(B_1) = 0$ and $\Lambda(B_0^*) = 0$, $\omega^*(B_1^*) = 0$. Thus the harmonic measure ω is concentrated on the set $S \cup B_0$ of σ-finite linear measure (see (6.5.3)) in the sense that $\omega(S \cup B_0) = 1$. In particular (Mak84), if $\Lambda(S \cup S^*) = 0$ then the harmonic measures ω and ω^* are concentrated on a set of zero linear measure.

Proposition 6.28. *If J is a quasicircle then*

$$(7) \qquad T \overset{\circ}{=} S \overset{\circ}{=} S^*, \quad f^{-1}(T) \overset{\circ}{=} \mathrm{Sect}(f).$$

Proof. Let A and A^* be the sets on T where f and f^* are conformal. Since $A \overset{\circ}{=} \mathrm{Sect}(f)$ by Proposition 6.22 we see from Theorem 6.24 that

$$T \subset S \overset{\circ}{=} f(A) \subset T;$$

the last inclusion is a consequence of Theorem 5.5 because a quasidisk is a John domain. Thus $T \overset{\circ}{=} S$ and similarly $T \overset{\circ}{=} S^*$. The second relation (7) now follows from (4). □

Proposition 6.29. *There is a starlike Jordan curve J such that*

$$(8) \qquad T = \emptyset, \quad \mathrm{Sect}(f) \overset{\circ}{=} \mathrm{Sect}(f^*) \overset{\circ}{=} T.$$

Proof. Let $r(\vartheta)$ $(0 \le \vartheta \le 2\pi)$ be a continuous positive function with $r(0) = r(2\pi)$ that does not have any finite or infinite derivative at any point; see e.g. Fab07 for the existence of such a function. Then

$$J : r(\vartheta)e^{i\vartheta}, \quad 0 \le \vartheta \le 2\pi$$

is a starlike Jordan curve with $T = \emptyset$, and $\mathrm{Sect}(f) \overset{\circ}{=} T$ holds because f is conformal almost everywhere by Theorem 6.27, similarly $\mathrm{Sect}(f^*) \overset{\circ}{=} T$. □

Thus (7) may fail for non-quasicircles. The next result (BiCaGaJo89) holds for all Jordan curves.

Theorem 6.30. *There is a partition*

$$(9) \qquad J = T \cup B \cup B^* \quad \text{with} \quad \omega(B) = 0, \quad w^*(B^*) = 0.$$

Together with (5) this shows that ω and ω^* are mutually singular if and only if $\Lambda(T) = 0$. It follows (BrWe63) that the continuous functions in $\widehat{\mathbb{C}}$ that are analytic in $\widehat{\mathbb{C}} \setminus J$ form a Dirichlet algebra if and only if $\Lambda(T) = 0$. In the opposite direction (Bis88), there are quasicircles of very big size (Hausdorff dimension > 1) such that $1/M \leq \omega^*(E)/\omega(E) \leq M$ for $E \subset J$. See e.g. BiJo92 for further results.

The sense-preserving homeomorphism ("conformal welding")

$$(10) \qquad \qquad \varphi = f^{*-1} \circ f$$

of $\mathbb{T}$ onto $\mathbb{T}$ was studied in Sections 2.3 and 5.4. It follows from (9) and (5) that φ is singular if and only if $\Lambda(T) = 0$. The curve of Proposition 6.29 shows this may happen even if f and f^* are both conformal almost everywhere.

Proof. We modify the proof of Theorem 6.25. Let $M_1, \ldots$ denote constants. It follows from Corollary 9.18 (to be proved later) that

$$\Lambda(f^{-1}(J \cap D_{kj}))\Lambda(f^{*-1}(J \cap D_{kj})) \leq M_1 (\operatorname{diam} D_{kj})^2.$$

Since $B_{kj} \subset f^{-1}(D_{kj})$ and $\varphi(B_{kj}) \subset f^{*-1}(D_{kj})$, we conclude from (6.5.17) that

$$\Lambda(B_{kj})\Lambda(\varphi(B_{kj})) \leq M_2(1 - |\zeta_{kj}|)^2|f'(\zeta_{kj})|^2 \leq M_3 2^{-2k}(1 - |\zeta_{kj}|)^2$$

and thus from (6.5.18) that

$$\Lambda(\varphi(B_{kj})) \leq M_3 2^{-2k}(1 - |\zeta_{kj}|).$$

Hence, for $k = 1, 2, \ldots$,

$$\sum_j \Lambda(\varphi(B_{kj})) \leq M_3 2^{-2k} \sum_j (1 - |\zeta_{kj}|) \leq M_4 2^{-2k}$$

because $1 - |\zeta_{kj}| < \Lambda(J_{kj})$ where the arcs $J_{kj} \subset \mathbb{T}$ are essentially disjoint. Therefore, by (6.5.19),

$$\Lambda(\varphi(A_0)) \leq \Lambda(\varphi(Z_m)) \leq \sum_{k \geq m} \sum_j \Lambda(\varphi(B_{kj})) < 2M_4 2^{-2m}$$

so that $\Lambda(\varphi(A_0)) = 0.$ ·

Using partitions as in Theorem 6.25 we define

$$B = [f^*(A_0^*) \cup f(A_1) \cup (S \cap S^*)] \setminus T,$$
$$B^* = [f(A_0) \cup f^*(A_1^*) \cup (S \cap S^*)] \setminus T.$$

Since $\Lambda(\varphi(A_0)) = 0$ and since $\Lambda(A_1^*) = 0$ by Theorem 6.25, we see that

$$\Lambda(f^{*-1}(B^*)) \leq \Lambda(\varphi(A_0)) + \Lambda(A_1^*) + \Lambda(f^{*-1}([S \cap S^*] \setminus T)) = 0$$

because $(S \cap S^*) \setminus T$ has linear measure zero and thus also its preimage, by (4). Hence $\omega^*(B^*) = 0$ and similarly $\omega(B) = 0$. Since

$$B \cup B^* \cup T = f(A_0) \cup f(A_1) \cup f^*(A_0^*) \cup f^*(A_1^*) \cup (S \cap S^*) = J$$

by (6.5.8), our assertion therefore follows if we replace B by $B \setminus (B \cap B^*)$. □

Exercises 6.6

1. If J is Dini-smooth show that $1/M \leq \omega^*(E)/\omega(E) \leq M$ for $E \subset J$ where M is a constant.

2. Show that $\varphi = f^{*-1} \circ f$ is absolutely continuous if and only if $\omega^*(T) = 1$.

3. If J is a quasicircle show that φ is absolutely continuous if and only if f is conformal almost everywhere.

4. Suppose that $\Lambda(T) = 0$ and extend f to $\mathbb{D}^*$ by defining $f(r\zeta) = f^*(r\varphi(\zeta))$ for $r \geq 1$ and $\zeta \in \mathbb{T}$. Show that f is a homeomorphism of $\widehat{\mathbb{C}}$ onto $\widehat{\mathbb{C}}$ that satisfies $|\partial f/\partial z| = |\partial f/\partial \bar{z}|$ for almost all $z \in \mathbb{D}^*$ ("almost nowhere quasiconformal").

5. Construct a Jordan curve J of σ-finite linear measure and a closed set $E \subset J$ such that $\Lambda(E) = 0$ and $\omega(E) > 0$.

Chapter 7. Smirnov and Lavrentiev Domains

7.1 An Overview

We now consider domains with rectifiable boundaries in more detail. Let f map $\mathbb{D}$ conformally onto the inner domain G of the Jordan curve J. We have shown in Section 6.3 that J is rectifiable if and only if f' belongs to the Hardy space H^1.

We call G a Smirnov domain if

$$(1) \qquad \log|f'(z)| = \frac{1}{2\pi} \int_0^{2\pi} \frac{1-|z|^2}{|\zeta-z|^2} \log|f'(\zeta)| \, |d\zeta| \qquad \text{for} \quad z \in \mathbb{D}.$$

In the language of Hardy spaces this means that f' is an "outer function". This is not always true, and the Smirnov condition has not been completely understood from the geometrical point of view. See e.g. Dur70, p. 173 for its significance for approximation theory.

Zinsmeister has shown (Theorem 7.6) that G is a Smirnov domain if J is Ahlfors-regular, i.e. if

$$(2) \qquad \Lambda(\{z \in J : |z-w| < r\}) \le M_1 r \qquad \text{for} \quad w \in \mathbb{C}, \quad 0 < r < \infty$$

for some constant M_1. The condition of Ahlfors-regularity appears in many different contexts, for instance in G. David's work on the Cauchy integral operator.

We call G a Lavrentiev domain if

$$(3) \qquad \Lambda(J(a,b)) \le M_2 |a-b| \qquad \text{for} \quad a,b \in J,$$

where $J(a,b)$ is the shorter arc of J between a and b. This is the quasicircle condition (5.4.1) with diameter replaced by length, and Lavrentiev domains are the Ahlfors-regular quasidisks (Proposition 7.7). Tukia has proved (Theorem 7.9) that Lavrentiev curves can be characterized as the bilipschitz images of circles.

A different characterization was given by Jerison and Kenig (Theorem 7.11): A domain is a Lavrentiev domain if and only if it is a Smirnov quasidisk such that $|f'|$ satisfies the Coifman-Fefferman (A_∞) condition. This is closely connected to the Muckenhoupt (A_p) and (B_q) conditions, also to the space BMOA of analytic functions of bounded mean oscillation, the dual space of H^1 by a famous result of Fefferman. We shall only quote some results of this rich theory (see e.g. Gar81) which has close connections to stochastic processes (see e.g. Durr84).

Let again $I(re^{it}) = \{e^{i\vartheta} : |\vartheta - t| \leq \pi(1 - r)\}$. The condition

$$(4) \qquad \frac{1}{\Lambda(I(z))} \int_{I(z)} |f'(\zeta)|^q |d\zeta| \leq M_3 |f'(z)|^q \quad (z \in \mathbb{D})$$

is related to the (B_q) condition and will appear repeatedly in this chapter:

a) If (4) holds for some $q > 0$ then G is a Smirnov domain (Proposition 7.5) and even $\log f' \in \text{BMOA}$ (Proposition 7.13).

b) If J is Ahlfors-regular then (4) is satisfied with $q < 1/3$ (Theorem 7.6).

c) The quasidisk G is a Lavrentiev domain if and only if (4) is true for $q = 1$ (Theorem 7.8). Then (4) automatically holds for some $q > 1$ (Corollary 7.12) by a result of Gehring.

7.2 Integrals of Bloch Functions

It will be convenient first to consider integrals

$$(1) \qquad h(z) = \int_0^z (g(\zeta) - g(0))\zeta^{-1}\, d\zeta \quad (z \in \mathbb{D})$$

of Bloch functions. These will also be needed in Section 10.4. Let $K_1, K_2, \ldots$ denote suitable absolute constants and $M_1, M_2, \ldots$ other positive constants.

Proposition 7.1. *If* $g \in \mathcal{B}$ *then* h *is continuous in* $\overline{\mathbb{D}}$ *and, for* $|z| \leq 1$, $0 < |t| \leq \pi$,

$$(2) \qquad \left| \frac{h(e^{it}z) - h(z)}{it} - g\left(\left(1 - \frac{|t|}{\pi}\right)z\right) + g(0) \right| \leq K_1 \|g\|_{\mathcal{B}}.$$

Proof. It follows from (4.2.4) that $h'(z) = O((1 - |z|)^{-1/2})$ as $|z| \to 1$. Hence h is (Hölder-) continuous in $\overline{\mathbb{D}}$; see Dur70, p. 74. In order to prove (2) we may assume that $g(0) = 0$, $1/2 < |z| < 1$ and $0 < |t| \leq \pi/2$; see (4.2.4). Now let $0 < t \leq \pi/2$ and $|\tau| \leq t$. We choose σ such that $t = \pi(1 - e^{-\sigma})$. Integration by parts shows that

$$\int_{e^{-\sigma}z}^{e^{i\tau}z} g'(\zeta) \log \frac{e^{i\tau}z}{\zeta}\, d\zeta = h(e^{i\tau}z) - h(e^{-\sigma}z)(\sigma + i\tau)g(e^{-\sigma}z).$$

It follows from (4.2.1) and from

$$\left| \log \frac{e^{i\tau}z}{\zeta} \right| \leq \log \frac{1}{|\zeta|} + \left| \arg \frac{e^{i\tau}z}{\zeta} \right| \leq K_2(1 - |\zeta|) \quad \text{for} \quad \zeta \in [e^{-\sigma}z, e^{i\tau}z]$$

that the integrand is bounded by $K_2 \|g\|_{\mathcal{B}}$. Hence

(3) $$|h(e^{i\tau}z) - h(e^{-\sigma}z) - (\sigma+i\tau)g(e^{-\sigma}z)| \le K_3\sigma\|g\|_{\mathcal{B}}$$

and we obtain (2) if we choose $\tau = \pm t$ and $\tau = 0$ and then subtract. $\square$

The next result (Zyg45) characterizes the integrals h of Bloch functions defined in (1).

Theorem 7.2. *Let g be analytic in $\mathbb{D}$. Then $g \in \mathcal{B}$ if and only if h is continuous in $\overline{\mathbb{D}}$ and*

(4) $$\max_{|z|=1} |h(e^{it}z) + h(e^{-it}z) - 2h(z)| \le M_1 t \quad for \ \ t > 0.$$

We say that a continuous function $h : \mathbb{T} \to \mathbb{C}$ belongs to the *Zygmund class* Λ_* if (4) is satisfied. We easily obtain from Theorem 7.2 that

(5) $$h \text{ analytic in } \mathbb{D}, h \in \Lambda_* \quad \Leftrightarrow \quad h' \in \mathcal{B}.$$

Proof. First let $g \in \mathcal{B}$. If we use (2) with t and $-t$ and then subtract we obtain that

$$|h(ze^{it}) + h(ze^{-it}) - 2h(z)| \le 2K_1 t\|g\|_{\mathcal{B}}.$$

Conversely let (4) hold. Since h is analytic in $\mathbb{D}$ and continuous in $\mathbb{D} \cup \mathbb{T}$ the Poisson integral formula shows that, for $0 < r < 1$,

$$h(re^{is}) - h(e^{i\vartheta}) = \frac{1}{2\pi} \int_0^{2\pi} \frac{1-r^2}{1-2r\cos(s-\tau)+r^2} [h(e^{i\tau}) - h(e^{i\vartheta})] \, d\tau.$$

Differentiating twice with respect to s and then substituting $s = \vartheta$, $\tau = \vartheta \pm t$, we obtain from (1) that

$$e^{i\vartheta} g'(re^{i\vartheta}) = \frac{1-r^2}{\pi} \int_0^\pi \frac{(1+r^2)\cos t - 2r - 2r\sin^2 t}{(1-2r\cos t + r^2)^3}$$
$$\times [h(e^{i(\vartheta+t)}) + h(e^{i(\vartheta-t)}) - 2h(e^{i\vartheta})] \, dt.$$

Using (3.1.6) it can be shown that the inner quotient is bounded by $K_4/[t(1-r)(1-2r\cos t + r^2)]$. Hence it follows from (4) that

$$(1-r^2)|g'(re^{i\vartheta})| \le \frac{2K_4 M_1}{\pi} \int_0^\pi \frac{1-r^2}{1-2r\cos t + r^2} \, dt = 2K_4 M_1$$

so that $g \in \mathcal{B}$. $\square$

Remark. If we only assume (4) with h replaced by $\text{Im}\, h$, then we obtain that $(1-r^2)|\text{Im}[e^{i\vartheta}g'(re^{i\vartheta})]| \le K_5 M_1$. This is enough to conclude that $g \in \mathcal{B}$; see Exercise 4.2.6.

The non-negative finite measure μ on $\mathbb{T}$ is called a *Zygmund measure* if

(6) $$|\mu(I) - \mu(I')| \le M_2 \Lambda(I)$$

where I and I' are adjacent arcs of equal length. Hence the corresponding increasing function $\mu^*(t) = \mu(\{e^{i\tau} : 0 \le \tau \le t\})$ is continuous and (6) can be written as

$$(7) \qquad |\mu^*(\vartheta + t) + \mu^*(\vartheta - t) - 2\mu^*(\vartheta)| \le M_2 t \quad \text{for} \quad t > 0.$$

These measures can be used to characterize Bloch functions of positive real part (DuShSh66):

Proposition 7.3. *The function g belongs to $\mathcal{B}$ and satisfies $\operatorname{Re} g(z) > 0$ $(z \in \mathbb{D})$ if and only if it can be represented as*

$$(8) \qquad g(z) = ib + \int_{\mathbb{T}} \frac{\zeta + z}{\zeta - z} \, d\mu(\zeta) \quad (z \in \mathbb{D})$$

where μ is a Zygmund measure and $b \in \mathbb{R}$.

Proof. Every analytic function g of positive real part can be represented in the form (8) where μ is a non-negative finite measure, and vice versa (Dur70, p. 3). We may therefore assume that $\operatorname{Re} g(z) > 0$ and $g(0) = 1$.

Let $f(z) = z \exp h(z)$ where h is defined by (1). Then

$$z f'(z)/f(z) = 1 + z h'(z) = g(z) \quad (z \in \mathbb{D})$$

has positive real part so that f is univalent and starlike in $\mathbb{D}$; see (3.6.5). It follows from (3.6.7) and (3.6.10) that, for some constant a,

$$a + 2\pi\mu^*(t) = \lim_{r \to 1-} \arg f(re^{it}) = t + \operatorname{Im} h(e^{it}).$$

Thus (7) is equivalent to (4) with h replaced by $\operatorname{Im} h$. Hence our assertion follows from Theorem 7.2 and the remark after its proof. $\qquad\qquad \square$

Exercises 7.2

1. Let $g(z) = \sum_0^\infty b_n z^n$ be a Bloch function. Show that

$$\left| \sum_{n=1}^{\infty} n^{-1} b_n (\sin nt)^2 z^n \right| \le M_3 t \quad \text{for} \quad |z| \le 1, \quad t > 0.$$

2. Deduce (DuShSh66) that $\sum_{n=1}^m n^2 |b_n|^2 = O(m^2)$ as $m \to \infty$.

3. Let $\mathcal{B}_0$ be defined as in (4.2.22). Show that (Zyg45)

$$g \in \mathcal{B}_0 \quad \Leftrightarrow \quad \max_{|z|=1} |h(e^{it} z) + h(e^{-it} z) - 2h(z)| = o(t) \quad (t \to 0).$$

7.3 Smirnov Domains and Ahlfors-Regularity

Let f map $\mathbb{D}$ conformally onto $G \subset \mathbb{C}$ and let the linear measure $\Lambda(\partial G)$ be finite. We do not assume that G is a Jordan domain. Then $f' \in H^1$ by (6.3.10) so that f has a finite nonzero angular derivative $f'(\zeta)$ for almost all $\zeta \in \mathbb{T}$. Since f' has no zeros there is a representation (Dur70, p. 24)

$$(1) \qquad \log f'(z) = ib + \frac{1}{2\pi} \int_{\mathbb{T}} \frac{\zeta + z}{\zeta - z} \log |f'(\zeta)| \, |d\zeta| - \frac{1}{2\pi} \int_{\mathbb{T}} \frac{\zeta + z}{\zeta - z} \, d\mu(\zeta)$$

for $z \in \mathbb{D}$, where $b = \arg f'(0)$ and where μ is a non-negative singular measure, i.e. the corresponding increasing function has derivative zero almost everywhere. It follows from (1) that

$$(2) \qquad \log |f'(z)| = \int_{\mathbb{T}} p(z, \zeta) \log |f'(\zeta)| \, |d\zeta| - \int_{\mathbb{T}} p(z, \zeta) \, d\mu(\zeta)$$

for $z \in \mathbb{D}$ where p is again the Poisson kernel.

We say that f satisfies the *Smirnov condition* if

$$(3) \qquad \log |f'(z)| = \int_{\mathbb{T}} p(z, \zeta) \log |f'(\zeta)| \, |d\zeta| \quad \text{for} \quad z \in \mathbb{D},$$

i.e. if the singular measure μ is zero. Then (Gar81, p. 15)

$$(4) \qquad \int_{\mathbb{T}} \big| \log |f'(\rho z)| - \log |f'(z)| \big| \, |dz| \to 0 \quad \text{as} \quad \rho \to 1 - \,.$$

We call G a *Smirnov domain* if $\Lambda(\partial G) < \infty$ and if (3) is satisfied.

The Smirnov condition holds if $\log f' \in H^1$ (but not vice versa, see Dur72). If $\log f' \in H^1$ then (3) is satisfied even with $|f'|$ replaced by f'.

We now construct an example (DuShSh66) to show that the Smirnov condition is not always satisfied (KeLa37).

There exists (Pir66) a strictly increasing continuous function μ^* that is singular and satisfies

$$(5) \qquad \sup_{0 \le t \le 2\pi} |\mu^*(t + \tau) + \mu^*(t - \tau) - 2\mu^*(t)| = o(\tau) \quad \text{as} \quad \tau \to 0$$

("Zygmund smooth"). The corresponding measure μ on $\mathbb{T}$ is thus a Zygmund measure so that (7.2.8) defines a Bloch function with $\operatorname{Re} g(z) > 0$ for $z \in \mathbb{D}$. We define

$$(6) \qquad f(z) = \int_0^z \exp(-ag(\zeta)) \, d\zeta \quad \text{for} \quad z \in \mathbb{D}$$

where $a = (2\|g\|_B)^{-1}$. Then

$$(1 - |z|^2)|f''(z)/f'(z)| = a(1 - |z|^2)|g'(z)| \le 1/2 \quad \text{for} \quad z \in \mathbb{D}$$

so that f maps $\mathbb{D}$ conformally onto a quasidisk G by Corollary 5.23. Since $\mu^{*\prime}(t) = 0$ for almost all t it follows (Dur70, p. 4) that $\operatorname{Re} g(re^{it}) \to 0$ as $r \to 1-$ and thus, by (6),

(7) $|f'(\zeta)| = 1$ for almost all $z \in \mathbb{T}$, $|f'(z)| < 1$ for $z \in \mathbb{D}$;

the last fact follows from $\operatorname{Re} g(z) > 0$. Hence $f' \in H^1$ and thus $\Lambda(\partial G) < \infty$. But the Smirnov condition (3) is not satisfied because of (7).

Let $K_1, K_2, \ldots$ denote suitable absolute constants and $M_1, M_2, \ldots$ constants that may depend on f and other parameters. We write again

(8) $I(z) = \{e^{it} : |t - \arg z| \le \pi(1 - |z|)\}$ for $z \in \mathbb{D}$.

Proposition 7.4. *If f maps $\mathbb{D}$ conformally onto a Smirnov domain then, for $z \in \mathbb{D}$,*

(9) $$\left| \frac{1}{\Lambda(I(z))} \int_{I(z)} \log |f'(\zeta)| \, |d\zeta| - \log |f'(z)| \right| \le K_1 .$$

Proof. We may assume that $f'(0) = 1$. We consider the Bloch function $g = \log f'$ and define h by (7.2.1); note that $g(0) = 0$. Let $|z| = r < 1$. It follows from Proposition 7.1 that, with $0 < \rho < 1$ and $t = \pi(1 - r)$,

$$\left| \frac{1}{2it} \left(h\left(\frac{\rho}{r} z e^{it}\right) - h\left(\frac{\rho}{r} z e^{-it}\right) \right) - g(\rho z) \right| \le K_2 \|g\|_B \le 6K_2 ;$$

see Proposition 4.1. Taking the real part we obtain from (7.2.1) that

$$\left| \frac{1}{2t} \int_{I(z)} \log |f'(\rho\zeta)| \, |d\zeta| - \log |f'(\rho z)| \right| \le 6K_2 .$$

It follows from (4) that, as $\rho \to 1-$, the integral converges to the corresponding integral with $\rho = 1$. Hence (9) holds because $2t = \Lambda(I(z))$. $\square$

No geometric characterization of the Smirnov condition is known. We shall give two sufficient conditions.

Proposition 7.5. *Let f map $\mathbb{D}$ conformally onto G with $\Lambda(\partial G) < \infty$. Suppose that $q > 0$ and*

(10) $$\frac{1}{\Lambda(I(z))} \int_{I(z)} |f'(\zeta)|^q |d\zeta| \le M_1 |f'(z)|^q   \text{for} z \in \mathbb{D}.$$

Then G is a Smirnov domain. If $0 < q < 1$ then furthermore

(11) $$\int_{\mathbb{T}} p(z, \zeta) |f'(\zeta)|^q |d\zeta| \le M_2 |f'(z)|^q   \text{for} z \in \mathbb{D};$$

if G is in addition linearly connected then (10) implies (11) for $0 < q \le 1$.

The univalent function $f(z) = (1-z)^2$ satisfies (10) for all $q > 0$ but not (11) for $q = 1$ so that the assumption about linear connectedness cannot be deleted.

Proof. We first prove (11) where we may assume that $z = r > 0$. Let $q \leq 1$ if G is linearly connected and $q < 1$ otherwise. We define $I_{-1} = \emptyset$ and

$$m = \left[\log \frac{1}{1-r}\right], \quad r_n = 1 - (1-r)e^n, \quad I_n = I(r_n) \quad (n = 0, 1, \ldots, m)$$

so that $r_0 = r$ and $0 \leq r_m < 1 - e^{-1}$, furthermore $r_{m+1} = 0$ and $I_{m+1} = I(0) = \mathbb{T}$. It follows from (10) that

$$\int_{\mathbb{T}} p(r,\zeta)|f'(\zeta)|^q |d\zeta| = \frac{1}{2\pi} \sum_{n=0}^{m+1} \int_{I_n \backslash I_{n-1}} \frac{1 - r^2}{|\zeta - r|^2} |f'(\zeta)|^q |d\zeta|$$

$$\leq K_3 \sum_{n=0}^{m+1} \frac{1-r}{1-r_n} M_1 |f'(r_n)|^q .$$

Since $0 \leq r_n \leq r$ we see from (5.3.3) that

$$|f'(r_n)| \leq 8|f'(r)|[(1-r_n)/(1-r)]^{\beta-1} ,$$

where $\beta < 2$ if G is linearly connected and $\beta = 2$ in the general case; see Theorem 5.7 and the remark following its proof. Since $1 - r_n \geq e^{n-1}(1-r)$ and $(\beta - 1)q < 1$ we conclude that

$$\int_{\mathbb{T}} p(r,\zeta)|f'(\zeta)|^q |d\zeta| \leq K_4 M_1 |f'(r)|^q \sum_{n=0}^{\infty} e^{-n(1-(\beta-1)q)}$$

and the sum converges.

Now let $q > 0$. It follows from Hölder's inequality that if (10) holds for $q > 1/2$ then it also holds for $q = 1/2$. Hence (11) holds for some $q > 0$. The inequality (6.3.14) between the geometric and arithmetic mean therefore shows that

$$\int_{\mathbb{T}} p(z,\zeta) \log|f'(\zeta)| \, |d\zeta| \leq \frac{1}{q} \log \left(\int_{\mathbb{T}} p(z,\zeta)|f'(\zeta)|^q \, |d\zeta| \right)$$

$$\leq \frac{1}{q} \log M_2 + \log|f'(z)| .$$

Hence we see from (2) that

$$\int_{\mathbb{T}} p(z,\zeta) \, d\mu(\zeta) \leq \frac{1}{q} \log M_2 \quad \text{for} \quad z \in \mathbb{D}$$

and since μ is a singular measure it follows (Gar81, p. 76) that μ is zero. Hence f satisfies the Smirnov condition. $\quad\square$

Let again $M_1, M_2, \ldots$ denote suitable constants. We say that the curve $C \subset \mathbb{C}$ is *Ahlfors-regular* if

$$(12) \qquad \Lambda(C \cap D(w, r)) \leq M_1 r \qquad \text{for} \quad w \in \mathbb{C}, \quad 0 < r < \infty.$$

We can write (12) as $\Lambda(\tau(C) \cap \mathbb{D}) \leq M_1$ where $\tau(z) = az + b$ ($z \in \mathbb{C}$). It is more difficult to show (Meyer, see Dav84) that C is Ahlfors-regular if and only if

$$(13) \qquad \Lambda(\tau(C)) \leq M_2 \operatorname{diam} \tau(C) \qquad \text{for all} \quad \tau \in \text{Möb} .$$

If E is bounded then (12) implies $\Lambda(E) < \infty$ but not vice versa. For example, if

$$E = \{x + ix^{3/2}\sin(1/x) : 0 < x \leq 1\}$$

then $\Lambda(E \cap D(0, r)) > \sqrt{r}/M_3$.

We call G an *Ahlfors-regular domain* if ∂G is (locally connected and) Ahlfors-regular. This is the weakest known geometric condition for Smirnov domains (Zin84).

Theorem 7.6. *Let f map $\mathbb{D}$ conformally onto an Ahlfors-regular domain. Then (10) holds for $q < 1/3$ so that G is a Smirnov domain.*

Proof. We restrict ourselves to prove (10) for $q = 1/4$. Hölder's inequality shows that, with $z \in \mathbb{D}$ and $I = I(z)$,

$$(14) \qquad \int_I |f'(\zeta)|^{1/4}|d\zeta| \leq \left(\int_I \frac{|f'(\zeta)|\,|d\zeta|}{|f(\zeta) - f(z)|^{6/5}} \right)^{1/4} \left(\int_I |f(\zeta) - f(z)|^{2/5}|d\zeta| \right)^{3/4}.$$

We split the second integral into the contributions from

$$\{\zeta \in I : e^n\delta \leq |f(\zeta) - f(z)| < e^{n+1}\delta\} \quad (n = 0, 1, \ldots)$$

where $\delta = d_f(z) = \operatorname{dist}(f(z), \partial G)$. The Ahlfors-regularity condition (12) shows that

$$(15) \qquad \int_I \frac{|f'(\zeta)|\,|d\zeta|}{|f(\zeta) - f(z)|^{6/5}} \leq \sum_{n=0}^{\infty} \frac{M_1 e^{n+1}\delta}{(e^n\delta)^{6/5}} = \frac{M_4}{\delta^{1/5}} .$$

Let $\varphi(s) = (s + z)/(1 + \bar{z}s)$ for $s \in \mathbb{D}$. Then $|\varphi'(s)| = |1 - \bar{z}\varphi(s)|^2/(1 - |z|^2) \leq K_5(1 - |z|^2)$ for $\varphi(s) \in I$ and thus, by Theorem 1.7,

$$\int_I |f(\zeta) - f(z)|^{2/5}|d\zeta| \leq K_5(1 - |z|^2) \int_{\varphi^{-1}(I)} |f(\varphi(s)) - f(\varphi(0))|^{2/5}|ds|$$

$$\leq K_6(1 - |z|^2)^{7/5}|f'(z)|^{2/5} .$$

Therefore we see from (14) and (15) that

$$\int_I |f'(\zeta)|^{1/4}|d\zeta| \leq M_5\delta^{-1/20}(1 - |z|^2)^{21/20}|f'(z)|^{6/20} \leq M_1\Lambda(I)|f'(z)|^{1/4}$$

because $\delta \geq (1 - |z|^2)|f'(z)|/4$ by Corollary 1.4. $\qquad \square$

A problem that has been extensively studied is the following: For what curves L in $\mathbb{C}$ is it true that

$$(16) \qquad \Lambda\left(f^{-1}(L \cap f(\mathbb{D}))\right) < \infty$$

for all conformal maps f of $\mathbb{D}$ into $\mathbb{C}$?

Partial answers were given in HayWu81, FeHa87, FeZi87 and FeHeMa89, see also Bae89. The complete solution (BiJo90) is that (16) holds if and only if L is Ahlfors-regular.

Ahlfors-regularity also plays a central role in the study of singular integral operators such as the Cauchy integral operator (Dav84); see also e.g. Jo90.

Exercises 7.3

1. Show that the definition of a Smirnov domain G does not depend on the choice of the function f mapping $\mathbb{D}$ conformally onto G.

2. Let φ map $\mathbb{D}$ conformally onto a Smirnov domain in $\mathbb{D}$. If f maps $\mathbb{D}$ conformally onto the inner domain of a Dini-smooth curve, show that $f \circ \varphi(\mathbb{D})$ is a Smirnov domain.

3. Show that the inner domain of a rectifiable starlike curve is a Smirnov domain, also every bounded regulated domain (Section 3.5).

4. Show that the spiral $w(t) = e^{-t + \imath t^\alpha}$, $0 \le t < \infty$, is Ahlfors-regular if $\alpha = 1$ but rectifiable and nonregular if $\alpha > 1$.

5. Let f be a conformal map such that $f' \in H^1$ and $1/f' \in H^p$ for some $p > 0$. Show that $\log f' \in H^2$ so that $f(\mathbb{D})$ is a Smirnov domain.

6. Let f map $\mathbb{D}$ conformally onto an Ahlfors-regular domain. Show (Dav84, Zin85) that

$$\sup_{z \in \mathbb{D}} (1 - |z|) \int_{\mathbb{T}} \frac{|f'(z) f'(\zeta)|}{|f(z) - f(\zeta)|^2} |d\zeta| < \infty .$$

7.4 Lavrentiev Domains

Let $M_1, M_2, \ldots$ again denote positive constants. A *Lavrentiev curve* is a rectifiable Jordan curve J such that

$$(1) \qquad \Lambda(J(a,b)) \le M_1 |a - b| \qquad \text{for} \quad a, b \in J$$

where $J(a, b)$ is the shorter arc of J between a and b; see Fig. 7.1. Lavrentiev curves are also called "quasismooth" or "chord-arc-curves". The inner domain of a Lavrentiev curve is called a *Lavrentiev domain*.

For example, if $\varphi(x)$ $(0 \le x \le 1)$ is positive and satisfies a Lipschitz condition then $\{x + iy : 0 < x < 1, 0 < y < \varphi(x)\}$ is a Lavrentiev domain.

Proposition 7.7. *A domain is a Lavrentiev domain if and only if it is an Ahlfors-regular quasidisk.*

Proof. If G is an Ahlfors-regular quasidisk then (1) follows at once from (5.4.1) and (7.3.12). Conversely let G be a Lavrentiev domain. It is trivial that G is a quasidisk.

To show that J is Ahlfors-regular, let $w \in \mathbb{C}$ and first assume that $r \leq r_0 = \operatorname{diam} J/(2M_1 + 3)$. Then there exists $c \in J$ with $|c - w| > (M_1 + 1)r$. If $J \cap D(w,r) = \emptyset$ there is nothing to prove. Otherwise there exist $a, b \in J \cap \partial D(w,r)$ such that the arc J' of J between a and b containing c lies outside $D(w,r)$; see Fig. 7.1. Since $\Lambda(J') \geq 2(|c - w| - r) > 2M_1 r \geq M_1|a - b|$ it follows for the other arc J'' of J between a and b that

$$\Lambda(J \cap D(w,r)) \leq \Lambda(J'') \leq M_1|a - b| \leq 2M_1 r .$$

Now let $r > r_0$. Then $\Lambda(J \cap D(w,r)) \leq \Lambda(J) < [\Lambda(J)/r_0]r.$ $\qquad\square$

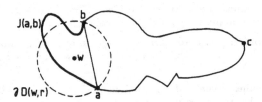

Fig. 7.1. A Lavrentiev domain and the proof of its Ahlfors-regularity

We now give an analytic characterization (Zin85). Let $I(z)$ be defined as in (7.3.8).

Theorem 7.8. *Let f map $\mathbb{D}$ conformally onto the inner domain G of a rectifiable Jordan curve. Then G is a Lavrentiev domain if and only if G is linearly connected and*

$$(2) \qquad \frac{1}{\Lambda(I(z))} \int_{I(z)} |f'(\zeta)|\,|d\zeta| \leq M_2|f'(z)| \qquad for \quad z \in \mathbb{D}.$$

The condition that G is linearly connected (Section 5.3) cannot be deleted as again the example $f(z) = (1 - z)^2$ shows. It can however be proved (Pom82, Zin85) that G is an Ahlfors-regular John domain (Section 5.2) if (2) holds.

Proof. Let G be a Lavrentiev domain. Then G is linearly connected by Theorem 5.9 (ii). It follows from Theorem 5.2 (ii) that, with $I = I(z)$,

$$\operatorname{diam} f(I) \leq M_3 d_f(z) \leq M_3(1 - |z|^2)|f'(z)|$$

so that $f(I)$ lies in a disk of this radius. Hence it follows from the Ahlfors-regularity condition (7.3.12) that

$$\int_I |f'(\zeta)|\,|d\zeta| = \Lambda(f(I)) \leq M_4(1 - |z|^2)|f'(z)| < M_4\Lambda(I)|f'(z)| .$$

Conversely let G be linearly connected and let (2) hold (Fig. 7.2). Given $a, b \in J$ we choose $z \in \mathbb{D}$ such that $J(a, b) = f(I)$ where $I = I(z)$. Then

$$\Lambda(J(a, b)) = \int_I |f'(\zeta)| \, |d\zeta| \leq 2\pi (1 - |z|) M_2 |f'(z)|$$
$$\leq K_1 M_2 d_f(z) \leq K_2 M_2 d_f(z^*)$$

by Proposition 1.6. If S is the non-euclidean segment connecting the endpoints of I then $d_f(z^*) \leq \operatorname{diam} f(S)$ so that (1) follows from Proposition 5.6.   □

Remark. The opposite inequality

$$(3) \qquad |f'(z)| \leq \frac{K_3}{\Lambda(I(z))} \int_{I(z)} |f'(\zeta)| \, |d\zeta| \quad (z \in \mathbb{D})$$

holds for all conformal maps because $\Lambda(I(z))|f'(z)| \leq K_3 \operatorname{diam} f(I(z))$ by Proposition 4.19.

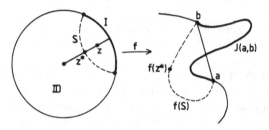

Fig. 7.2

We say that h is a *bilipschitz map* of A into $\mathbb{C}$ if

$$(4) \quad M_5^{-1}|z_1 - z_2| \leq |h(z_1) - h(z_2)| \leq M_5 |z_1 - z_2| \quad \text{for} \quad z_1, z_2 \in A.$$

A sense-preserving bilipschitz map of $\mathbb{C}$ onto $\mathbb{C}$ is quasiconformal but not vice versa. For example, a conformal map onto a quasidisk has a quasiconformal extension to $\mathbb{C}$ (Section 5.1) but no bilipschitz extension if it has unbounded derivative.

We give a characterization (Tuk80, Tuk81, JeKe82b) of Lavrentiev domains that is analogous to the characterization of quasicircles as quasiconformal images of circles.

Theorem 7.9. *The curve J is a Lavrentiev curve if and only if there is a bilipschitz map h of $\mathbb{C}$ onto $\mathbb{C}$ such that $J = h(\mathbb{T})$.*

Proof. Let h be a bilipschitz map with $J = h(\mathbb{T})$ and let $h(\zeta_1), h(\zeta_2) \in J$ be given. We divide the shorter arc of $\mathbb{T}$ between ζ_1 and ζ_2 into subarcs by points z_k $(k = 1, \ldots, n)$ where $z_0 = \zeta_1$ and $z_n = \zeta_2$. It follows from (4) that

$$\sum_{k=1}^{n} |h(z_k) - h(z_{k-1})| \le M_5 \sum_{k=1}^{n} |z_k - z_{k-1}| \le \pi M_5 |\zeta_1 - \zeta_2|$$

and thus, again by (4),

$$\Lambda(J(h(\zeta_1), h(\zeta_2))) \le \pi M_5^2 |h(\zeta_1) - h(\zeta_2)|.$$

Conversely let J be a Lavrentiev curve. We may assume that $\Lambda(J) = 2\pi$. The arc length parametrization $J : h(e^{is})$, $0 \le s \le 2\pi$ satisfies

$$|h(e^{is_1}) - h(e^{is_2})| \le |s_1 - s_2| \le \pi |e^{is_1} - e^{is_2}|$$

for $|s_1 - s_2| \le \pi$ and furthermore, by (1),

$$|e^{is_1} - e^{is_2}| \le |s_1 - s_2| \le M_1 |h(e^{is_1}) - h(e^{is_2})|.$$

Hence h is a bilipschitz map from $\mathbb{T}$ into $\mathbb{C}$ which can be extended to a bilipschitz map of $\mathbb{C}$ onto $\mathbb{C}$ by the following result (Tuk80, Tuk81):   □

Theorem 7.10. *Every bilipschitz map h of $\mathbb{T}$ into $\mathbb{C}$ can be extended to a bilipschitz map of $\mathbb{C}$ onto $\mathbb{C}$.*

Proof (see Fig. 7.3). The function h is quasisymmetric on $\mathbb{T}$; indeed (5.4.4) is satisfied with $\lambda(x) \le M_5^2 x$ $(x > 0)$ by (4). Hence $J = h(\mathbb{T})$ is a quasicircle (Proposition 5.10) which we may assume to be positively oriented.

As in the proof of Theorem 5.16, let f and g map $\mathbb{D}$ and $\mathbb{D}^*$ conformally onto the inner and outer domain of J respectively; we assume that $g(\infty) = \infty$. Then $\varphi = f^{-1} \circ h$ and $\psi = g^{-1} \circ h$ are quasisymmetric by Proposition 5.12. Let $\widetilde{\varphi}$ and $\widetilde{\psi}$ denote the Douady-Earle extensions of φ and ψ; for simplicity we shall write φ, ψ instead of $\widetilde{\varphi}$ and $\widetilde{\psi}$. We extend h to a homeomorphism of $\mathbb{C}$ onto $\mathbb{C}$ by defining

$$(5) \qquad h(z) = f(\varphi(z)) \quad \text{for} \quad |z| \le 1, \quad = g(\psi(z)) \quad \text{for} \quad |z| \ge 1$$

as in (5.5.6).

Let $z \in \mathbb{D}$ be given and write $\omega = e^{2\pi i/3}$. We choose $\tau \in \text{Möb}(\mathbb{D})$ such that $\tau(0) = z$, $\tau(\omega^2) = z/|z|$ and define $z_\nu = \tau(\omega^\nu)$ for $\nu = 1, 2, 3$. Finally we choose $\sigma \in \text{Möb}(\mathbb{D})$ such that

$$(6) \qquad \sigma(\omega^\nu) = \varphi(z_\nu) \equiv \zeta_\nu \quad \text{for} \quad n = 1, 2, 3.$$

Then $\varphi^* = \sigma^{-1} \circ \varphi \circ \tau$ gives a sense-preserving quasisymmetric map of $\mathbb{T}$ onto $\mathbb{T}$ satisfying $\varphi^*(\omega^\nu) = \omega^\nu$. Using the Möbius-invariant condition (5.4.8) of Proposition 5.14, we conclude from Proposition 5.21 that $|\varphi^{*\prime}(0)| \le r_0 < 1$ where r_0 does not depend on z.

It follows from (5) that $h \circ \tau = f \circ \varphi \circ \tau = f \circ \sigma \circ \varphi^*$ in $\mathbb{D}$ and therefore, with $\zeta = \sigma(\varphi^*(0)) = \varphi(z)$ and $h_z = \partial h/\partial z$, that

(7) $\qquad |h_z(z)| = \dfrac{|\sigma'(\varphi^*(0))f'(\zeta)\varphi_z^*(0)|}{|\tau'(0)|} = \dfrac{(1-|\zeta|^2)|f'(\zeta)|\,|\varphi_z^*(0)|}{(1-|z|^2)(1-|\varphi^*(0)|^2)}\,.$

By our choice of τ, all three distances $|z_1 - z_2|$, $|z_2 - z_3|$, $|z_3 - z_1|$ are comparable to $1 - |z|$, and since $\sigma \in \text{Möb}(\mathbb{D})$ and $|\varphi^*(0)| \leq r_0$, at least one of the distances $|\zeta_1 - \zeta_2|$, $|\zeta_2 - \zeta_3|$, $|\zeta_3 - \zeta_1|$ is comparable to $1 - |\zeta|$. Hence we may assume that

(8) $\qquad \dfrac{1}{M_6} \leq \dfrac{|z_1 - z_2|}{1 - |z|} \leq M_6\,, \qquad \dfrac{1}{M_7} \leq \dfrac{|\zeta_1 - \zeta_2|}{1 - |\zeta|} \leq M_7\,.$

Since $f(\zeta_\nu) = h(z_\nu)$ by (5) and (6), we can write

$$\frac{(1 - |\zeta|^2)|f'(\zeta)|}{1 - |z|^2} = \frac{(1 - |\zeta|^2)|f'(\zeta)|}{|f(\zeta_1) - f(\zeta_2)|}\,\frac{|h(z_1) - h(z_2)|}{|z_1 - z_2|}\,\frac{|z_1 - z_2|}{1 - |z|^2}\,.$$

All factors on the right-hand side are bounded from above and below by positive constants: The first factor because J is a quasicircle and in view of (8) (see Theorems 5.2(ii) and 5.7), the second factor because h is bilipschitz on $\mathbb{T}$, and the third factor because of (8). If we now write $\chi(z) = (\varphi^*(z) - \varphi^*(0))/(1 - \overline{\varphi^*(0)}\varphi^*(z))$ then $\chi(0) = 0$ and we conclude from (7) and the analogous identity for $|h_{\overline{z}}(z)|$ that

(9) $\quad |h_z(z)| \leq \dfrac{M_8|\varphi_z^*(0)|}{1 - |\varphi^*(0)|^2} = M_8|\chi_z(0)| \leq M_9\,, \quad |h_{\overline{z}}(z)| \leq M_8|\chi_{\overline{z}}(0)| \leq M_9\,,$

(10) $\qquad |h_z(z)|^2 - |h_{\overline{z}}(z)|^2 \geq \big(|\chi_z(0)|^2 - |\chi_{\overline{z}}(0)|^2\big)/M_{10} \geq 1/M_{11}\,;$

the last estimates follow from (5.5.21) and (5.5.22) as in part (b) of the proof of Theorem 5.15 using that $\chi = \gamma \circ \varphi \circ \tau$ with $\gamma, \tau \in \text{Möb}(\mathbb{D})$.

It follows from (9) and (10) that h and its inverse function h^{-1} have bounded partial derivatives for $z \in \mathbb{D}$ and similarly for $z \in \mathbb{D}^*$. The right-hand inequality (4) follows at once by integration over $[z_1, z_2]$ because h is continuous on $\mathbb{T}$.

To prove the left-hand inequality (4) we consider $w_1, w_2 \in \mathbb{C}$. If w_1 and w_2 lie in the closure of the same component of $\mathbb{C} \setminus J$ then we integrate over the hyperbolic geodesic from w_1 to w_2. In the other case, we choose $w_0 \in J \cap [w_1, w_2]$ and integrate over the hyperbolic geodesics from w_1 to w_0 and from w_0 to w_2. Since J is a quasicircle, these geodesics have lengths comparable to the distance of their endpoints (Exercise 5.4.4). Hence we obtain

$$|h^{-1}(w_1) - h^{-1}(w_2)| \leq M_{12}|w_1 - w_2|\,.$$

Thus (4) holds so that h is a bilipschitz map. $\qquad\qquad\qquad\qquad\square$

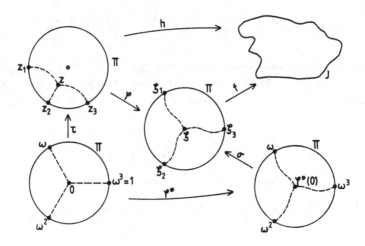

Fig. 7.3. Note that $h = f \circ \varphi$ and $\sigma \circ \varphi^* = \varphi \circ \tau$

Exercises 7.4

Let f map $\mathbb{D}$ conformally onto $G \subset \mathbb{C}$ and let $M_1, \ldots$ denote suitable constants.

1. Let G be a Lavrentiev domain. Show that

$$\int_I |f'(\zeta)|\,|d\zeta| \leq M_1 |f'(0)| \quad \text{for some arc } I \subset \mathbb{T} \text{ with } \Lambda(I) = \pi$$

where M_1 depends only on G and not on f.

2. Let G be linearly connected. Use Proposition 7.5 to show that G is a Lavrentiev domain if and only if

$$|f'(z)| \leq \int_{\mathbb{T}} p(z, \zeta)|f'(\zeta)|\,|d\zeta| \leq M_2 |f'(z)| \quad \text{for} \quad z \in \mathbb{D};$$

the left-hand inequality is always true.

3. Show that the bilipschitz image of a Lavrentiev domain is again a Lavrentiev domain.

4. Deduce Proposition 7.7 from Theorem 7.9.

5. Let G be a quasidisk. Show that f can be extended to a bilipschitz map of $\mathbb{C}$ onto $\mathbb{C}$ if and only if $1/M_3 < |f'(z)| < M_3$ ($z \in \mathbb{D}$). (This is in contrast to the case of quasiconformal extension.)

7.5 The Muckenhoupt Conditions and BMOA

1. We introduce several rather technical but important conditions on "weights", see ReRy75 or Gar81, Sect. VI, 6. Let φ be a nonnegative measurable function defined on $\mathbb{T}$. We say that φ satisfies the *Coifman-Fefferman* (A_∞) *condition* (CoFe74; ReRy75, p. 52) if one of the following three equivalent conditions holds:

(i) For all arcs $I \subset \mathbb{T}$,

$$\frac{1}{\Lambda(I)} \int_I \varphi(\zeta)|d\zeta| \leq M_1 \exp\left(\frac{1}{\Lambda(I)} \int_I \log \varphi(\zeta)|d\zeta|\right).$$

(ii) There exists $\alpha > 0$ such that

$$\int_E \varphi|d\zeta| \Big/ \int_I \varphi|d\zeta| \leq M_2 \Lambda(E)^\alpha / \Lambda(I)^\alpha$$

for all arcs $I \subset \mathbb{T}$ and measurable sets $E \subset I$.

(iii) There exists $\beta > 0$ such that, for arcs $I \subset \mathbb{T}$ and measurable $E \subset I$,

$$\int_E \varphi|d\zeta| \Big/ \int_I \varphi|d\zeta| < \beta \quad \Rightarrow \quad \Lambda(E)/\Lambda(I) < 1/2.$$

Here $M_1, M_2, \ldots$ again denote suitable positive constants. Condition (i) is a reversed geometric-arithmetic mean inequality, while (ii) is a strong form of condition (iv') for John domains (Section 5.2). We now prove another characterization of Lavrentiev domains (Lav36, JeKe82b).

Theorem 7.11. *Let f map $\mathbb{D}$ conformally onto G. Then G is a Lavrentiev domain if and only if*

(1) *G is linearly connected,*
(2) *G is a Smirnov domain,*
(3) *$|f'|$ satisfies the (A_∞) condition.*

It is clear that every Lavrentiev domain satisfies (1) and (2); see Proposition 7.7 and Theorem 7.6. In spite of this fact it is not possible to delete either condition. The function defined in (7.3.6) maps $\mathbb{D}$ conformally onto a non-Smirnov quasidisk with $|f'(\zeta)| = 1$ for almost all $\zeta \in \mathbb{T}$ so that $|f'|$ satisfies the (A_∞) condition (i).

A domain is an Ahlfors-regular John domain if and only if (2) and (3) are satisfied (Pom82). This holds in particular (FiLe86) if G is a Jordan domain such that every $w \in \partial G$ is vertex of an open triangle in G which is congruent to some fixed triangle.

Proof. First let G be a Lavrentiev domain. It follows from Theorem 7.8 and Proposition 7.5 that G is a Smirnov domain. If I is an arc on $\mathbb{T}$ we choose $z \in \mathbb{D}$ such that $I = I(z)$ and obtain from (7.4.2) and Proposition 7.4 that

$$\frac{1}{\Lambda(I)} \int_I |f'||d\zeta| \leq M_3|f'(z)| \leq M_3 \exp\left(K_1 + \frac{1}{\Lambda(I)} \int_I \log|f'||d\zeta|\right).$$

Hence (i) holds with $\varphi = |f'|$.

Conversely let conditions (1), (2) and (3) be satisfied. It follows from (i) and Proposition 7.4 that

$$\frac{1}{\Lambda(I)} \int_I |f'|\,|d\zeta| \le M_1 \exp\left(\frac{1}{\Lambda(I)} \int_I \log|f'|\,|d\zeta|\right)$$

$$\le M_1 \exp(K_1 + \log|f'(z)|) = M_1 e^{K_1}|f'(z)|$$

so that G is a Lavrentiev domain by Theorem 7.8. $\qquad\square$

There are further conditions related to reversed Hölder inequalities. If $1 < p < \infty$ we say that $\varphi \ge 0$ satisfies the *Muckenhoupt* (A_p) *condition* if

$$(4) \qquad \frac{1}{\Lambda(I)} \int_I \varphi\,|d\zeta| \le M_4 \left(\frac{1}{\Lambda(I)} \int_I \varphi^{-1/(p-1)}|d\zeta|\right)^{-(p-1)}$$

for all arcs $I \subset \mathbb{T}$. If $q > 1$ then $\varphi \ge 0$ satisfies the (B_q) *condition* if

$$(5) \qquad \left(\frac{1}{\Lambda(I)} \int_I \varphi(\zeta)^q |d\zeta|\right)^{1/q} \le \frac{M_5}{\Lambda(I)} \int_I \varphi(\zeta)|d\zeta|.$$

It is remarkable that (Muc72; Gar81, p. 262)

$$(6) \qquad \varphi \text{ satisfies } (A_p) \quad\Rightarrow\quad \begin{cases} \varphi \text{ satisfies } (A_{p'}) \text{ for some } p' < p, \\ \varphi \text{ satisfies } (B_q) \text{ for some } q > 1, \end{cases}$$

and (Geh73; Gar81, p. 260)

$$(7) \qquad \varphi \text{ satisfies } (B_q) \quad\Rightarrow\quad \varphi \text{ satisfies } (B_{q'}) \text{ for some } q' > q.$$

The following conditions are equivalent (CoFe74; ReRy75, p. 50) to (i), (ii) and (iii):

(iv) φ satisfies (A_p) for some $p < \infty$,
(v) φ satisfies (B_q) for some $q > 1$.

These facts can be used to sharpen Theorem 7.8.

Corollary 7.12. *If f maps $\mathbb{D}$ conformally onto a Lavrentiev domain then there are $\delta > 0$ and $q > 1$ such that*

$$(8) \qquad \frac{1}{\Lambda(I(z))} \int_{I(z)} |f'(\zeta)|^q |d\zeta| \le M_6 |f'(z)|^q \qquad \text{for } z \in \mathbb{D},$$

$$(9) \qquad \frac{1}{\Lambda(I(z))} \int_{I(z)} \frac{|d\zeta|}{|f'(\zeta)|^\delta} \le \frac{M_7}{|f'(z)|^\delta} \qquad \text{for } z \in \mathbb{D}.$$

The first assertion follows at once from Theorems 7.8 and 7.11 because (v) is equivalent to (i). The second assertion follows from (iv) and (7.4.3).

It is not possible to choose the exponent δ independent of f. For every $\delta_0 > 0$ there exists (Zin85) a conformal map onto a Lavrentiev domain such that (9) becomes false for $\delta = \delta_0$. The corresponding fact holds for $q > 1$ (JoZi82).

The *Helson-Szegö theorem* (Gar81, p. 149, p. 255) implies for Smirnov domains that $|f'|$ satisfies the (A_2) condition if and only if

(10) $f' = e^{g+h}$ with $\sup_{z \in \mathbb{D}} |\operatorname{Re} g(z)| < \infty$, $\sup_{z \in \mathbb{D}} |\operatorname{Im} h(z)| < \dfrac{\pi}{2}$.

2. Let BMO($\mathbb{T}$) denote the space of all integrable functions g on $\mathbb{T}$ such that

(11) $\sup_I \dfrac{1}{\Lambda(I)} \displaystyle\int_I |g(\zeta) - a_I| \, |d\zeta| < \infty$, $a_I = \dfrac{1}{\Lambda(I)} \displaystyle\int_I g(\zeta) |d\zeta|$

where the supremum is taken over all arcs $I \subset \mathbb{T}$. These are the functions of *bounded mean oscillation*. It is a basic fact (CoFe74; Gar81, pp. 235, 258) that

(12) $g \in \text{BMO}(\mathbb{T}) \;\Leftrightarrow\; g^* \in \text{BMO}(\mathbb{T}) \;\Leftrightarrow\; e^{qg}$ satisfies (A_∞) for some $q > 0$,

where g is real-valued and where g^* is the conjugate function of g.

We say that $g \in \text{BMOA}$ if $g \in H^1$ and if the boundary values of g on $\mathbb{T}$ belong to BMO($\mathbb{T}$).

Proposition 7.13. *Let f map $\mathbb{D}$ conformally into $\mathbb{C}$ and let $f' \in H^p$ for some $p > 0$. Then $\log f' \in \text{BMOA}$ if and only if there exists $q > 0$ such that*

(13) $\dfrac{1}{\Lambda(I(z))} \displaystyle\int_{I(z)} |f'(\zeta)|^q |d\zeta| \le M_8 |f'(z)|^q$ *for* $z \in \mathbb{D}$.

In particular, it follows from Theorem 7.6 that $\log f' \in \text{BMOA}$ if $f(\mathbb{D})$ is Ahlfors-regular (Zin84).

Proof. Let f_q be defined by $f_q'(z) = f'(z)^q$ for $z \in \mathbb{D}$. If $0 < q < 1/6$ then, by Proposition 4.1,

$$(1 - |z|^2)|f_q''(z)/f_q'(z)| = q(1 - |z|^2)|f''(z)/f'(z)| \le 6q < 1$$

for $z \in \mathbb{D}$ so that f_q maps $\mathbb{D}$ conformally onto a quasidisk by Corollary 5.23.

If $\log f' \in \text{BMOA}$ then $|f_q'| = \exp(q \operatorname{Re} \log f')$ satisfies (A_∞) if $q \in (0, 1/6)$ is chosen small enough; see (12). Since $\log f_q' \in H^1$ the Smirnov condition is satisfied. Hence $f_q(\mathbb{D})$ is a Lavrentiev domain by Theorem 7.11, and (13) follows from Theorem 7.8.

Conversely suppose that (13) holds for some $q > 0$; we may assume that $q < p$ and $q < 1/6$. Then $f_q(\mathbb{D})$ is a Lavrentiev domain by Theorem 7.8, and Theorem 7.11 shows that $|f_q'|$ satisfies (A_∞). By Theorem 7.6, $\log f_q'$ can be represented by its Poisson integral and it follows (Dur70, p. 34) that $\log f_q'$ belongs to H^1. Hence we see from (12) that $\log f_q'$ and thus $\log f'$ belong to BMOA. $\qquad\square$

A function g in H^2 belongs to BMOA if and only if (see e.g. Bae80)

(14) $\|g\|_* = |g(0)| + \sup_{z \in \mathbb{D}} \left(\displaystyle\int_{\mathbb{T}} |g(\zeta) - g(z)|^2 p(z, \zeta) |d\zeta| \right)^{1/2} < \infty$,

and BMOA is a Banach space with this norm. The last integral can be rewritten as

(15) $$\frac{1}{2\pi}\int_{\mathbb{T}}|g_z(\zeta)|^2|d\zeta| \quad \text{with} \quad g_z(\zeta) \equiv g\left(\frac{\zeta+z}{1+\bar{z}\zeta}\right) - g(z)$$

which is $\geq |g_z'(0)|^2 = \left(1-|z|^2\right)^2|g'(z)|^2$. It follows that

(16) $$\text{BMOA} \subset \mathcal{B}.$$

The subspace VMOA ("vanishing mean oscillation", Sar75) consists of all g such that

(17) $$\int_{\mathbb{T}}|g(\zeta)-g(z)|^2 p(z,\zeta)|d\zeta| \to 0 \quad \text{as} \quad |z| \to 1.$$

It is easy to see that VMOA $\subset \mathcal{B}_0$. There is the deep duality (Fef 71)

(18) $$\text{VMOA}^* \cong H^1, \quad H^{1*} \cong \text{BMOA}$$

where $\cong$ denotes topological isomorphism.

If f is a conformal map of $\mathbb{D}$ onto the inner domain of a Jordan curve J then $\log f' \in$ VMOA if and only if

(19) $$\Lambda(J(a,b))/|a-b| \to 1 \quad \text{as} \quad |a-b| \to 0, \quad a,b \in J;$$

compare (7.4.1). Such curves are called *asymptotically smooth* (Pom78), and are, in particular, asymptotically conformal (Section 11.2).

See e.g. Zin84, Sem88, Zin89, AsZi91 and BiJo92 for further results connecting BMOA and conformal mapping.

Exercises 7.5

Let f map $\mathbb{D}$ conformally onto G.

1. If G is a Lavrentiev domain show that $f' \in H^q$ for some $q > 1$. Show (Zin85) furthermore that, for each $q > 1$, there exists a Lavrentiev domain such that $f' \notin H^q$.

2. If f satisfies (8), show that (compare (ii))
$$\Lambda(f(E))/\Lambda(f(I)) \leq M_9[\Lambda(E)/\Lambda(I)]^{1-1/q}$$
for all arcs $I \subset \mathbb{T}$ and measurable $E \subset I$.

3. Let G be a Lavrentiev domain and h a conformal map of G onto $\mathbb{D}$. Show that there exists $p > 1$ such that
$$\left(\frac{1}{\Lambda(C)}\int_C |h'(w)|^p|dw|\right)^{1/p} \leq \frac{M_{10}}{\Lambda(C)}\int_C |h'(w)|\,|dw|$$
for all arcs $C \subset \partial G$.

4. If $\arg f'$ is bounded in $\mathbb{D}$ show that $\log f' \in$ BMOA.

5. If $|\arg[zf'(z)/f(z)]| \leq \pi\alpha/2$ for $z \in \mathbb{D}$ with $\alpha < 1$, show that G is a Lavrentiev domain and that $|f'|$ satisfies (A$_2$).

Chapter 8. Integral Means

8.1 An Overview

We first consider the classical problem how the integral means

$$\int_0^{2\pi} |f(re^{it})|^p \, dt \quad (0 < r < 1)$$

of the conformal map f grow as $r \to 1-$. Next we consider the integral means of the derivative which are more important for our purposes. Let $p \in \mathbb{R}$ and let $\beta_f(p)$ be the smallest number such that

(1) $$\int_0^{2\pi} |f'(re^{it})|^p \, dt = O\big((1 - r)^{-\beta_f(p)-\varepsilon}\big) \quad \text{as} \quad r \to 1-$$

for every $\varepsilon > 0$. We shall show that

(2) $$\sup_f \beta_f(p) = 3p - 1 \quad \text{for} \quad p \geq 2/5,$$

(3) $$0.117p^2 < \sup_f \beta_f(p) < 3.001p^2 \quad \text{for small } p;$$

these are results of Feng-MacGregor, Clunie, Makarov and others. An interesting open problem is the *Brennan conjecture* that

(4) $$\beta_f(-2) \leq 1.$$

It is only known that $\beta_f(-2) < 1.601$. The proofs use differential inequalities.

The case $p = 1$ is connected with the growth of the coefficients. Not even the exact order of growth is known for bounded univalent functions or for functions in Σ. We shall give a brief survey of results.

In Section 8.5 we consider integral means of Bloch functions g and deduce:

Makarov Law of the Iterated Logarithm. *If $g \in \mathcal{B}$ then*

(5) $$\limsup_{r \to 1} \frac{|g(r\zeta)|}{\sqrt{\log \frac{1}{1-r} \log \log \log \frac{1}{1-r}}} \leq \|g\|_{\mathcal{B}} \quad \text{for almost all } \zeta \in \mathbb{T}.$$

In the last section we consider lacunary series. The typical case is

$$(6) \qquad g(z) = \sum_{k=1}^{\infty} b_k z^{q^k}, \qquad q = 2, 3, \dots .$$

If (b_k) is bounded then $g \in \mathcal{B}$. The theory of Bloch functions and lacunary series has many connections to probability theory. We shall list some of the remarkable properties of lacunary series.

The lower bound in (3) comes from a function such that $\log f'$ has the form (6), and lacunary series show that (5) is essentially best possible.

In Chapter 10, we shall apply the results of this section to the problem how Hausdorff measure behaves under conformal mapping.

8.2 Univalent Functions

Let g be analytic in $\mathbb{D}$ and $p \in \mathbb{R}$. We write $z = re^{it}$, where $0 \le r < 1$. A calculation shows that

$$(1) \qquad \left(r \frac{\partial}{\partial r} \right)^2 |g(z)|^p + \left(\frac{\partial}{\partial t} \right)^2 |g(z)|^p = p^2 |g(z)|^p \left| z \frac{g'(z)}{g(z)} \right|^2 .$$

If we integrate over $0 \le t \le 2\pi$ the second integral vanishes by periodicity and we obtain *Hardy's identity*

$$(2) \qquad \frac{d}{dr} \left(r \frac{d}{dr} \int_0^{2\pi} |g(re^{it})|^p \, dt \right) = p^2 r \int_0^{2\pi} |g(re^{it})|^{p-2} |g'(re^{it})|^2 \, dt .$$

Proposition 8.1. *Let f map $\mathbb{D}$ conformally into $\mathbb{C}$ and let*

$$(3) \qquad M(r) = \max_{|z|=r} |f(z)| \quad \text{for} \quad 0 \le r < 1 .$$

If $f(0) = 0$ and $p > 0$ then

$$(4) \qquad \int_0^{2\pi} |f(re^{it})|^p \, dt \le 2\pi p \int_0^r M(\rho)^p \rho^{-1} \, d\rho$$

and, for $1/2 \le r < 1$,

$$(5) \qquad \int_0^{2\pi} |f(re^{it})|^{p-2} |f'(re^{it})|^2 \, dt \le K(p) \frac{M(r)^p}{1-r} \le \frac{K(p)|f'(0)|^p}{(1-r)^{2p+1}}$$

where $K(p)$ depends only on p.

Proof. Integrating Hardy's identity (2) (with $g = f$) we see that

$$r \frac{d}{dr} \int_0^{2\pi} |f(re^{it})|^p \, dt = p^2 \iint_{|z| \le r} |f(z)|^{p-2} |f'(z)|^2 \, dx \, dy .$$

We substitute $w = f(z)$. Since f is univalent we see from (3) that the last expression is

$$\leq p^2 \iint_{|w| \leq M(r)} |w|^{p-2} \, du \, dv = 2\pi p^2 \int_0^{M(r)} s^{p-1} \, ds = 2\pi p M(r)^p,$$

and (4) follows by a further integration because $M(0) = |f(0)| = 0$.
If $z = \rho e^{it}$, $1 - 3(1-r)/2 \leq \rho \leq r$, then

$$|f(re^{it})|^{p-2} |f'(re^{it})|^2 \leq K_0(p)|f(z)|^{p-2}|f'(z)|^2$$

by Corollary 1.6. We integrate over ρ and see as in the first part that

$$p^2 \frac{1-r}{2} \int_0^{2\pi} |f(re^{it})|^{p-2} |f'(re^{it})|^2 \, dt \leq K_0(p)8\pi p M(r)^p$$

which implies (5) because $M(r) \leq |f'(0)|(1-r)^{-2}$ by Theorem 1.3.   □

Theorem 8.2. *Let f map $\mathbb{D}$ conformally into $\mathbb{C}$ and suppose that*

(6) $$f(z) = O((1 - |z|)^{-\alpha}) \quad as \quad |z| \to 1$$

where $0 \leq \alpha \leq 2$. If $p > 1/\alpha$ then

(7) $$\int_0^{2\pi} |f(re^{it})|^p \, dt = O\left(\frac{1}{(1-r)^{\alpha p - 1}}\right) \quad as \quad r \to 1;$$

if $0 < p < 1/\alpha$ then f belongs to the Hardy space H^p.

This result (Pra27) follows at once from (4). See Hay58b for generalizations to multivalent functions. Since (6) with $\alpha = 2$ holds for all conformal maps, we conclude that (compare Theorem 1.7)

(8) $$\int_0^{2\pi} |f(re^{it})|^p \, dt = O\left(\frac{1}{(1-r)^{2p-1}}\right) \quad as \quad r \to 1$$

for $p > 1/2$ while $f \in H^p$ for $p < 1/2$. We remark that the integral is maximized in S by the Koebe function (Bae74; Dur83, p. 215; Hay89, p. 674).
If P_α denotes the Legendre function then (MaObSo66, p. 184)

(9) $$\frac{1}{2\pi} \int_0^{2\pi} \frac{dt}{|1 - re^{it}|^q} = \frac{1}{(1-r^2)^{q/2}} P_{-q/2}\left(\frac{1+r^2}{1-r^2}\right).$$

As $r \to 1 - 0$ this is (MaObSo66, p. 157, formula 10)

(10)
$$\sim 2^{-q}\pi^{-1/2}\Gamma\left(\frac{1-q}{2}\right)\Gamma\left(1 - \frac{q}{2}\right)^{-1} \quad \text{if} \quad 0 \leq q < 1,$$
$$\sim 2^{-1}\pi^{-1/2}\Gamma\left(\frac{q-1}{2}\right)\Gamma\left(\frac{q}{2}\right)(1-r)^{-q+1} \quad \text{if} \quad q > 1.$$

The conformal map $(1 - z)^{-\alpha}$ $(z \in \mathbb{D})$, $0 < \alpha \le 2$ therefore shows that Theorem 8.2 is best possible.

Exercises 8.2

1. Use Hardy's identity to show that $\int_{\mathbb{T}} |g(r\zeta)|^p |d\zeta|$ is increasing in $0 < r < 1$ if $p \ge 0$.

2. Let f map $\mathbb{D}$ conformally onto a domain symmetric with respect to 0. Show that

$$\int_0^{2\pi} |f(re^{it})| \, dt = O\left(\log \frac{1}{1-r}\right) \quad \text{as} \quad r \to 1-.$$

3. Let f map $\mathbb{D}$ conformally into $\mathbb{C}$. Use Hilbert's inequality to show that

$$f \in H^2 \quad \Leftrightarrow \quad \int_0^1 M(r)^2 \, dr < \infty.$$

4. Use (3.1.6) to prove directly that, for $q > 1$,

$$\int_0^{2\pi} |1 - re^{it}|^{-q} \, dt = O\left((1-r)^{-q+1}\right) \quad \text{as} \quad r \to 1.$$

8.3 Derivatives of Univalent Functions

Let f be a conformal map of $\mathbb{D}$ into $\mathbb{C}$. For $p \in \mathbb{R}$ we define

$$(1) \qquad \beta(p) = \beta_f(p) = \limsup_{r \to 1-} \left(\log \int_{\mathbb{T}} |f'(r\zeta)|^p |d\zeta|\right) / \log \frac{1}{1-r}.$$

Thus $\beta(p)$ is the smallest number such that, for every $\varepsilon > 0$,

$$(2) \qquad \int_0^{2\pi} |f'(re^{it})|^p \, dt = O\left(\frac{1}{(1-r)^{\beta(p)+\varepsilon}}\right) \quad \text{as} \quad r \to 1-.$$

Proposition 8.3. *The function β_f is continuous and convex in $\mathbb{R}$ and satisfies*

$$(3) \qquad \beta(p+q) \le \begin{cases} \beta(p) + 3q & \text{if } q > 0, \\ \beta(p) + |q| & \text{if } q < 0. \end{cases}$$

Proof. The Hölder inequality shows that

$$\int_{\mathbb{T}} |f'|^{\lambda p + (1-\lambda)q} |d\zeta| \le \left(\int_{\mathbb{T}} |f'|^p |d\zeta|\right)^{\lambda} \left(\int_{\mathbb{T}} |f'|^q |d\zeta|\right)^{1-\lambda}$$

for $p, q \in \mathbb{R}$ and $0 \le \lambda \le 1$. Hence we see from (1) that

(4) $$\beta_f(\lambda p + (1-\lambda)q) \leq \lambda\beta_f(p) + (1-\lambda)\beta_f(q) \quad (0 \leq \lambda \leq 1)$$

so that β_f is convex and therefore continuous. Inequality (3) follows from (1) and the fact (Theorem 1.3) that

$$\frac{|f'(0)|}{8}(1-|z|) \leq |f'(z)| \leq 2|f'(0)|(1-|z|)^{-3} \quad \text{for} \quad z \in \mathbb{D}. \qquad \square$$

As an example, consider the conformal map

$$f_\alpha(z) = \left(\frac{1+z}{1-z}\right)^\alpha, \quad f'_\alpha(z) = 2\alpha\frac{(1+z)^{\alpha-1}}{(1-z)^{\alpha+1}},$$

where $1 < \alpha \leq 2$. We easily see from (8.2.10) that

(5) $$\beta_{f_\alpha}(p) = \begin{cases} (\alpha+1)p - 1 & \text{if } 1/(\alpha+1) < p < +\infty, \\ (\alpha-1)|p| - 1 & \text{if } -\infty < p < -1/(\alpha-1), \\ 0 & \text{otherwise} \end{cases}$$

Sharp estimates (FeMG76) are known only for the range $p > 2/5$ and for close-to-convex functions (e.g. starlike functions, see Section 3.6).

Theorem 8.4. *Let f map $\mathbb{D}$ conformally into $\mathbb{C}$. Then*

(6) $$\beta_f(p) \leq 3p - 1 \quad \text{for} \quad 2/5 \leq p < +\infty.$$

If f is close-to-convex then

(7) $$\beta_f(p) \leq \begin{cases} |p| - 1 & \text{for } -\infty < p < -1, \\ 0 & \text{for } -1 \leq p \leq 1/3, \\ 3p - 1 & \text{for } 1/3 < p < \infty. \end{cases}$$

Proof. We may assume that $f \in S$. First let $2/5 < p < 2$ and choose $(2-p)/4 < x < p$. Hölder's inequality shows that

$$\int_\mathbb{T} |f'(r\zeta)|^p |d\zeta| \leq \left(\int_\mathbb{T} \frac{|f'(r\zeta)|^2}{|f(r\zeta)|^{2x/p}}|d\zeta|\right)^{\frac{p}{2}} \left(\int_\mathbb{T} |f(r\zeta)|^{\frac{2x}{2-p}}|d\zeta|\right)^{1-\frac{p}{2}}.$$

Estimating the second integral by (8.2.5) and the third integral by (8.2.8), we see that the right-hand side is bounded by a constant times

$$[(1-r)^{-2(2-2x/p)-1}]^{p/2}[(1-r)^{-2\cdot 2x/(2-p)+1}]^{1-p/2} = (1-r)^{-3p+1}.$$

Hence (6) follows from (1) for $2/5 < p < 2$ and from (3) and continuity for $2/5 \leq p < \infty$.

Now let f be close-to-convex. We first consider $p \geq 0$. By (3.6.14) and (3.6.5) we can write

(8) $$zf'(z) = g(z)q(z) \quad \text{with } g \text{ starlike}, \quad \text{Re } q(z) > 0$$

for $z \in \mathbb{D}$ and it follows from Hölder's inequality that

$$(9) \qquad \int_{\mathbb{T}} |rf'(r\zeta)|^p |d\zeta| \le \left(\int_{\mathbb{T}} |g(r\zeta)|^{3p/2} |d\zeta| \right)^{2/3} \left(\int_{\mathbb{T}} |q(r\zeta)|^{3p} |d\zeta| \right)^{1/3}.$$

Since g is univalent the second integral is $O(1)$ if $0 \le p < 1/3$ and $O((1-r)^{-3p+1})$ if $p > 1/3$; see (8.2.8).

Since q has positive real part we can furthermore write

$$q(z) = \frac{q(0) + \overline{q(0)}\varphi(z)}{1 - \varphi(z)} \qquad \text{with} \quad |\varphi(z)| \le 1, \quad \varphi(0) = 0,$$

so that, by Littlewood's subordination theorem (Dur70, p. 10),

$$(10) \qquad \int_{\mathbb{T}} |q(r\zeta)|^{3p} |d\zeta| \le \int_{\mathbb{T}} \left| \frac{q(0) + \overline{q(0)}\zeta}{1 - \zeta} \right|^{3p} |d\zeta| \quad (0 \le r < 1).$$

By (8.2.9) the last integral is $O(1)$ if $0 \le p < 1/3$ and $O((1-r)^{-3p+1})$ if $p > 1/3$. Hence (7) follows from (9) for $p \ge 0$; the case $p = 1/3$ follows from the continuity of β_f.

Finally if $p < 0$ we write

$$f'(z)^{-1} = \frac{z}{g(z)} \frac{1}{q(z)} \qquad \text{for} \quad z \in \mathbb{D}.$$

The first factor is bounded because g is univalent and $g(0) = 0$. The second factor has positive real part. Hence (7) follows as in (10) but with $|p|$ instead of $3p$. $\qquad \Box$

The next result (ClPo67, Pom85) covers the other ranges of p in the general case but is not sharp.

Theorem 8.5. *If f maps $\mathbb{D}$ conformally into $\mathbb{C}$ then*

$$(11) \qquad \beta_f(p) \le p - \frac{1}{2} + \left(4p^2 - p + \frac{1}{4} \right)^{1/2}$$

for $p \in \mathbb{R}$, furthermore

$$(12) \qquad \beta_f(-1) < 0.601.$$

The right-hand side of (11) is $< 3p^2$ for $p < 0$ and $< 3p^2 + 7p^3$ for $p > 0$. A miracle happens for $p = -1$ which allows us to give a much better estimate; this estimate is in fact true even if f has a pole in $\mathbb{D}$. In the opposite direction we have (Mak86, Roh89):

Theorem 8.6. *There exists a conformal map f of $\mathbb{D}$ onto a quasidisk such that $\beta_f(-1) > 0.109$ and*

$$(13) \qquad \beta_f(p) \ge 0.117p^2 \quad \text{for small } |p|.$$

In particular it follows that (6) need not hold for $p = 1/3$. We shall prove this theorem in Section 8.6 using lacunary series.

The *Brennan conjecture* (Bre78) states that

(14)
$$\beta_f(p) \leq |p| - 1 \quad \text{for} \quad -\infty < p \leq -2;$$

this would be sharp by (5) with $\alpha = 2$. In view of (3) the Brennan conjecture is equivalent to $\beta_f(-2) \leq 1$. It is only known that

(15)
$$\beta_f(p) < |p| - 0.399 \quad \text{for} \quad -\infty < p \leq -1$$

which follows from (12) and (3). In Fig. 8.1, the upper curve gives the best known bounds for $\beta_f(p)$ and comes from Theorems 8.4 and 8.5 together with convexity. The lower curve comes from the example in Theorem 8.6 and from (5) with $\alpha = 2$.

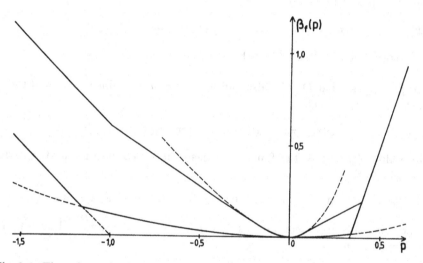

Fig. 8.1. The values of $\sup_f \beta_f(p)$ that are possible according to present knowledge lie between the two curves

If $\beta_f(p) < 1$ then, with $\beta_f(p) < \alpha < 1$,

$$\int_0^1 \int_0^{2\pi} |f'(re^{it})|^p \, r \, dt \, dr = \int_0^1 O\big((1-r)^{-\alpha}\big) \, dr < \infty.$$

Hence it follows from (15) and (6) that

(16)
$$f' \in L^p(\mathbb{D}) \quad \text{for} \quad -1.399 \leq p < 2/3.$$

The Brennan conjecture would imply that $f' \in L^p(\mathbb{D})$ for $-2 < p < 2/3$.

For the proof of Theorem 8.5 we need a result about *differential inequalities* (see e.g. BeBe65, p. 139; Her72).

Proposition 8.7. *Let $n \geq 2$ and let $a_k(r)$ $(k = 0, \ldots, n-1)$ be continuous in $r_0 \leq r < 1$. Assume that*

$$(17) \qquad a_k(r) \geq 0 \quad \text{for} \quad k = 0, \ldots, n-2.$$

Let u and v be n times differentiable in $[r_0, 1)$ and

$$(18) \qquad u^{(n)} < \sum_{k=0}^{n-1} a_k u^{(k)}, \quad v^{(n)} = \sum_{k=0}^{n-1} a_k v^{(k)}.$$

If $u^{(k)}(r_0) < v^{(k)}(r_0)$ for $k = 0, \ldots, n-1$ then

$$(19) \qquad u(r) < v(r) \quad \text{for} \quad r_0 \leq r < 1.$$

Thus u satisfies a differential inequality while v satisfies the corresponding differential equation. Note that a_{n-1} may have any sign. In the case $n = 2$ the assumptions reduce to

$$(20) \qquad u'' < a_1 u' + a_0 u, \quad v'' = a_1 v' + a_0 v, \quad a_0 \geq 0$$

and $u(r_0) < v(r_0)$, $u'(r_0) < v'(r_0)$.

Proof. Suppose that (19) is false and write $\varphi = u - v$. Since $\varphi(r_0) < 0$ there exists $r_n \in (r_0, 1)$ such that

$$\varphi(r_n) = 0, \quad \varphi(r) < 0 \quad \text{for} \quad r_0 < r < r_n.$$

Since also $\varphi'(r_0) < 0$, the function φ has a first local minimum at r_{n-1} in (r_0, r_n), so that

$$\varphi'(r_{n-1}) = 0, \quad \varphi''(r_{n-1}) \geq 0, \quad \varphi'(r) \leq 0 \quad \text{for} \quad r_0 < r < r_{n-1}.$$

Continuing this argument we find $r_0 < r_1 < \cdots < r_{n-1} < r_n < 1$ such that, for $k = 0, \ldots, n-1$,

$$\varphi^{(k)}(r_{n-k}) = 0, \quad \varphi^{(k+1)}(r_{n-k}) \geq 0, \quad \varphi^{(k)}(r) \leq 0 \quad \text{for} \quad r_0 < r < r_{n-k}.$$

Hence $\varphi^{(k)}(r_1) \leq 0$ $(k = 0, \ldots, n-2)$, $\varphi^{(n-1)}(r_1) = 0$ and $\varphi^{(n)}(r_1) \geq 0$. Therefore (17) and (18) show that

$$\varphi^{(n)}(r_1) < \sum_{k=0}^{n-2} a_k(r_1) \varphi^{(k)}(r_1) \leq 0$$

which contradicts $\varphi^{(n)}(r_1) \geq 0$. $\qquad \square$

Proof of Theorem 8.5. (a) We write

$$(21) \qquad u(r) = \int_{\mathbb{T}} |f'(r\zeta)|^p |d\zeta| \quad \text{for} \quad 0 < r < 1.$$

Differentiating under the integral sign we obtain

$$(22) \qquad u'(r) = p \int_{\mathbb{T}} |f'|^p \, \mathrm{Re}[\zeta f''(r\zeta)/f'(r\zeta)]|d\zeta|,$$

and Hardy's identity (8.2.2) shows that

$$(23) \qquad u''(r) + r^{-1}u'(r) = \frac{1}{r}\frac{d}{dr}[ru'(r)] = p^2 \int_{\mathbb{T}} |f'(r\zeta)|^p \left| \zeta \frac{f''(r\zeta)}{f'(r\zeta)} \right|^2 |d\zeta|.$$

Since $u'(r) \geq 0$ by Exercise 8.2.1, we conclude from (21), (22) and (23) that

$$u''(r) - \frac{4rp}{1-r^2}u'(r) + \frac{4r^2p^2}{(1-r^2)^2}u(r) \leq p^2 \int_{\mathbb{T}} |f'|^p \left| \zeta \frac{f''(r\zeta)}{f'(r\zeta)} - \frac{2r}{1-r^2} \right|^2 |d\zeta|$$

and this is bounded by $16p^2(1-r^2)^{-2}u(r)$ because of Proposition 1.2. Hence

$$u''(r) \leq \frac{4rp}{1-r^2}u'(r) + \frac{(16-4r^2)p^2}{(1-r^2)^2}u(r) < \frac{2p+\varepsilon}{1-r}u'(r) + \frac{3p^2+\varepsilon}{(1-r)^2}u(r)$$

if $r_0(\varepsilon) \leq r < 1$. The corresponding differential equation

$$(24) \qquad v''(r) = \frac{2p+\varepsilon}{1-r}v'(r) + \frac{3p^2+\varepsilon}{(1-r)^2}v(r) \quad (r_0 \leq r < 1)$$

has the solution $c(1-r)^{-\lambda}$ where $\lambda = \lambda(\varepsilon)$ is the positive root of

$$(25) \qquad \lambda(\lambda+1) = (2p+\varepsilon)\lambda + 3p^2 + \varepsilon.$$

Since (20) is satisfied we conclude from Proposition 8.7 that

$$\int_{\mathbb{T}} |f'(r\zeta)|^p |d\zeta| = u(r) < \frac{c(\varepsilon)}{(1-r)^{\lambda(\varepsilon)}} \qquad \text{for} \quad r_0(\varepsilon) \leq r < 1$$

if $c(\varepsilon)$ is chosen large enough. Hence (11) follows from (2) because $\lambda(\varepsilon)$ tends to the expression on the right-hand side of (11) as $\varepsilon \to 0$.

(b) Now consider

$$(26) \qquad g(z) = f'(z)^{-1/2} = \sum_{n=0}^{\infty} b_n z^n.$$

Parseval's formula shows that

$$(27) \qquad u(r) \equiv \frac{1}{2\pi} \int_{\mathbb{T}} \frac{|d\zeta|}{|f'(r\zeta)|} = \frac{1}{2\pi} \int_{\mathbb{T}} |g(r\zeta)|^2 |d\zeta| = \sum_{n=0}^{\infty} |b_n|^2 r^{2n}$$

for $0 \leq r < 1$, hence

$$u^{(4)}(r) = \sum_{n=2}^{\infty} 2n(2n-1)(2n-2)(2n-3)|b_n|^2 r^{2n-4}$$

$$\leq 16 \sum_{n=2}^{\infty} n^2(n-1)^2 |b_n|^2 r^{2n-4} = \frac{16}{2\pi} \int_{\mathbb{T}} |g''(r\zeta)|^2 |d\zeta|.$$

Computation shows that $g'' = -S_f g/2$ where S_f denotes the Schwarzian derivative. Since $|S_f(r\zeta)| \leq 6(1-r^2)^{-2}$ by Exercise 1.3.6 we see from (27) that

$$u^{(4)}(r) \leq \frac{4}{2\pi} \int_{\mathbb{T}} |S_f(r\zeta)|^2 |g(r\zeta)|^2 |d\zeta|$$

$$\leq \frac{144}{(1-r^2)^4} u(r) < \frac{9.01}{(1-r)^4} u(r) \quad (r_0 \leq r < 1).$$

The corresponding differential equation

$$v^{(4)}(r) = \frac{9.01}{(1-r)^4} v(r) \quad (r_0 \leq r < 1)$$

has the solution $c(1-r)^{-\lambda}$ where λ is the positive root of $\lambda(\lambda+1)(\lambda+2)(\lambda+3)$ $= 9.01$; calculation shows that $0.600 < \lambda < 0.601$. It follows from Proposition 8.7 that

$$\frac{1}{2\pi} \int_{\mathbb{T}} |f'(r\zeta)|^{-1} |d\zeta| = u(r) < \frac{c}{(1-r)^\lambda} \quad \text{for} \quad r_0 \leq r < 1$$

if c is chosen sufficiently large, and we conclude that $\beta_f(-1) \leq \lambda < 0.601$. $\square$

Exercises 8.3

1. Show that $b = \lim_{x \to +\infty} x^{-1}\beta(x)$ exists and that $\beta(p+q) \leq \beta(p) + bq$ for $p, q > 0$.

2. Let f map $\mathbb{D}$ conformally into $\mathbb{C}$. Show that

$$\beta_f(p) < \begin{cases} 0.649|p| - 0.048 & \text{for } -1 \leq p \leq -0.2, \\ 0.545p - 0.018 & \text{for } 0.1 \leq p \leq 0.4. \end{cases}$$

3. Show that the Brennan conjecture is equivalent to the following assertion: If g maps $G \subset \mathbb{C}$ conformally onto $\mathbb{D}$ then $g' \in L^q(G)$ for $4/3 < q < 4$.

4. Suppose that $(1 - |z|^2)|zf''(z)/f'(z)| \leq \kappa$ for $z \in \mathbb{D}$; compare Corollary 5.23. Show that $\beta_f(p) \leq \kappa^2 p^2/4$ for $p \in \mathbb{R}$.

5. Let f lie in a linearly invariant family of order α where $1 \leq \alpha < \infty$; see (5.3.13). Show that

$$\beta_f(p) \leq p - \frac{1}{2} + \sqrt{\alpha^2 p^2 - p + \frac{1}{4}} \sim (\alpha^2 - 1)p^2 \quad (p \to 0).$$

8.4 Coefficient Problems

We now consider the power series expansion

(1)
$$f(z) = \sum_{n=0}^{\infty} a_n z^n \quad \text{for } |z| < 1.$$

If $n = 2, 3, \ldots$ and $r_n = 1 - 1/n$ then

(2) $\quad n|a_n| = \dfrac{1}{2\pi r_n^{n-1}} \left| \displaystyle\int_0^{2\pi} f'(r_n e^{it}) e^{-i(n-1)t}\, dt \right| \leq \dfrac{e}{2\pi} \displaystyle\int_0^{2\pi} |f'(r_n e^{it})|\, dt .$

This allows us to apply the results of Section 8.3 to estimate $|a_n|$.

Theorem 8.8. *Let f map $\mathbb{D}$ conformally into $\mathbb{C}$ and suppose that*

(3)
$$f(z) = O((1 - |z|)^{-\alpha}) \quad \text{as } |z| \to 1.$$

If $1/2 < \alpha \leq 2$ then, as $r \to 1$ and $n \to \infty$,

(4)
$$\int_0^{2\pi} |f'(re^{it})|\, dt = O\left(\frac{1}{(1-r)^\alpha} \right), \quad a_n = O(n^{\alpha-1}).$$

If f is bounded, i.e. if $\alpha = 0$, then

(5)
$$\int_0^{2\pi} |f'(re^{it})|\, dt = O\left(\frac{1}{(1-r)^{0.491}} \right), \quad a_n = O(n^{0.491-1}).$$

The first part (LiPa32) is sharp as the functions $(1 - z)^{-\alpha}$ show. For functions of large growth, the size of the coefficients is mainly determined by the growth of $M(r)$. These results also hold for multivalent functions (see e.g. Hay58b, p. 46).

For bounded functions however, the coefficient size is mainly determined by the complexity of the boundary; see also Corollary 10.19. The result for $\alpha = 0$ (ClPo67, Pom85) is not sharp.

The situation is less clear for the range $0 < \alpha \leq 1/2$. The estimate (4) still holds (Bae86) at least for $\alpha \geq 0.497$. See e.g. Pom75, p. 131, p. 146 for further results.

Proof. We may assume that $f(0) = 0$. First let $1/2 < \alpha \leq 2$ and choose $1/\alpha < q < 2$. We see from the Schwarz inequality that

$$\left(\int_0^{2\pi} |f'(re^{it})|\, dt \right)^2 \leq \int_0^{2\pi} |f|^{-q}|f'|^2\, dt \int_0^{2\pi} |f|^q\, dt$$

$$= O((1-r)^{-(2-q)\alpha-1}) O((1-r)^{-\alpha q+1}) = O\left((1-r)^{-2\alpha}\right)$$

by (3), Proposition 8.1 and Theorem 8.2. This implies (4); the second estimate follows from (2).

Now let f be bounded. We see from (8.2.5) for $p = 2$ that

(6)
$$\int_0^{2\pi} |f'(re^{it})|^2 \, dt = O\left(\frac{1}{1-r}\right) \qquad \text{as} \quad r \to 1.$$

Hence $\beta_f(2) \le 1$ using the notation (8.3.1), and it follows from Theorem 8.5 by the convexity of β_f that, for $0 < p < 1$,

$$\beta_f(1) \le \frac{1}{2-p}\beta_f(p) + \frac{1-p}{2-p}\beta_f(2) \le \frac{1}{2-p}\left(\frac{1}{2} + \left(4p^2 - p + \frac{1}{4}\right)^{1/2}\right).$$

If we choose $p = 0.07$ we conclude that $\beta_f(1) < 0.491$ which implies (5). We remark that $\beta_f(2 - \delta) \le 1 - \delta + O(\delta^2)$ as $\delta \to 0$; see JoMa92. $\qquad\square$

It might appear that one loses much by applying the triangle inequality in (2). But this is not the case because (CaJo91)

(7)
$$\gamma \equiv \sup_f \beta_f(1) = \sup_f \limsup_{n\to\infty} \frac{\log(n|a_n|)}{\log n}$$

where the supremum is taken over all bounded univalent functions. A direct construction (Pom75, p. 133; Dur83, p. 238) shows that $\gamma > 0.17$. Numerical calculations using (7) and Julia sets show (CoJo91) that $\gamma > 0.23$ so that, together with (5),

(8)
$$0.23 < \gamma < 0.491.$$

Carleson and Jones have conjectured that $\gamma = 1/4$.

Closely related is the coefficient problem for the class Σ of the univalent functions

(9)
$$g(\zeta) = \zeta + \sum_{n=0}^{\infty} b_n \zeta^{-n} \qquad \text{for} \quad |\zeta| > 1.$$

If γ is defined by (7) then (CaJo91)

(10)
$$\gamma = \sup_{g\in\Sigma} \limsup_{n\to\infty} \frac{\log(n|b_n|)}{\log n}$$

so that γ is the smallest number such that, for all $g \in \Sigma$,

(11)
$$b_n = O(n^{\gamma+\varepsilon-1}) \quad (n \to \infty) \quad \text{for each } \varepsilon > 0.$$

The method of integral means will not give sharp coefficient bounds. It is often important to get sharp estimates for the coefficients and their combinations because these estimates can be transferred to bounds for other functionals via the Koebe transform (1.3.12).

The class S consists of all analytic univalent functions $f(z) = z + a_2 z^2 + \dots$ in $\mathbb{D}$. It is now known (deB85) that

(12)
$$|a_n| \le n \qquad \text{for} \quad f \in S, \quad n = 1, 2, \dots.$$

This Bieberbach conjecture had been open for a long time; see (1.3.11) and
e.g. Dur83. The bound $|a_2| \leq 2$ leads to the Koebe distortion theorem as we
have seen in Section 1.3.

We state some further sharp bounds without proof. Let $f \in S$. If $0 \leq \vartheta \leq 2\pi$ and $0 < \lambda \leq 4$ then (Jen60, p. 173)

$$(13) \qquad \mathrm{Re}[(a_3 - a_2^2)e^{2i\vartheta} + \lambda a_2 e^{i\vartheta}] \leq 1 + \frac{3}{8}\lambda^2 + \frac{1}{4}\lambda^2 \log \frac{4}{\lambda},$$

furthermore (FeSz33; Dur83, p. 104)

$$(14) \qquad |a_3 - \alpha a_2^2| \leq 1 + 2\exp[-2\alpha/(1-\alpha)] \quad \text{for} \quad 0 \leq \alpha < 1.$$

If we write $\log[f(z)/z] = \Sigma c_n z^n$ then (deB85)

$$(15) \qquad \sum_{k=1}^{n} k(n+1-k)|c_k|^2 \leq 4 \sum_{k=1}^{n} \frac{n+1-k}{k}$$

which implies (12) by the Lebedev-Milin inequalities (Dur83, p. 143).

Now let $g \in \Sigma$ and $0 \leq \vartheta \leq 2\pi$. Then (Jen60, p. 184)

$$(16) \qquad \mathrm{Re}[b_2 e^{3i\vartheta} + \lambda b_1 e^{2i\vartheta}] \leq \frac{2}{3} + \frac{1}{2}\lambda^2 - \frac{1}{12}\lambda^3 \quad \text{for} \quad -\frac{2}{3} \leq \lambda \leq 2,$$

in particular $|b_2| \leq 2/3$ (Schi38). Furthermore (Jen60, p. 191; Pom75, p. 121)

$$(17) \qquad \mathrm{Re}\left[\left(b_3 + \frac{1}{2}b_1^2 \right) e^{2i\vartheta} + \lambda b_1 e^{i\vartheta} \right] \geq -\frac{1}{2} - \frac{3}{16}\lambda^2 - \frac{1}{8}\lambda^2 \log \frac{4}{\lambda}$$

for $0 < \lambda \leq 4$, in particular $|b_3| \leq 1/2 + e^{-6}$ (GaSch55b).

Exercises 8.4

1. Let $f \in S$ and write $[z^{-1}f(z)]^\lambda = \sum_0^\infty c_n z^n$. Show that $c_n = O(n^{2\lambda-1})$ as $n \to \infty$
   if $\lambda > 1/4$.
   (See HayHu86 for more precise results; the Koebe function is not extremal for
   $0.25 < \lambda < 0.4998$.)

2. Suppose that $f(z) = a_1 z + a_2 z^2 + \ldots$ maps $\mathbb{D}$ conformally into $\mathbb{D}$. Show that
   $|a_2| \leq 2|a_1|(1 - |a_1|)$ by considering $f(z)/(1 - e^{i\vartheta}f(z))^2$ (Pic17; see e.g. Tam78
   or Pom75, p. 98 for further results.)

8.5 The Growth of Bloch Functions

We now consider the class $\mathcal{B}$ of Bloch functions, i.e. of functions g analytic in
$\mathbb{D}$ such that

$$(1) \qquad \|g\|_{\mathcal{B}} = \sup_{z \in \mathbb{D}}(1 - |z|^2)|g'(z)| < \infty;$$

see Section 4.2. There are non-trivial bounds of their integral means (Mak85):

Theorem 8.9. *If $g \in \mathcal{B}$ and $g(0) = 0$ then*

(2)
$$\frac{1}{2\pi} \int_{\mathbb{T}} |g(r\zeta)|^{2n} |d\zeta| \le n! \|g\|_{\mathcal{B}}^{2n} \left(\log \frac{1}{1 - r^2} \right)^n$$

for $0 < r < 1$ and $n = 0, 1, \ldots$.

Proof. The case $n = 0$ is trivial. Suppose that (2) holds for some n. Hardy's identity (8.2.2) shows that

$$\frac{d}{dr} \left(r \frac{d}{dr} \right) \left(\frac{1}{2\pi} \int_{\mathbb{T}} |g(r\zeta)|^{2n+2} |d\zeta| \right) = \frac{4(n+1)^2 r}{2\pi} \int_{\mathbb{T}} |g(r\zeta)|^{2n} |g'(r\zeta)|^2 |d\zeta|.$$

Writing $\lambda(r) = \log[1/(1 - r^2)]$ we see from (2) and (1) that this is

$$\le 4(n+1)^2 r \, n! \|g\|_{\mathcal{B}}^{2n} \lambda(r)^n \cdot (1 - r^2)^{-2} \|g\|_{\mathcal{B}}^2$$

$$\le (n+1)! \|g\|_{\mathcal{B}}^{2n+2} \frac{d}{dr} \left[r \frac{d}{dr} \lambda(r)^{n+1} \right].$$

Hence we obtain by integration that

$$\frac{d}{dr} \left(\frac{1}{2\pi} \int_{\mathbb{T}} |g(r\zeta)|^{2(n+1)} |d\zeta| \right) \le (n+1)! \|g\|_{\mathcal{B}}^{2(n+1)} \frac{d}{dr} \left(\lambda(r)^{n+1} \right)$$

and (2) for $n + 1$ follows by another integration because both sides vanish for $r = 0$. $\quad\square$

We deduce the *Makarov law of the iterated logarithm* (Mak85).

Theorem 8.10. *If $g \in \mathcal{B}$ then, for almost all $\zeta \in \mathbb{T}$,*

(3)
$$\limsup_{r \to 1} \frac{|g(r\zeta)|}{\sqrt{\log \frac{1}{1-r} \log \log \log \frac{1}{1-r}}} \le \|g\|_{\mathcal{B}}.$$

There exists $g \in \mathcal{B}$ such that this limes superior is $> 0.685\|g\|_{\mathcal{B}}$ for almost all $\zeta \in \mathbb{T}$.

Here is a probabilistic interpretation: Consider the probability measure $(2\pi)^{-1} \Lambda$ on $\mathbb{T}$. Then

(4)
$$Z_t(\zeta) = g((1 - e^{-t})\zeta) \quad (z \in \mathbb{T}), \quad 0 \le t < \infty$$

defines a complex stochastic process with expectation $g(0)$. Its variance satisfies

(5)
$$\sigma(t)^2 \le \|g\|_{\mathcal{B}}^2 t \quad (0 \le t < \infty)$$

by (2) with $n = 1$. We can write (3) as

(6)
$$\limsup_{t \to \infty} \frac{|Z_t|}{\|g\|_B \sqrt{t \log \log t}} \leq 1 .$$

This is therefore a one-sided form of the law of the iterated logarithm; see e.g. Bin86 for a survey. The last assertion of Theorem 8.10 shows that (3) and thus (6) are not far from being best possible. See Bañ86 and Lyo90 for an approach using Brownian motion and Mak89d using martingales.

Proof. We may assume that $\|g\|_B = 1$. Consider the maximal function

(7)
$$g^*(s, \zeta) = \max_{0 \leq r \leq 1 - e^{-s}} |g(r\zeta)| \quad \text{for} \quad e \leq s < \infty, \quad \zeta \in \mathbb{T} .$$

The Hardy-Littlewood maximal theorem (Dur70, p. 12) applied to the analytic function g^{2n} and Theorem 8.9 show that

$$\int_{\mathbb{T}} g^*(s, \zeta)^{2n} |d\zeta| \leq \frac{K}{2\pi} \int_{\mathbb{T}} |g((1 - e^{-s})\zeta)|^{2n} |d\zeta| \leq K \, n! \, s^n$$

where K is an absolute constant. We multiply by $s^{-n} \psi_n(s)$ where

$$\psi_n(s) = -n \frac{d}{ds} \left[(\log s)^{-1/n} \right] = s^{-1} (\log s)^{-1-1/n} > 0 .$$

Integrating we obtain from Fubini's theorem that

$$\int_{\mathbb{T}} \left(\int_e^{\infty} g^*(s, \zeta)^{2n} s^{-n} \psi_n(s) \, ds \right) |d\zeta| \leq K \, n! \int_e^{\infty} \psi_n(s) \, ds = K \, n! \, n .$$

Hence there are sets $A_n \subset \mathbb{T}$ $(n = 1, 2, ...)$ with $\Lambda(A_n) > 2\pi - K/n^2$ such that

(8)
$$\int_e^{\infty} g^*(s, \zeta)^{2n} s^{-n} \psi_n(s) \, ds \leq n! \, n^3 \quad \text{for} \quad \zeta \in A_n .$$

Since $-\frac{d}{ds} \left[s^{-n} (\log s)^{-1-1/n} \right] \leq 3ns^{-n} \psi_n(s)$ we therefore conclude that, for $\zeta \in A_n$ and $e \leq \sigma < \infty$,

(9) $\quad g^*(\sigma, \zeta)^{2n} \sigma^{-n} (\log \sigma)^{-1-1/n} \leq 3n \int_{\sigma}^{\infty} g^*(s, \zeta)^{2n} s^{-n} \psi_n(s) \, ds \leq 3 \, n! \, n^4 .$

The set $A = \bigcup_{k=1}^{\infty} \bigcap_{n=k}^{\infty} A_n$ satisfies $\Lambda(A) = 2\pi$ because $\Lambda(A_n) \geq 2\pi - K/n^2$. Now let $\zeta \in A$. Then $\zeta \in A_n$ for $n \geq k$ and suitable $k = k(\zeta)$. If $r < 1$ is sufficiently close to 1 then

$$n = \lfloor \log \log \sigma \rfloor \geq k \quad \text{where} \quad \sigma = \log 1/(1 - r) .$$

Since $\log \log \sigma \geq n$ and $\log \sigma < e^{n+1}$ we therefore see from (7) and (9) that

$$\frac{|g(r\zeta)|^2}{\sigma \log \log \sigma} \leq \frac{g^*(\sigma, \zeta)^2}{\sigma \log \log \sigma} \leq \left(\frac{3 \, n! \, n^4}{n^n} \right)^{1/n} e^{(n+1)^2/n^2}$$

which tends $\to 1$ as $n \to \infty$ by Stirling's formula. This proves (3). We postpone the proof of the second assertion to the end of the next section. $\qquad\square$

Corollary 8.11. *If f maps $\mathbb{D}$ conformally into $\mathbb{C}$ then*

$$(10) \qquad \limsup_{r \to 1} \frac{|\log f'(r\zeta)|}{\sqrt{\log \frac{1}{1-r} \log\log\log \frac{1}{1-r}}} \leq 6 \quad \text{for almost all } \zeta \in \mathbb{T},$$

in particular $f'(r\zeta) = O((1-r)^{-\varepsilon})$ as $r \to 1$ for $\varepsilon > 0$ and almost all ζ.

This result (Mak85) follows at once from Theorem 8.10 and the fact that $\|\log f'\|_{\mathcal{B}} \leq 6$ by Proposition 4.1.

Exercises 8.5

1. Let $g(z) = \sum_0^\infty b_n z^n$. Show that (b_n) is bounded if $g \in \mathcal{B}$ and that $b_n \to 0$ if $g \in \mathcal{B}_0$. (See Fern84 for deeper results.)

2. If $g \in \mathcal{B}$, $g(0) = 0$ and $0 < \alpha \|g\|_{\mathcal{B}} < 1$, show that

$$\int_{\mathbb{T}} \exp\left[\alpha^2 |g(r\zeta)|^2 / \log\frac{1}{1-r^2}\right] |d\zeta| \leq \frac{2\pi}{1 - \alpha^2\|g\|_{\mathcal{B}}^2} \quad (0 \leq r < 1).$$

3. Let $g \in \mathcal{B}$, $g(0) = 0$. Use (2) to prove that, with some constant K,

$$\Lambda\left(\left\{\zeta \in \mathbb{T} : \frac{|g(r\zeta)|}{\|g\|_{\mathcal{B}}\sqrt{\log 1/(1-r^2)}} \geq x\right\}\right) \leq Kxe^{-x^2} \quad \text{for } 0 < r < 1, \quad x \geq 1.$$

This is related to the central limit theorem.

4. If f maps $\mathbb{D}$ conformally into $\mathbb{C}$ and if $\alpha < 1$, show that $|f(\zeta) - f(r\zeta)| = O((1-r)^\alpha)$ as $r \to 1$ for almost all $\zeta \in \mathbb{T}$.

8.6 Lacunary Series

1. We say that the function g has *Hadamard gaps* if

$$(1) \qquad g(z) = b_0 + \sum_{k=1}^\infty b_k z^{n_k}, \quad \frac{n_{k+1}}{n_k} \geq q \quad (k = 1, 2, \ldots)$$

for some real $q > 1$.

Proposition 8.12. *If g has Hadamard gaps then*

$$g \in \mathcal{B} \quad \Leftrightarrow \quad \sup |b_k| < \infty,$$
$$g \in \mathcal{B}_0 \quad \Leftrightarrow \quad b_k \to 0 \quad \text{as } k \to \infty.$$

Proof. The implications $\Rightarrow$ hold for all Bloch functions; see Exercise 8.5.1. Now let $|b_k| \leq M$ and $|z| = r < 1$. It follows from (1) that

$$\frac{r|g'(z)|}{1-r} \leq \frac{M}{1-r} \sum_{k=1}^{\infty} n_k r^{n_k} = M \sum_{m=1}^{\infty} \left(\sum_{n_k \leq m} n_k \right) r^m ,$$

furthermore that

(2) $\qquad n_k \leq q^{k-j} n_j \quad$ for $\quad k \leq j, \quad n_k \geq q^{k-j} n_j \quad$ for $\quad k \geq j.$

We conclude that

$$\frac{r|g'(z)|}{1-r} \leq M \sum_{m=1}^{\infty} \frac{qm}{q-1} r^m = \frac{Mq}{q-1} \frac{r}{(1-r)^2}$$

so that $(1-r^2)|g'(z)| \leq 2Mq/(q-1)$. Hence $g \in \mathcal{B}$. If $b_k \to 0$ then we choose N so large that $|b_k| < \varepsilon$ for $k \geq N$ and write $g(z) = p(z) + \sum_{k=N}^{\infty} b_k z^{n_k}$ where p is a polynomial. As above we see that

$$\limsup_{r \to 1} (1-r^2)|g'(z)| \leq \lim_{r \to 1} (1-r^2)|p'(z)| + 2\varepsilon q/(q-1)$$

for every $\varepsilon > 0$. Hence the limes superior is $= 0$ and thus $g \in \mathcal{B}_0$. $\qquad \square$

Proposition 8.13. *Let the lacunary series g be given by (1). If $r_j = 1 - 1/n_j$ for $j = 1, 2, \ldots$ then*

(3) $\qquad \left| b_0 + \sum_{k=1}^{j} b_k \zeta^{n_k} - g(r\zeta) \right| \leq K(q) \sup |b_k| \quad (\zeta \in \mathbb{T})$

for $r_j \leq r \leq r_{j+1}$ where $K(q)$ depends only on $q > 1$.

Proof. We may assume that $|b_k| \leq 1$. It follows from (1) that

$$\left| b_0 + \sum_{k=1}^{j} b_k \zeta^{n_k} - g(r\zeta) \right| \leq \sum_{k=1}^{j} (1 - r^{n_k}) + \sum_{k=j+1}^{\infty} r^{n_k} .$$

Since $1 - r^{n_k} \leq n_k(1-r_j) = n_k/n_j$ and $r^{n_k} \leq \exp(-n_k/n_{j+1})$, we see from (2) that the right-hand side is

$$\leq \sum_{k=1}^{j} q^{k-j} + \sum_{k=j+1}^{\infty} \exp(-q^{k-j-1}) < \frac{q}{q-1} + \sum_{\nu=0}^{\infty} \exp(-q^\nu) . \qquad \square$$

Lacunary series are useful for constructing "pathological" conformal maps (see e.g. GnHPo86).

Proposition 8.14. *We define*

(4) $\qquad B_q = \max_{0 \leq x \leq 1} 2 \sum_{n=-\infty}^{\infty} q^{n+x} \exp(-q^{n+x})$

for $q = 2, 3, \ldots$. If

(5)
$$\log f'(z) = \sum_{k=0}^{\infty} b_k z^{q^k}, \quad \sup_k |b_k| < 1/B_q$$

then f maps $\mathbb{D}$ conformally onto a quasidisk.

q	Upper bound for B_q	q	Upper bound for B_q
2	2.8913	12	0.9266
3	1.8291	13	0.9104
4	1.4614	14	0.8968
5	1.2754	15	0.8851
6	1.1642	20	0.8455
7	1.0907	30	0.8074
8	1.0386	40	0.7890
9	0.9998	50	0.7782
10	0.9699	80	0.7621
11	0.9460	∞	0.7358

Proof. The sum in (4) is continuous in x and has the period 1. Hence the maximum B_q exists. Let $0 < |z| < 1$. We write

$$|z| = \exp(-q^{-j+x}) \quad \text{with} \quad j \in \mathbb{Z}, \quad 0 \le x < 1.$$

If $|b_k| \le M < 1/B_q$ then, by (5),

$$\left(1 - |z|^2\right) \left| z \frac{f''(z)}{f'(z)} \right| \le 2 \log \frac{1}{|z|} \sum_{k=0}^{\infty} |b_k| q^k |z|^{q^k}$$

$$\le 2M \sum_{k=0}^{\infty} q^{k-j+x} \exp(-q^{k-j+x})$$

which is $< MB_q < 1$ by (4). Hence f maps $\mathbb{D}$ conformally onto a quasidisk by the Becker univalence criterion (Corollary 5.23). The upper bounds are obtained by splitting $[0, 1]$ into 128 subintervals and using the fact that te^{-t} increases for $0 < t \le 1$ and then decreases. $\square$

It follows from Proposition 8.14 that

(6)
$$f(z) = \int_0^z \exp\left(b \sum_{k=1}^{\infty} \zeta^{q^k} \right) d\zeta = z + \ldots, \quad q = 2, 3, \ldots$$

maps $\mathbb{D}$ conformally onto a quasidisk if $|b| < 1/B_q$. The *modified Bessel functions* are defined by

(7)
$$I_n(x) = \left(\frac{x}{2}\right)^n \sum_{\nu=0}^{\infty} \frac{x^{2\nu}}{2^{2\nu} \nu! (\nu + n)!} \quad (n = 0, 1, \ldots);$$

see e.g. MaObSo66, p. 66, p. 70. They appear in the Fourier series

$$(8) \qquad \exp(x\cos\vartheta) = I_0(x) + 2\sum_{n=1}^{\infty} I_n(x)\cos n\vartheta\,.$$

We give a lower estimate (Roh89) of the integral means of f'.

Proposition 8.15. *Let f be the conformal map defined by (6) with q odd and $0 < b < 1/B_q$. If $p \in \mathbb{R}$ then*

$$(9) \qquad \int_{\mathbb{T}} |f'(r\zeta)|^p |d\zeta| \geq c(1-r)^{-a} \quad (0 < r < 1)$$

where $c > 0$ and $a = \log I_0(bp)/\log q$.

Proof. First we prove by induction on $m = 0, 1, \ldots$ that

$$(10) \qquad \int_0^{2\pi} \prod_{k=1}^{m} \cos(n_k t)\, dt \geq 0 \quad \text{for} \quad n_1, \ldots, n_m \in \mathbb{Z}.$$

The cases $m = 0$ and $m = 1$ are trivial. Suppose that $m \geq 2$ and that (10) holds for $m - 1$. Using $\cos\alpha\cos\beta = [\cos(\alpha+\beta)+\cos(\alpha-\beta)]/2$ we can write our integral as

$$\frac{1}{2}\int_0^{2\pi} \left(\prod_{k=1}^{m-2}\cos n_k t\right)[\cos(n_m+n_{m-1})t + \cos(n_m - n_{m-1})t]\,dt$$

which is the sum of two integrals with $m-1$ factors and thus ≥ 0. Furthermore, if $n_1 + \cdots + n_m$ is odd then the integral in (10) changes sign if we replace t by $t + \pi$ and is thus $= 0$.

If $s_m(t) = \sum_{k=1}^{m}\cos(q^k t)$ then

$$\int_0^{2\pi} \exp[bp s_m(t)]\,dt = \int_0^{2\pi} \prod_{k=1}^{m} \exp[bp\cos(q^k t)]\,dt\,,$$

and by (8) this is

$$= \int_0^{2\pi} \prod_{k=1}^{m} \left[I_0(bp) + 2\sum_{n=1}^{\infty} I_n(bp)\cos(nq^k t)\right]dt\,.$$

We multiply out and integrate term-by-term. We see from (7) that $I_{n_1}(bp)\cdots I_{n_m}(bp) \geq 0$ if $n_1 + \cdots + n_m$ is even so that the result is $\geq 2\pi I_0(bp)^m$. Hence we obtain from Proposition 8.13 that, for $1 - q^{-m} \leq r \leq 1 - q^{-m-1}$,

$$\int_{\mathbb{T}} |f'(r\zeta)|^p |d\zeta| \geq c_1 \int_0^{2\pi} \exp[bp s_m(t)]\,dt \geq 2\pi c_1 I_0(bp)^m$$

and (9) follows because $(m+1)\log q \geq \log[1/(1-r)]$. $\qquad\square$

Proof of Theorem 8.6. We choose $q = 15$ and $b = 1.129 < 1/B_{15}$. Then

$$\beta(-1) \geq \log I_0(1.129) / \log 15 > 0.109 \,.$$

Furthermore $\log I_0(x) \sim x^2/4$ as $x \to 0$ and thus $\beta(p) \geq 0.117p^2$ for small $|p|$ by Proposition 8.15. $\qquad\square$

2. We now list without proof some of the surprising properties of lacunary series. We assume throughout that g is given by (1) so that g has Hadamard gaps.

If g has a finite asymptotic value (see (4.2.14)) at some point of $\mathbb{T}$ then (HaLi26, Binm69)

$$(11) \qquad\qquad b_k \to 0 \quad\text{as}\quad k \to \infty$$

and thus $g \in \mathcal{B}_0$. Conversely every function in $\mathcal{B}_0$ (lacunary or not) has finite radial limits on a set of Hausdorff dimension 1; see (11.2.7).

If however g has finite radial limits on a set of positive measure then (Zyg68, I, p. 203)

$$(12) \qquad\qquad \sum_k |b_k|^2 < \infty \,.$$

Conversely if (12) holds then $g \in H^p$ for every $p < \infty$, even (Zyg68, I, p. 215)

$$(13) \qquad \sup_r \int_{\mathbb{T}} \exp\left(\alpha |g(r\zeta)|^2\right) |d\zeta| < \infty \quad\text{for small } \alpha > 0 \,.$$

Furthermore (12) implies $g \in \text{VMOA}$; see e.g. HaSa90.

If $g \in H^\infty$ then more is true, namely (Sid27; Zyg68, I, p. 247)

$$(14) \qquad\qquad \sum_k |b_k| < \infty$$

so that g is continuous in $\overline{\mathbb{D}}$. Furthermore (14) holds if (GnPo83)

$$(15) \qquad \int_\Gamma |f'(z)|\,|dz| < \infty \quad\text{for some curve } \Gamma \text{ ending on } \mathbb{T} \,.$$

It is remarkable that (14) already holds (Mur81) if $g(\mathbb{D}) \neq \mathbb{C}$. It is an open problem what $g(\mathbb{D}) \neq \mathbb{C}$ implies under weaker gap assumptions. An interesting example is the (non-Hadamard) series

$$(16) \qquad \sum_{k=0}^\infty (-1)^k (2k+1) z^{k(k+1)/2} = \prod_{n=1}^\infty (1 - z^n)^3 \neq 0$$

which comes from the theory of theta functions.

If $g \notin \mathcal{B}$, that is if

$$(17) \qquad\qquad \sup |b_k| = \infty \,,$$

then g has the asymptotic value ∞ at every point of $\mathbb{T}$; see Mur83. Note that g has the angular limit ∞ almost nowhere on $\mathbb{T}$ by the Privalov uniqueness theorem.

There is a close connection of lacunary series g to probability theory; see Kah85 for the relation to random power series. The functions $g(r\zeta)(\zeta \in \mathbb{T})$ behave like a family of weakly dependent random variables on the probability space $(\mathbb{T}, \text{Borel}, (2\pi)^{-1}\Lambda)$.

To be more precise, let (b_k) be bounded but

$$(18) \qquad b(r)^2 \equiv \sum_{k=1}^{\infty} |b_k|^2 r^{2n_k} \to \infty \qquad \text{as} \quad r \to 1-$$

so that $g \in \mathcal{B} \setminus H^2$. We define r_t by $2b(r_t)^2 = t$ for $0 \le t < \infty$. Then (PhSt75, p.60) there is a probability space $(\Omega, \mathcal{A}, P)$ and a stochastic process $S_t(0 \le t < \infty)$ with

$$(19) \qquad P(S_t \le x) = (2\pi)^{-1}\Lambda(\{\zeta \in \mathbb{T} : \operatorname{Re} g(r_t\zeta) \le x\})$$

for $x \in \mathbb{R}$ such that

$$(20) \qquad S_t = X_t + O(t^{0.47}) \quad (t \to \infty) \text{ almost surely},$$

where X_t is the standard Brownian motion (of expectation 0 and variance t). Our continuous version is equivalent to the discrete version in PhSt75 by Proposition 8.13.

There are several important consequences. We formulate them for the modulus instead of the real part. Let $b_0 = 0$. The *central limit theorem* (Zyg68, II, p.264) states that, for each set $E \subset \mathbb{T}$ with $\Lambda(E) > 0$ and $x \in \mathbb{R}$,

$$(21) \qquad \Lambda\left(\{\zeta \in E : |g(r\zeta)| \le xb(r)\}\right) \to \frac{\Lambda(E)}{\sqrt{2\pi}} \int_{-\infty}^{x} e^{-\xi^2/2}\, d\xi \qquad \text{as} \quad r \to 1.$$

The *law of the iterated logarithm* (Wei59, p.469) states that

$$(22) \qquad \limsup_{r \to 1} \frac{|g(r\zeta)|}{b(r)\sqrt{\log\log b(r)}} = 1 \quad \text{for almost all } \zeta \in \mathbb{T}.$$

3. Using these properties we finally prove two results stated earlier.

Proof of Proposition 4.12. Let f be defined by $f(0) = 0$ and

$$(23) \qquad \log f'(z) = \frac{i}{3} \sum_{k=0}^{\infty} z^{2^k} \qquad \text{for} \quad z \in \mathbb{D}.$$

It follows from Proposition 8.14 that f maps $\mathbb{D}$ conformally onto a quasidisk. Now suppose that f is isogonal at some point $e^{i\vartheta}$. Then

$$\arg f'(re^{i\vartheta}) = \frac{1}{3} \sum_{k=0}^{\infty} r^{2^k} \cos(2^k\vartheta) \to \gamma \qquad \text{as} \quad r \to 1-$$

for some γ by Proposition 4.11. Hence $\sum_k \cos(2^k \vartheta) z^{2^k}$ has a finite radial limit at 1 and it follows from (11) that $\cos(2^k \vartheta) \to 0$ as $k \to \infty$. This is impossible because $\cos(2^{k+1} \vartheta) = 2 \cos^2(2^k \vartheta) - 1$. $\qquad\square$

Proof of Theorem 8.10 (Conclusion). We now prove the second assertion. Let $q = 15$ and consider the lacunary series

$$g(z) = \sum_{k=m}^{\infty} z^{q^k} \quad \text{for} \quad z \in \mathbb{D}.$$

Then $(1 - |z|^2)|z g'(z)| \le B_q < 0.886$ as in the proof of Proposition 8.14. It is easy to see that this implies $\|g\|_B < 0.886$ if m is chosen large enough. With $b(r)$ defined by (18) we have

$$\frac{b(r)^2}{1 - r^2} = \frac{1}{1 - r^2} \sum_{k=m}^{\infty} r^{2q^k} = \sum_{n=1}^{\infty} \left(\sum_{q^k \le n, k \ge m} 1 \right) r^{2n}$$

$$\ge \sum_{n=1}^{\infty} \left(\frac{\log n}{\log q} - m \right) r^{2n} \ge \frac{1}{\log q} \sum_{n=1}^{\infty} \left(1 + \cdots + \frac{1}{n} \right) r^{2n} - \frac{M}{1 - r^2}$$

for some constant M. It follows that $b(r)^2 \ge \log[1/(1 - r^2)]/\log q - M$ and thus from (22) that, for almost all $\zeta \in \mathbb{T}$,

$$\limsup_{r \to 1} \frac{|g(r\zeta)|}{\sqrt{\log \frac{1}{1-r} \log \log \log \frac{1}{1-r}}} \ge \frac{1}{\sqrt{\log q}} > 0.685 \|g\|_B. \qquad\square$$

Exercises 8.6

1. Use (6) to construct a bounded univalent function the coefficients of which are not $O(n^{0.064 - 1})$; compare (8.4.8).

2. Construct a univalent function such that the coefficients of $1/f'$ are not $O(n^{0.064})$.

3. Let $g \in \mathcal{B}_0$ have Hadamard gaps. Show that the left-hand side of (3) tends $\to 0$ uniformly in $\mathbb{T}$ for $r_j \le r \le r_{j+1}$, $j \to \infty$.

4. Let g have Hadamard gaps. If g has a finite radial limit at $\zeta \in \mathbb{T}$, show that the series converges at ζ (HaLi26).

5. Use the fact that (15) implies (14) to prove Murai's theorem that $g(\mathbb{D}) \ne \mathbb{C}$ implies (14).

6. Let the univalent function $f(z) = \sum_0^{\infty} a_n z^n$ have Hadamard gaps. Show that $\sum n|a_n| < \infty$.

7. Let f map $\mathbb{D}$ conformally onto the inner domain of the Jordan curve J and let $\log f'$ have Hadamard gaps. If $|f'|$ is bounded show that f' is continuous and nonzero in $\overline{\mathbb{D}}$. If J has tangents on a set of positive linear measure show that J is rectifiable. (Use also Proposition 6.22 and Theorem 6.24).

Chapter 9. Curve Families and Capacity

9.1 An Overview

We introduce two important concepts. Let Γ be a family of curves in the Borel set $B \subset \mathbb{C}$. The *module* of Γ is defined by

$$(1) \qquad \operatorname{mod} \Gamma = \inf_{\rho} \iint_B \rho(z)^2 \, dx \, dy$$

where the infimum is taken over the functions $\rho \geq 0$ with

$$(2) \qquad \int_C \rho(z)|dz| \geq 1 \qquad \text{for all} \quad C \in \Gamma.$$

The quantity $1/\operatorname{mod}\Gamma$ is called the extremal length of Γ. The module concept has its root in the length-area method used by Grötzsch and later e.g. by Warschawski. It was introduced by Beurling and Ahlfors (AhBeu50) and has many applications in conformal mapping, in particular in connection with quadratic differentials (see e.g. Jen65, Kuz80, Str84). It plays a key role in the theory of quasiconformal maps (LeVi73), in particular in $\mathbb{R}^n$ (see e.g. Väi71, Vuo88).

For a compact set $E \subset \mathbb{C}$, following Fekete, we define the logarithmic *capacity* by

$$(3) \qquad \operatorname{cap} E = \lim_{n \to \infty} \max_{(z_k)} \prod \prod_{j \neq k} |z_j - z_k|^{1/n(n-1)}$$

where the maximum is taken over all point systems $z_1, \ldots, z_n \in E$. An alternative definition is through potentials; see e.g. Lan72, HayKe76. The concept of capacity can be generalized to Borel (and even Souslin) sets.

Apart from the linear measure Λ, the capacity is the most important tool to describe the size of a set in conformal mapping. Exceptional sets are often of zero capacity (Car67), and we say that a property holds *nearly everywhere* if it holds outside a set of zero capacity. If $E \subset \mathbb{T}$ then $\operatorname{cap} E \geq \sin[\Lambda(E)/4]$ with equality for an arc. More generally

$$(4) \qquad E \subset \mathbb{C}, \quad \operatorname{cap} E = 0 \quad \Rightarrow \quad \Lambda(E) = 0$$

so that "nearly everywhere" implies "almost everywhere". The converse does not hold; for example the classical Cantor set has positive capacity but zero measure. See e.g. Nev70 and Section 10.2 where we study the relation to Hausdorff measure.

Pfluger's Theorem. *If $\Gamma_E(r)$ is the family of curves that connect $\{|z| = r\}$ with the Borel set $E \subset \mathbb{T}$ then*

$$(5) \qquad \operatorname{cap} E = \lim_{r \to 0} \frac{1}{\sqrt{r}} \exp\left(-\frac{\pi}{\operatorname{mod} \Gamma_E(r)}\right).$$

This result can be used to establish many important connections between capacity and conformal mapping which are more precise than some of the estimates proved in the first chapters. For instance, if f is a conformal map of $\mathbb{D}$ then (Beu40)

$$(6) \qquad \int_0^1 |f'(r\zeta)|\, |d\zeta| < \infty \quad \text{for nearly all } \zeta \in \mathbb{T},$$

and if $I = \{e^{it} : |t - \arg \zeta| \leq \pi(1 - r)\}$ with $\zeta \in \mathbb{T}$, $0 \leq r < 1$ then

$$(7) \qquad |f(z) - f(r\zeta)| \leq R(1 - r^2)|f'(r\zeta)| \quad \text{for } z \in I \setminus E$$

where $\operatorname{cap} E < 11(1 - r)/\sqrt{R}$; see Corollary 9.20. If f maps $\mathbb{D}$ conformally into $\mathbb{D}$ with $f(0) = 0$ and if $A \subset \mathbb{T}$, $f(A) \subset \mathbb{T}$ then (Theorem 9.21)

$$(8) \qquad \operatorname{cap} f(A) \geq |f'(0)|^{-1/2} \operatorname{cap} A \geq \operatorname{cap} A.$$

This is an analogue of Löwner's lemma that $\Lambda(f(A)) \geq \Lambda(A)$.

9.2 The Module of a Curve Family

Let B be a Borel set in $\mathbb{C}$. A *curve family* Γ in B consists of open, halfopen or closed arcs or curves in B. The function ρ in B is called *admissible* for Γ if it is non-negative and Borel measurable in B and if

$$(1) \qquad \int_C \rho(z)|dz| \geq 1 \quad \text{for all } C \in \Gamma.$$

Thus the curves C in Γ are required to have length ≥ 1 with respect to the metric $\rho(z)|dz|$; the length may be infinite.

We define the *module* of the curve family Γ by

$$(2) \qquad \operatorname{mod} \Gamma = \inf_\rho \iint_B \rho(z)^2 \, dx \, dy,$$

where the infimum is taken over all functions ρ admissible for Γ. Note that the integral is the area of B with respect to $\rho(z)|dz|$. If $B \subset B'$ then Γ is also a curve family in B' with the same module; we set $\rho(z) = 0$ for $z \in B' \setminus B$.

Example 1. Let Γ be the family of all curves C in the rectangle

(3)
$$G = \{x + iy : 0 < x < a, 0 < y < b\}$$

that connect its two horizontal sides; see Fig. 9.1 (i). Then $\rho_0(z) = 1/b$ $(z \in G)$ is admissible because every curve in Γ has length $\geq b$. It follows that

(4)
$$\mod \Gamma \leq \iint_G \rho_0(z)^2 \, dx \, dy = \frac{a}{b}.$$

Now let ρ be any admissible function. Since the vertical segments $(x, x + ib)$ belong to Γ for $0 < x < a$ we obtain from (1) by Schwarz's inequality that

$$1 \leq \left(\int_0^b \rho(x + iy) \, dy \right)^2 \leq b \int_0^b \rho(x + iy)^2 \, dy$$

and thus

$$\frac{a}{b} \leq \int_0^a \left(\int_0^b \rho(x + iy)^2 \, dy \right) dx = \iint_G \rho(z)^2 \, dx \, dy.$$

Hence we conclude from (2) and (4) that

(5)
$$\mod \Gamma = a/b.$$

Example 2. We consider the annulus

(6)
$$H = \{z : r_1 < |z| < r_2\} \quad (0 \leq r_1 < r_2 < \infty).$$

Let Γ be the family of curves in H that connect the two boundary components and Γ' the family of curves in H that separate them, see Fig. 9.1 (ii). Then (see e.g. Jen65, p. 18)

(7)
$$\mod \Gamma = 2\pi / \log \frac{r_2}{r_1}, \quad \mod \Gamma' = \frac{1}{2\pi} \log \frac{r_2}{r_1};$$

if $r_1 = 0$ then $\mod \Gamma = 0$.

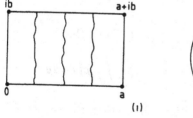
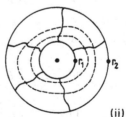

Fig. 9.1. The families of Examples 9.1 and 9.2

An important property of the module is its *conformal invariance.* The image $f(\Gamma)$ consists of all image curves $f(C)$ where $C \in \Gamma$.

Proposition 9.1. *Let Γ be a curve family in the domain G. If f maps G conformally into $\mathbb{C}$ then*

$$(8) \qquad\qquad \mod f(\Gamma) = \mod \Gamma.$$

Proof. Let ρ^* be admissible for $f(\Gamma)$. We define $\rho(z) = \rho^*(f(z))|f'(z)|$ for $z \in G$. If $C \in \Gamma$ then

$$(9) \qquad \int_C \rho(z)|dz| = \int_C \rho^*(f(z))|f'(z)|\,|dz| = \int_{f(C)} \rho^*(w)|dw| \geq 1$$

so that ρ is admissible for Γ. Since f is a conformal map we have

$$(10) \quad \iint_G \rho(z)^2\, dx\, dy = \iint_G \rho^*(f(z))^2 |f'(z)|^2\, dx\, dy = \iint_{f(G)} \rho^*(w)^2\, du\, dv$$

and therefore $\mod \Gamma \leq \mod f(\Gamma)$. The reverse inequality is obtained by considering f^{-1}. $\square$

The module is monotone and subadditive.

Proposition 9.2. *If $\Gamma \subset \Gamma'$ or if, for every $C \in \Gamma$, there exists $C' \in \Gamma'$ with $C' \subset C$, then $\mod \Gamma \leq \mod \Gamma'$. Furthermore*

$$(11) \qquad\qquad \mod\left(\bigcup_k \Gamma_k\right) \leq \sum_k \mod \Gamma_k.$$

In particular $\mod \Gamma_0 = 0 \Rightarrow \mod(\Gamma \cup \Gamma_0) = \mod \Gamma$.

Proof. If ρ is admissible for Γ' then ρ is also admissible for Γ because we can find $C' \in \Gamma'$ with $C' \subset C$. Hence $\mod \Gamma \leq \mod \Gamma'$ follows from (2). To prove (11), let ρ_k be admissible for Γ_k. We set $\rho_k(z) = 0$ where it is undefined. Then

$$\rho \equiv \left(\sum_j \rho_j^2\right)^{1/2} \geq \rho_k \qquad \text{for all} \quad k$$

so that ρ is admissible for $\Gamma = \bigcup \Gamma_k$, and (2) shows that

$$\mod \Gamma \leq \iint \rho^2\, dx\, dy = \sum_k \iint \rho_k^2\, dx\, dy$$

which implies (11). The final assertion follows from the fact that

$$\mod \Gamma \leq \mod(\Gamma \cup \Gamma_0) \leq \mod \Gamma + \mod \Gamma_0.$$ $\square$

Proposition 9.3 *Let Γ_k be curve families in disjoint Borel sets B_k. If $\bigcup_k \Gamma_k \subset \Gamma$ then*

(12)
$$\sum_k \operatorname{mod} \Gamma_k \le \operatorname{mod} \Gamma.$$

If Γ is a curve family in B and if each $C \in \Gamma$ contains some $C_k \in \Gamma_k$ for every k, then

(13)
$$\frac{1}{\operatorname{mod} \Gamma} \ge \sum_k \frac{1}{\operatorname{mod} \Gamma_k}.$$

The quantity $1/\operatorname{mod} \Gamma$ is called the *extremal length* of Γ and is often used in preference to the module; see e.g. Ahl73.

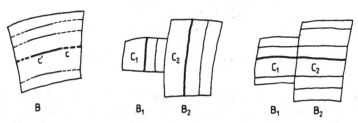

Fig. 9.2. The situations considered in Proposition 9.2 and Proposition 9.3

Proof. We restrict ourselves to prove the second assertion and write $a_k = 1/\operatorname{mod}\Gamma_k$, $a = \sum_j a_j$. We may assume that $a < \infty$; the case $a = \infty$ is obtained as a limit. If ρ_k is admissible for Γ_k we define

$$\rho(z) = a^{-1} a_k \rho_k(z) \quad \text{for} \quad z \in B_k, \quad \rho(z) = 0 \quad \text{for} \quad z \in B \setminus \bigcup B_j.$$

For $C \in \Gamma$ we have $C \supset \bigcup C_k$ with $C_k \in \Gamma_k$ and thus

$$\int_C \rho |dz| \ge \sum_k a^{-1} a_k \int_{C_k} \rho_k |dz| \ge \sum_k a^{-1} a_k = 1.$$

Thus ρ is admissible for Γ. Furthermore

$$\iint_B \rho^2 \, dx \, dy \le \sum_k \iint_{B_k} \rho^2 \, dx \, dy = a^{-2} \sum_k a_k^2 \iint_{B_k} \rho_k^2 \, dx \, dy$$

and thus, by (2),

$$\operatorname{mod} \Gamma \le a^{-2} \sum_k a_k^2 \operatorname{mod} \Gamma_k = a^{-2} \sum_k a_k = a^{-1}$$

which implies (13).   □

Next we consider modules and symmetry; see Fig. 9.3.

Proposition 9.4. *Let the domain G be symmetric with respect to $\mathbb{R}$ and let $G^+ = \{\operatorname{Im} z > 0\} \cap G$, $A^+ \subset \{\operatorname{Im} z > 0\} \cap \partial G$, furthermore G^-, A^- the reflected sets in $\{\operatorname{Im} z < 0\}$. If*

$$\Gamma = \{curves\ C\ in\ G\ from\ A^-\ to\ A^+\}$$
$$\Gamma^\pm = \{curves\ C\ in\ G^\pm\ from\ \mathbb{R}\ to\ A^\pm\}$$

then $\mathrm{mod}\,\Gamma^+ = \mathrm{mod}\,\Gamma^- = 2\,\mathrm{mod}\,\Gamma$.

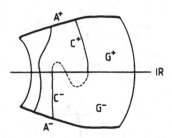

Fig. 9.3. A domain G symmetric to the real axis

Proof. It is easy to see that $\mathrm{mod}\,\Gamma^+ = \mathrm{mod}\,\Gamma^-$. Since every $C \in \Gamma$ satisfies $C \supset C^+ \cup C^-$ with $C^\pm \in \Gamma^\pm$, we conclude from (13) that $\mathrm{mod}\,\Gamma^+ \geq 2\,\mathrm{mod}\,\Gamma$.

Conversely let ρ be admissible for Γ and define $\rho^+(z) = \rho(z) + \rho(\bar{z})$ for $z \in G^+$. Let $C^+ \in \Gamma^+$ and let C^- be the reflected curve in G^-. Then $C^+ \cup (\overline{C}^+ \cap \overline{C}^-) \cup C^- \in \Gamma$ and thus

$$\int_{C^+} \rho^+ |dz| = \int_{C^+} \rho(z)|dz| + \int_{C^-} \rho(z)|dz| \geq 1$$

so that ρ^+ is admissible for Γ^+. Furthermore

$$\iint_{G^+} \rho^+(z)^2\,dx\,dy = \iint_{G^+} \left[\rho(z)^2 + 2\rho(z)\rho(\bar{z}) + \rho(\bar{z})^2\right]\,dx\,dy$$
$$\leq 2 \iint_{G^+} \rho(z)^2\,dx\,dy + 2 \iint_{G^-} \rho(z)^2\,dx\,dy$$
$$= 2 \iint_G \rho^2\,dx\,dy$$

and therefore $\mathrm{mod}\,\Gamma^+ \leq 2\,\mathrm{mod}\,\Gamma$. $\qquad\square$

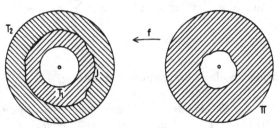

Fig. 9.4

We now prove one of the *Teichmüller module theorems* (Tei38); see Fig. 9.4 and also (18), (19) below.

Proposition 9.5. *Let* $0 < r_1 < r_2$ *and* $T_j = \{|z| = r_j\}$ *for* $j = 1, 2$. *Let the Jordan curve* J *separate* T_1 *and* T_2 *and let* Γ_j *denote the family of Jordan curves that separate* T_j *and* J. *If*

(14) $\operatorname{mod} \Gamma_1 + \operatorname{mod} \Gamma_2 > (2\pi)^{-1} \log(r_2/r_1) - \varepsilon \quad (0 < \varepsilon < 1/36)$

then, with an absolute constant K,

(15) $$1 \le \max_{z \in J} |z| / \min_{z \in J} |z| < 1 + K \sqrt{\varepsilon \log \frac{1}{\varepsilon}}\,.$$

It follows from (7) and (12) that always

(16) $$(2\pi)^{-1} \log(r_2/r_1) \ge \operatorname{mod} \Gamma_1 + \operatorname{mod} \Gamma_2\,,$$

and our proposition states that approximate equality implies that J is approximately a concentric circle.

Proof. Let the functions

$$f(\zeta) = \sum_{n=1}^{\infty} a_n \zeta^n\,, \quad g(\zeta) = b\zeta^{-1} + \sum_{n=0}^{\infty} b_n \zeta^n\,, \quad a_1 > 0\,, \quad b > 0$$

map $\mathbb{D}$ conformally onto the inner and outer domain of J. For $r > r_2$ let $\Gamma(r)$ denote the family of Jordan curves separating J and $\{|z| = r\}$. Since the preimage of $\{|z| = r\}$ is approximately $\{|\zeta| = b/r\}$ for large r, we easily obtain from (7), Proposition 9.3 and Proposition 9.1 that

$$2\pi \operatorname{mod} \Gamma_2 - \log r_2 \le \lim_{r \to \infty} (2\pi \operatorname{mod} \Gamma(r) - \log r) = -\log b$$

with a similar estimate for $\operatorname{mod} \Gamma_1$. Hence we see from (14) that

(17) $\log(b/a_1) \le \log(r_2/r_1) - 2\pi(\operatorname{mod} \Gamma_1 + \operatorname{mod} \Gamma_2) < 2\pi\varepsilon\,.$

The area of the inner domain of J is $\pi \sum_1^{\infty} n|a_n|^2$ and also equal to $\pi(b^2 - \sum_1^{\infty} n|b_n|^2)$. Hence

$$\sum_{n=2}^{\infty} n|a_n|^2 + \sum_{n=1}^{\infty} n|b_n|^2 = b^2 - a_1^2$$

and we see from Schwarz's inequality that, for $|\zeta| = \rho = a_1/b$,

$$\left| g(\zeta) - b\zeta^{-1} - b_0 \right|^2 \le (b^2 - a_1^2) \sum_{n=1}^{\infty} \frac{\rho^{2n}}{n} = b^2(1 - \rho^2) \log \frac{1}{1 - \rho^2}\,.$$

Since $\rho > e^{-2\pi\varepsilon} > 0.83$ by (17) and thus $1 - \rho^2 < e^{-1}$, it follows that

$$|g(\zeta) - b\zeta^{-1} - b_0| \le b\eta \equiv bK_1\sqrt{\varepsilon \log \frac{1}{\varepsilon}},$$

similarly $|f(\zeta) - a_1\zeta| \le b\eta$. Hence (17) implies that J lies between a circle of radius $b\rho^{-1} + b\eta < be^{2\pi\varepsilon} + b\eta$ and a circle of radius $a_1\rho - b\eta > be^{-4\pi\varepsilon} - b\eta$ around 0. $\square$

There are also estimates in the opposite direction due to Grötzsch and Teichmüller; see e.g. LeVi73, p. 53–61 for the proofs.

Let G be a doubly connected domain and Γ the family of Jordan curves that separate the two boundary components. If $G \subset \mathbb{D}$ and if G separates $\{0, z_1\}$ from $\mathbb{T}$ then

(18) $$2\pi \operatorname{mod} \Gamma < \log(4/|z_1|).$$

If $G \subset \mathbb{C}$ separates $\{0, z_1\}$ from $\{z_2, \infty\}$ then

(19) $$2\pi \operatorname{mod} \Gamma < \log \frac{16(|z_1| + |z_2|)}{|z_1|}.$$

Exercises 9.2

1. Use conformal mapping onto a rectangle to prove Proposition 9.4 for the case of a Jordan domain.

2. Deduce the second formula (7) from (5) and Proposition 9.4.

3. Suppose that f maps $\mathbb{D}$ conformally onto the open kernel of

$$\bigcup_{k=1}^{n} \{k - 1 \le x \le k, \, 0 \le y \le y_k\}$$

such that the arc $\{e^{it} : 3\pi/4 \le t \le 5\pi/4\}$ is mapped onto $[0, iy_1]$ while $\{e^{it} : -\pi/4 \le t \le \pi/4\}$ is mapped onto $[n, n + iy_n]$. Give two proofs that

$$\sum_{k=1}^{n} 1/y_k \le 1.$$

4. Let the Jordan curve J be partitioned into four consecutive arcs A_1, A_2, A_3, A_4 and let G be the inner domain of J. Consider two curve families to prove that

$$\operatorname{dist}(A_1, A_3) \operatorname{dist}(A_2, A_4) \le \operatorname{area} G.$$

(See McM71 for much deeper related results.)

5. Let h be a κ-quasiconformal map from H to H^* where $0 \le \kappa < 1$. If Γ is a curve family in H show that

$$\operatorname{mod} h(\Gamma) \le [(1 + \kappa)/(1 - \kappa)] \operatorname{mod} \Gamma.$$

9.3 Capacity and Green's Function

1. Let E be a compact set in $\mathbb{C}$ and G its outer domain, i.e. the component of $\widehat{\mathbb{C}} \setminus E$ with $\infty \in G$. For $n = 2, 3, \ldots$ we consider

$$(1) \qquad \Delta_n = \Delta_n(E) = \max_{z_1,\ldots,z_n \in E} \prod_{\substack{k=1 \\ k \neq j}}^{n} \prod_{j=1}^{n} |z_k - z_j|.$$

The maximum is assumed for the *Fekete points*

$$(2) \qquad z_k = z_{nk} \in E \quad (k = 1, \ldots, n);$$

this system of points is not always uniquely determined, and the maximum principle shows that $z_{nk} \in \partial G$. We can write

$$(3) \qquad \Delta_n = \left| \det_{k=1,\ldots,n} \left(1 \; z_{nk} \; \ldots \; z_{nk}^{n-1}\right) \right|^2;$$

this is a Vandermonde determinant. We call

$$(4) \qquad q_n(z) = \prod_{k=1}^{n} (z - z_{nk})$$

the nth *Fekete polynomial* and write

$$(5) \qquad \gamma_n = \gamma_n(E) = \min_k |q_n'(z_{nk})| = \min_k \prod_{j \neq k} |z_{nk} - z_{nj}|.$$

If the minimum in (5) is assumed for $k = \nu$ then

$$\Delta_n = \prod_{k \neq \nu} |z_{n\nu} - z_{nk}|^2 \prod_{\substack{k \neq j \\ k,j \neq \nu}} |z_{nk} - z_{nj}| \leq \gamma_n^2 \Delta_{n-1}$$

by (1). Hence

$$(6) \qquad (\Delta_n / \Delta_{n-1})^{n/2} \leq \gamma_n^n \leq \Delta_n;$$

the second inequality follows from (5). We conclude that $\Delta_n^{1/[n(n-1)]}$ is decreasing and call

$$(7) \qquad \operatorname{cap} E = \lim_{n \to \infty} \Delta_n(E)^{\frac{1}{n(n-1)}}$$

the (logarithmic) *capacity* of E. It is also called the "transfinite diameter". See e.g. Lej61 for generalizations.

The next properties follow at once from the corresponding properties of the quantity Δ_n:

$$(8) \qquad \operatorname{cap} E = \operatorname{cap} \partial G;$$

(9) $$E_1 \subset E_2 \quad \Rightarrow \quad \operatorname{cap} E_1 \le \operatorname{cap} E_2 ;$$

(10) $$\operatorname{cap} \varphi(E) = |a| \operatorname{cap} E \quad \text{if} \quad \varphi(z) = az + b;$$

(11) $$\operatorname{cap} \varphi(E) \le \operatorname{cap} E \quad \text{if } \varphi \text{ is a contraction of } E.$$

Proposition 9.6. *If $E \subset \mathbb{C}$ is compact then*

(12) $$\operatorname{cap} E \le \max_{z \in E} |p(z)|^{1/N}$$

for every polynomial $p(z) = z^N + \dots$, furthermore

(13) $$\operatorname{cap} E = \lim_{n \to \infty} \gamma_n^{1/n} = \lim_{n \to \infty} \max_{z \in E} |q_n(z)|^{1/n} .$$

Proof. Let $|p(z)| \le M$ for $z \in E$ and let $m = 1, 2, \dots$. We write $n = mN$ and consider the polynomials

(14) $$p_{\mu N + \nu}(z) \equiv p(z)^{\mu} z^{\nu} = z^{\mu N + \nu} + \dots \quad \text{for} \quad 0 \le \mu < m, \quad 0 \le \nu < N.$$

Adding suitable multiples of the first $j - 1$ columns of the Vandermonde determinant in (3) to the jth column we see that

$$\Delta_n = \left| \det_{k=1,\dots,n} \left(1\, p_1(z_{nk}) \dots p_{n-1}(z_{nk}) \right) \right|^2 .$$

We now apply Hadamard's determinant inequality

(15) $$\left| \det_{k=1,\dots,n} (a_{k1}\, a_{k2} \dots a_{kn}) \right|^2 \le \prod_{j=1}^{n} \left(\sum_{k=1}^{n} |a_{kj}|^2 \right) .$$

If $E \subset D(0, r)$ we obtain from (14) that

$$\Delta_n \le \prod_{j=0}^{n-1} \left(\sum_{k=1}^{n} |p_j(z_{nk})|^2 \right)$$
$$\le n^n \prod_{\mu=0}^{m-1} \prod_{\nu=0}^{N-1} \left(r^{2\nu} M^{2\mu} \right) = n^n r^{N(N-1)m} M^{Nm(m-1)} .$$

Taking the $n(n-1)$-th root and using that $n = Nm$, we deduce for $m \to \infty$ that $\operatorname{cap} E \le M^{1/N}$ which proves (12).

Furthermore (1) and (4) show that, for $z \in E$,

$$|q_n(z)|^2 \Delta_n = \prod_{j=1}^{n} |z - z_{nj}|^2 \prod_{k \ne j} \prod |z_{nk} - z_{nj}| \le \Delta_{n+1} .$$

Hence we obtain from (6) and (12) that

(16) $$\operatorname{cap} E \le \max_{z \in E} |q_n(z)|^{\frac{1}{n}} \le \left(\frac{\Delta_{n+1}}{\Delta_n} \right)^{\frac{1}{2n}} \le \gamma_{n+1}^{\frac{1}{n}} \le \Delta_{n+1}^{\frac{1}{n(n+1)}}$$

and (13) follows from (7) for $n \to \infty$. $\square$

2. We now introduce the Green's function with pole at ∞ by the method of Fekete points (Myr33, Lej34). A probability measure μ on the compact set E is a measure with $\mu(E) = 1$ and $\mu(B) \geq 0$ for all Borel sets $B \subset E$.

Theorem 9.7. *Let G be a domain with $\infty \in G$ and let $E = \partial G$ and $\operatorname{cap} E > 0$. Then there is a probability measure μ on E with the following properties: The function*

$$(17) \quad g(z) = \int_E \log|z - \zeta|\, d\mu(\zeta) - \log \operatorname{cap} E = \log|z| - \log \operatorname{cap} E + O\left(\frac{1}{|z|}\right)$$

is subharmonic in $\mathbb{C}$ and harmonic in $G \setminus \{\infty\}$. It satisfies

$$(18) \quad g(z) \geq 0 \quad \text{for} \quad z \in \overline{G}, \quad g(z) = 0 \quad \text{for} \quad z \in \mathbb{C} \setminus \overline{G}.$$

If u is any positive harmonic function in G with $u(z) = \log|z| + O(1)$ as $z \to \infty$ then

$$(19) \quad u(z) \geq g(z) > 0 \quad \text{for} \quad z \in G,$$

and this property uniquely determines g in G.

We call μ an *equilibrium measure* of E and g the (extended) *Green's function* with pole at ∞. Since it is subharmonic and thus *upper semicontinuous* we have

$$(20) \quad g(\zeta) = \limsup_{z \to \zeta} g(z) \quad \text{for} \quad \zeta \in \mathbb{C}.$$

Proof. (a) For $n = 2, 3, \ldots$ let $z_{n1}, \ldots, z_{nn}$ be Fekete points of E and q_n the corresponding Fekete polynomial. We introduce

$$(21) \quad v_n(z) = \frac{1}{n} \log\left[|q_n(z)|/\gamma_n\right] \quad \text{for} \quad z \in \mathbb{C},$$

where $v_n(z_{nk}) = -\infty$. It follows from Proposition 9.6 by the maximum principle that

$$(22) \quad \max_{z \in \mathbb{C} \setminus G} v_n(z) = \log \max_{z \in E} |q_n(z)|^{1/n} - \log(\gamma_n^{1/n}) \to 0 \quad \text{as} \quad n \to \infty.$$

The Lagrange interpolation formula shows that

$$1 = \sum_{k=1}^{n} \frac{q_n(z)}{q_n'(z_{nk})(z - z_{nk})} \quad \text{for} \quad z \notin E$$

so that, by (5),

$$1 \leq n|q_n(z)|/[\gamma_n \operatorname{dist}(z, E)].$$

Hence we see from (21) that, for each $z \notin E$,

$$(23) \quad \liminf_{n \to \infty} v_n(z) \geq \liminf_{n \to \infty} \frac{1}{n}(\log \operatorname{dist}(z, E) - \log n) = 0.$$

(b) We define probability measures μ_n on E by assigning to each Fekete point $z_{n1}, \ldots, z_{nn}$ the mass $1/n$ so that, by (4) and (21),

$$(24) \qquad v_n(z) = \int_E \log |z - \zeta| \, d\mu_n(\zeta) - \frac{1}{n} \log \gamma_n \qquad \text{for} \quad z \in \mathbb{C}.$$

The space of all probability measures on E is weakly compact (e.g. HayKe76, p. 205), i.e. there exists a subsequence (μ_{n_j}) and a probability measure μ on E such that

$$(25) \qquad \int_E \psi(\zeta) \, d\mu_{n_j}(\zeta) \rightarrow \int_E \psi(\zeta) \, d\mu(\zeta) \qquad \text{as} \quad j \rightarrow \infty$$

for every function ψ continuous on E.

We now define g by (17). Since $\log |z - \zeta|$ is subharmonic and $d\mu(\zeta) \geq 0$ it follows that g is subharmonic in $\mathbb{C}$. It follows from (13), (24) and (25) that, as $n = n_j \rightarrow \infty$,

$$(26) \qquad v_n(z) \rightarrow g(z) \qquad \text{for} \quad z \in \mathbb{C} \setminus E$$

and therefore from (23) that $g(z) \geq 0$ holds for $z \in \mathbb{C} \setminus E$ and thus also for $z \in E = \partial G$ by (20). Hence we conclude from (22) and (26) that $g(z) = 0$ for $z \in (\mathbb{C} \setminus G) \setminus E = \mathbb{C} \setminus \overline{G}$. It is clear from (17) that g is harmonic in $G \setminus \{\infty\}$. Since g is nonnegative and nonconstant in the domain G it follows that $g(z) > 0$ for $z \in G$.

(c) Finally let u be positive and harmonic in G with $u(z) = \log |z| + O(1)$ as $z \rightarrow \infty$. We apply the maximum principle to the harmonic function $v_{n_j} - u$ in G. Since $v_{n_j} - u < v_{n_j}$, we conclude from (22) and (26) that $g - u \leq 0$ in G which proves (19).

If $\tilde{g}$ is another positive harmonic function in G with $\tilde{g}(z) = \log |z| + O(1)$ as $z \rightarrow \infty$ that satisfies (19), it follows that $\tilde{g} \leq g$ and also that $g \leq \tilde{g}$, hence $\tilde{g} = g$ in G. Consequently (26) holds for all $n \rightarrow \infty$; the case that $z \in \mathbb{C} \setminus \overline{G}$ is trivial by (18) and (22). $\qquad \square$

The next result is a partial converse.

Theorem 9.8. *Let u be positive and harmonic in the domain G and let*

$$(27) \qquad u(z) = \log |z| + a + O(|z|^{-1}) \qquad as \quad z \rightarrow \infty.$$

Then $\operatorname{cap} E \geq e^{-a} > 0$ where $E = \partial G$. If furthermore $u(z) \rightarrow 0$ as $z \rightarrow \partial G$ then $u = g$ and $\operatorname{cap} E = e^{-a}$.

Proof. As in part (c) of the foregoing proof we see that $\lim v_n \leq u$ in G and thus, by (24) and (27),

$$\frac{1}{n} \log \gamma_n = \lim_{z \to \infty} (\log |z| - v_n(z)) \geq \lim_{z \to \infty} (\log |z| - u(z)) = -a > -\infty.$$

Hence $\operatorname{cap} E \geq e^{-a} > 0$ by (5) and Proposition 9.6 so that g exists.

Now assume that $u(z) \to 0$ as $z \to \partial G$. It follows from (27) and Theorem 9.7 that $v \doteq u - g \geq 0$ in G. Since v is harmonic in G including ∞ and since $0 \leq v(z) \leq u(z) \to 0$ as $z \to \partial G$, we obtain from the maximum principle that $v = 0$ and thus $u = g$. Finally $a = -\log \operatorname{cap} E$ is a consequence of (17) and (27). $\qquad\square$

For simply connected domains the Green's function is closely related to conformal maps. Let again $\mathbb{D}^* = \{|z| > 1\} \cup \{\infty\}$.

Corollary 9.9. *If*

$$(28) \qquad h(\zeta) = c\zeta + c_0 + c_1\zeta^{-1} + \ldots$$

maps $\mathbb{D}^*$ conformally onto G then $\operatorname{cap} \partial G = |c|$ and

$$(29) \qquad g(z) = \log|h^{-1}(z)| \quad \text{for} \quad z \in G, \quad g(z) = 0 \quad \text{for} \quad z \in \mathbb{C} \setminus G.$$

This follows at once from Theorem 9.8 applied to the positive harmonic function $u = \log|h^{-1}|$. We consider two examples.

The function $h(\zeta) = \zeta + \zeta^{-1}$ maps $\mathbb{D}^*$ conformally onto $[-2, 2]$. Hence it follows from (10) and Corollary 9.9 that

$$(30) \qquad \operatorname{cap}[a, b] = |b - a|/4 \quad \text{for} \quad a, b \in \mathbb{C}.$$

Let $0 < \alpha \leq \pi$ and $a = \sin(\alpha/2)$. Then

$$h(\zeta) = \zeta(a\zeta + 1)/(\zeta + a) = a\zeta + (1 - a^2) + \ldots$$

maps $\mathbb{D}^*$ conformally onto the complement of a circular arc and we obtain from Corollary 9.9 that

$$(31) \qquad \operatorname{cap}\{e^{i\vartheta} : -\alpha \leq \vartheta \leq \alpha\} = \sin(\alpha/2) \quad (0 < \alpha \leq \pi).$$

In particular $\operatorname{cap} \overline{\mathbb{D}} = \operatorname{cap} \mathbb{T} = 1$.

These examples are extremal in the following result (Pól28, AhBeu50).

Corollary 9.10. *If P is the orthogonal projection of the compact set E onto some line then*

$$(32) \qquad \Lambda(P) \leq 4\operatorname{cap} P \leq 4\operatorname{cap} E.$$

If E is a compact set on $\mathbb{T}$ then

$$(33) \qquad \sin[\Lambda(E)/4] \leq \operatorname{cap} E.$$

To prove (32) we may assume that we project onto $\mathbb{R}$. Since $z \mapsto \operatorname{Re} z$ is a contraction of E onto P and since

$$\varphi(x) = \Lambda(P \cap (-\infty, x]) \quad (x \in P)$$

is a contraction of P onto a segment of length $\Lambda(P)$, we conclude from (11) and (30) that

$$\operatorname{cap} E \geq \operatorname{cap} P \geq \operatorname{cap} \varphi(P) = \Lambda(P)/4 .$$

See AhBeu50 or Pom75, p. 337 for the proof of (33).

Proposition 9.11. *Let $E \subset \mathbb{C}$ be compact and $\operatorname{cap} E > 0$. If $\operatorname{diam} E < b$ and if μ is an equilibrium measure of E then*

$$(34) \qquad \mu(A) \leq \frac{\log(b/\operatorname{cap} E)}{\log(b/\operatorname{cap} A)} \quad \text{for compact } A \subset E ,$$

in particular $\mu(A) = 0$ if $\operatorname{cap} A = 0$.

Note that $\operatorname{cap} A \leq \operatorname{cap} E \leq \Delta_2^{1/2} \leq \operatorname{diam} E < b$; see (7).

Proof. We may assume that $a = \mu(A) > 0$ and $b = 1$, by (10). The function

$$(35) \quad u(z) = \frac{1}{a} \int_A \log|z - \zeta| \, d\mu(\zeta) - \frac{\log \operatorname{cap} E}{a} = \log|z| - \frac{\log \operatorname{cap} E}{a} + O\left(\frac{1}{|z|}\right)$$

is harmonic in $\mathbb{C} \setminus A$. Since $A \subset E$ and $\operatorname{diam} E < 1$, we see from (17) and (18) that

$$\liminf_{z \to A} u(z) = \liminf_{z \to A} \left(\frac{1}{a} \int_{E \setminus A} \log \frac{1}{|z - \zeta|} \, d\mu(\zeta) + \frac{g(z)}{a}\right) \geq 0$$

and we conclude that $u(z) > 0$ in the outer domain H of A. Hence it follows from Theorem 9.8 applied to H and from (35) that $\operatorname{cap} A \geq \exp(a^{-1} \log \operatorname{cap} E)$ which is equivalent to (34). $\qquad \square$

3. Finally we consider arbitrary sets in $\mathbb{C}$. The (*inner*) *capacity* of E is defined by

$$(36) \qquad \operatorname{cap} E = \sup\{\operatorname{cap} A : A \text{ compact}, \ A \subset E\} .$$

It is clear that $E_1 \subset E_2$ implies $\operatorname{cap} E_1 \leq \operatorname{cap} E_2$.

Theorem 9.12. *If E is a Borel or Souslin set then, for every $\varepsilon > 0$, there is an open set H with*

$$(37) \qquad E \subset H, \quad \operatorname{cap} H < \operatorname{cap} E + \varepsilon .$$

This important result (Cho55) states that Souslin sets are "capacitable". See e.g. HayKe76 for the proof. We deduce a subadditivity property.

Corollary 9.13. *Let E_n be Borel sets and $E = \bigcup_{n=1}^{\infty} E_n$. If $\operatorname{diam} E < b$ then*

$$(38) \qquad 1/\log \frac{b}{\operatorname{cap} E} \leq \sum_n 1/\log \frac{b}{\operatorname{cap} E_n} .$$

In particular the union of countably many sets of capacity zero has again capacity zero.

Proof. By Theorem 9.12 there are open sets $H_n \supset E_n$ such that

$$(39) \qquad 1/\log \frac{b}{\operatorname{cap} H_n} < 1/\log \frac{b}{\operatorname{cap} E_n} + \frac{\varepsilon}{2^n} \quad (n = 1, 2, \dots).$$

Then $E \subset H = \bigcup_n H_n$ and we may assume that $\operatorname{diam} H < b$. Every compact subset A of H can be covered by finitely many disks D_k such that $\overline{D}_k \subset H_{n_k}$ for some n_k. The compact sets $A_n = A \cap \bigcup_{k:n_k=n} \overline{D}_k \subset H_n$ satisfy $A = A_1 \cup \cdots \cup A_m$ for some m. If μ is an equilibrium measure of A then $1 = \mu(A) \le \mu(A_1) + \cdots + \mu(A_m)$ and therefore

$$1/\log \frac{b}{\operatorname{cap} A} \le \sum_{n=1}^{m} 1/\log \frac{b}{\operatorname{cap} A_n} < \sum_{n=1}^{m} 1/\log \frac{b}{\operatorname{cap} E_n} + \varepsilon$$

by (34), (39) and because of $A_n \subset H_n$. Hence (38) follows from (36). The final statement is an immediate consequence. $\qquad\Box$

We say that a property holds for *nearly all* $\zeta \in E$ if it holds for all $\zeta \in E$ except for a set of zero capacity. To say that a property holds nearly everywhere is much stronger than to say that it holds almost everywhere because a set of capacity zero has linear measure zero and even α-dimensional Hausdorff measure zero for every $\alpha > 0$; see (10.1.3).

We now show (Fro35) that the Green's function g vanishes nearly everywhere on $E = \partial G$; compare also Exercise 7.

Theorem 9.14. *Let G be a domain with $\infty \in G$ and $\operatorname{cap} \partial G > 0$. Then*

$$(40) \qquad g(z) \to 0 \quad as \quad z \to \zeta, z \in G \quad for \quad \zeta \in \partial G \setminus X$$

where X is an F_σ-set with $\operatorname{cap} X = 0$.

Proof. It follows from (4) and Proposition 9.6 that

$$\int_E \frac{1}{n} \sum_{j=1}^{n} \log |z_{nj} - \zeta| \, d\mu(\zeta) \le \max_{\zeta \in E} \frac{1}{n} \log |q_n(\zeta)| \to \log \operatorname{cap} E$$

as $n \to \infty$ and therefore, by (17),

$$(41) \qquad 0 \le \inf_{z \in E} g(z) \le \liminf_{n \to \infty} \frac{1}{n} \sum_{j=1}^{n} g(z_{nj}) \le 0.$$

Furthermore the sets

$$(42) \qquad A_k = \left\{ \zeta \in E : \limsup_{z \to \zeta, z \in G} g(z) \ge \frac{1}{k} \right\} \quad (k = 1, 2 \dots)$$

are compact. Now suppose that $\operatorname{cap} A_k > 0$ for some k and let g_k be the Green's function of A_k. Since G lies in the outer domain of A_k where g_k is positive, it follows from (19) that $g_k \geq g$ in G. Since $A_k \subset \partial G$ and g_k is upper semicontinuous we deduce from (42) that, for each $\zeta \in A_k$,

$$g_k(\zeta) \geq \limsup_{z \to \zeta, z \in G} g_k(z) \geq \limsup_{z \to \zeta, z \in G} g(z) \geq \frac{1}{k} > 0$$

in contradiction to (41) applied to g_k. Thus $\operatorname{cap} A_k = 0$ for all k, and (42) shows that $g(z) \to 0$ as $z \to \zeta$, $z \in G$ for $\zeta \in E \setminus X$ where $X = \bigcup A_k$. $\quad\square$

There is a great number of further results on Green's function; see e.g. BrCh51, duP70, Lan72, HayKe76.

Exercises 9.3

1. Use the Hadamard determinant inequality to show that $\exp(i\alpha + 2\pi i k/n)$ gives a system of Fekete points of $\mathbb{T}$ for each α and deduce that $\Delta_n(\mathbb{T}) = n^n$ and $\gamma_n(\mathbb{T}) = n$.

2. Use the contraction property to prove (33) for the special case that E lies in one half of $\mathbb{T}$.

3. Prove Theorem 9.12 for the special case that E is compact.

4. Let (A_k) be an increasing sequence of compact sets. Show that $\operatorname{cap} A_k \to \operatorname{cap}\left(\bigcup_n A_n\right)$ as $k \to \infty$.

In the following problems, let G be a domain with $\infty \in G$ and $\operatorname{cap} E > 0$ where $E = \partial G$. Let g be the Green's function.

5. Show that $g(z) \leq \log(\operatorname{diam} E / \operatorname{cap} E)$ for $z \in E$.

6. If $A \subset E$ is a continuum show that $g(z) = 0$ for $z \in A$.

7. If $g(\zeta) > 0$ for some $\zeta \in E$ show that ζ can be enclosed by Jordan curves in G of arbitrarily small diameter. Deduce that $g(\zeta) = 0$ holds for nearly all $\zeta \in E$ (which is stronger than (40)).

9.4 Pfluger's Theorem

Now we consider sets on the unit circle $\mathbb{T}$. There is a close connection to the starlike functions studied in Section 3.6. We begin with compact sets.

Proposition 9.15. *Let $E \subset \mathbb{T}$ be compact and $\operatorname{cap} E > 0$. Then the equilibrium measure μ is uniquely determined. The function*

$$(1) \qquad h(z) = z(\operatorname{cap} E)^2 \exp\left[-2 \int_E \log(1 - \bar{\zeta}z)\, d\mu(\zeta)\right] \quad (z \in \mathbb{D})$$

maps $\mathbb{D}$ conformally onto a starlike domain

$$(2) \qquad h(\mathbb{D}) = \{se^{i\vartheta} : 0 \leq s < R(\vartheta), 0 \leq \vartheta \leq 2\pi\} \subset \mathbb{D},$$

where $R(\vartheta) = 1$ for almost all ϑ and $|h(\zeta)| = 1$ for nearly all $\zeta \in E$.

Proof. The function h is univalent and starlike by Theorem 3.18 because μ can trivially be extended to a probability measure on $\mathbb{T}$. It follows from Theorem 9.7 that

(3) $$|h(z)| = |z| \exp[-2g(z)] < 1 \quad \text{for} \quad z \in \mathbb{D}.$$

Since $h'(0) > 0$ and since the Green's function g is uniquely determined by $E \subset \mathbb{T}$, the same holds for h. Hence it follows from (1) and (3.6.7) that μ is uniquely determined. Furthermore $h(\zeta)$ exists for $\zeta \in \mathbb{T}$ and $|h(\zeta)| = 1$ for $\zeta \in E \setminus X$ where X is the union of countably many compact sets A_ν of capacity zero by Theorem 9.14. Hence it follows from (3.6.7) and Proposition 9.11 that

$$\Lambda(\{\vartheta = \arg h(\zeta) : \zeta \in (\mathbb{T} \setminus E) \cup X\})$$
$$= 2\pi \left[\mu(\mathbb{T} \setminus E) + \mu(X)\right] \leq 2\pi \sum_\nu \mu(A_\nu) = 0$$

so that $R(\vartheta) = 1$ for almost all ϑ. $\quad\square$

Next we consider small F_σ-sets and construct a starlike function related to the "Evans function".

Proposition 9.16. *If $E \subset \mathbb{T}$ is an F_σ-set of zero capacity then there is a starlike function $h(z) = z + \ldots$ ($z \in \mathbb{D}$) such that*

(4) $$|h(z)| \to \infty \quad \text{as} \quad z \to \zeta, z \in \mathbb{D} \quad \text{for each } \zeta \in E.$$

Proof. The F_σ-set E is the union of countably many compact sets A_ν of capacity zero and by Corollary 9.13 we may assume that $A_\nu \subset A_{\nu+1}$. Let $q_{\nu n}$ denote the nth Fekete polynomial of A_ν defined by (9.3.4). By Proposition 9.6 we can find n_ν such that

(5) $$|q_{\nu n_\nu}(z)|^{1/n_\nu} < \exp(-4^\nu) \quad \text{for} \quad z \in H_\nu \quad (\nu = 0, 1, \ldots)$$

where H_ν is some open set containing A_ν. Let $z_{\nu nk}$ be nth Fekete points of A_ν. Then

(6) $$h(z) = z \prod_{\nu=0}^{\infty} \prod_{k=1}^{n_\nu} (1 - \overline{z}_{\nu n_\nu k} z)^{-1/(n_\nu 2^\nu)} \quad (z \in \mathbb{D})$$

is starlike by Theorem 3.18 because the sum of all exponents is -2.

Let $\zeta \in E$. Then $\zeta \in A_\nu$ for large ν. Since $|1 - \overline{z}_{\nu nk} z| \leq 2$ for $z \in \mathbb{D}$ and $|z_{\nu nk}| = 1$ we see from (6) and (5) that, for $z \in \mathbb{D} \cap H_\nu$,

$$|h(z)| \geq \frac{|z|}{4} |q_{\nu n_\nu}(z)|^{-1/(n_\nu 2^\nu)} > \frac{|z|}{4} \exp(2^\nu)$$

which implies (4). $\quad\square$

Finally we consider any Borel set E on $\mathbb{T}$. It follows from (9.3.33) and (9.3.36) that

(7) $$\frac{\Lambda(E)}{2\pi} \leq \sin \frac{\Lambda(E)}{4} \leq \operatorname{cap} E.$$

We now prove *Pfluger's theorem* (Pfl55) which is very useful for estimating the size of sets (Mak87).

Theorem 9.17. *Let E be a Borel set on $\mathbb{T}$ and let $\Gamma_E(r)$ $(0 < r < 1)$ denote the family of all curves in $\{r < |z| < 1\}$ that connect E with $\{|z| = r\}$. Then*

$$(8) \qquad \frac{\sqrt{r}}{1+r} \operatorname{cap} E \le \exp\left(-\frac{\pi}{\operatorname{mod} \Gamma_E(r)}\right) \le \frac{\sqrt{r}}{1-r} \operatorname{cap} E$$

for $0 < r \le 1/3$ and thus

$$(9) \qquad \operatorname{cap} E = \lim_{r \to 0} \frac{1}{\sqrt{r}} \exp\left(-\frac{\pi}{\operatorname{mod} \Gamma_r(r)}\right).$$

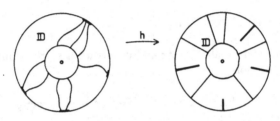

Fig. 9.5. Pfluger's theorem and its proof

Proof. (a) First we prove the lower estimate (8) assuming that E is compact. Let h be the starlike function of Proposition 9.15. Then

$$(10) \qquad |h(z)| \ge R_0 \equiv r(1+r)^{-2}(\operatorname{cap} E)^2 \quad \text{for} \quad |z| = r < 1$$

by (1) and Theorem 1.3. If $R(\vartheta) = 1$ in (2) then the preimage of $[0, e^{i\vartheta}]$ under h is a curve in $\mathbb{D}$ connecting 0 with E as we see from (3) and Theorem 9.7. If ρ is admissible for the family $h(\Gamma_E(r))$ and is zero where it was undefined, it follows that

$$1 \le \left(\int_{R_0}^1 \rho(se^{i\vartheta})\, ds\right)^2 \le \int_{R_0}^1 \frac{ds}{s} \int_{R_0}^1 \rho(se^{i\vartheta})^2 s\, ds$$

by Schwarz's inequality. Since $R(\vartheta) = 1$ holds for almost all ϑ we conclude that

$$2\pi \le \log \frac{1}{R_0} \iint_{R_0 < |w| < 1} \rho(w)^2\, du\, dv$$

and thus, by (9.2.2) and Proposition 9.1,

$$2\pi \le \operatorname{mod} h(\Gamma_E(r)) \log \frac{1}{R_0} = 2 \operatorname{mod} \Gamma_E(r) \log \frac{1+r}{\sqrt{r}\, \operatorname{cap} E}$$

because of (10). This gives the lower inequality (8) for the case that E is compact.

If E is any set on $\mathbb{T}$ then we can find compact sets $A_n \subset E$ with $\operatorname{cap} A_n \to \operatorname{cap} E$ by (9.3.36). Since $\Gamma_{A_n}(r) \subset \Gamma_E(r)$ and thus $\operatorname{mod} \Gamma_{A_n}(r) \le \operatorname{mod} \Gamma_E(r)$ by Proposition 9.2, we obtain the lower estimate in (8) for E by letting $n \to \infty$ in the estimate for A_n.

(b) Now we prove the upper estimate (8) assuming that E is an F_σ-set, thus

$$E = \bigcup_n A_n \quad \text{with } A_n \text{ compact}, \quad A_n \subset A_{n+1} \quad (n = 1, 2, \dots).$$

Let h_n be the starlike functions of Proposition 9.15 for the compact sets A_n. The union X of all the exceptional sets is again an F_σ-set of zero capacity. Let $h_0(z) = z + \dots$ be the starlike function associated with X according to Proposition 9.16. It satisfies $|h_0(\zeta)| \ge 1/4$ for $\zeta \in \mathbb{T}$ by Theorem 1.3.

Since $A_n \subset A_{n+1}$ we see from (3) and Theorem 9.7 that $|h_n| \le |h_{n+1}|$ in $\mathbb{D}$. Since $h_n'(0) > 0$ it follows that $h_n(z)$ converges to a function $h(z) = (\operatorname{cap} E)^2 z + \dots$ $(z \in \mathbb{D})$; see Exercise 9.3.4. If $\zeta \in E \setminus X$ then $\zeta \in A_n \setminus X$ for some n and thus $|h(\zeta)| \ge |h_n(\zeta)| = 1$.

Let $0 < \delta < 1 - \log 3 / \log 4$. The function

$$f(z) = h_0(z)^\delta h(z)^{1-\delta} = (\operatorname{cap} E)^{2-2\delta} z + \dots \quad (z \in \mathbb{D})$$

is again starlike and satisfies $|f(\zeta)| \ge 4^{-\delta}$ for $\zeta \in E \setminus X$ and $f(\zeta) = \infty$ for $\zeta \in X$, furthermore

$$(11) \quad |f(z)| \le R \equiv r(1-r)^{-2}(\operatorname{cap} E)^{2-2\delta} \le \frac{3}{4} < 4^{-\delta} \quad \text{for} \quad |z| = r \le \frac{1}{3}.$$

Let C be a curve in $\Gamma_E(r)$ ending at $\zeta \in E$. Since the angular limit $f(\zeta)$ belongs to the cluster set of f along C by (2.5.6), we see that

$$\liminf_{z \to \zeta, z \in C} |f(z)| \ge |f(\zeta)| \ge 4^{-\delta}.$$

Together with (11) this shows that every curve in $f(\Gamma_E(r))$ contains a curve connecting the two boundary circles of $\{R < |w| < 4^{-\delta}\}$. Hence we obtain from Proposition 9.2 and from (9.2.7) that

$$\operatorname{mod} \Gamma_E(r) = \operatorname{mod} f(\Gamma_E(r)) \le 2\pi / \log \frac{4^{-\delta}}{R} = \pi / \log \frac{2^{-\delta}(1-r)}{\sqrt{r}(\operatorname{cap} E)^{1-\delta}}$$

by (11), and the upper estimate (8) follows for $\delta \to 0$.

Finally let E be any Borel set on $\mathbb{T}$ and let $\varepsilon > 0$. By Theorem 9.12 there is an open set H on $\mathbb{T}$ such that $E \subset H$ and $\operatorname{cap} H < \operatorname{cap} E + \varepsilon$. Since H is an F_σ-set it follows that

$$\exp\left(-\frac{\pi}{\operatorname{mod} \Gamma_E(r)}\right) \le \exp\left(-\frac{\pi}{\operatorname{mod} \Gamma_H(r)}\right) \le \frac{\sqrt{r}}{1-r}(\operatorname{cap} E + \varepsilon). \qquad \square$$

The following result (BiCaGaJo89, Roh91) was used in the proof of Theorem 6.30; see Fig. 9.6.

Corollary 9.18. *Let f_1 and f_2 map $\mathbb{D}$ onto the inner and outer domain of the Jordan curve $J \subset \mathbb{C}$ and let*

(12) $$A_j = f_j^{-1}(J \cap D(a, R)) \quad (a \in J, R > 0)$$

for $j = 1, 2$. Then

(13) $$\Lambda(A_1)\Lambda(A_2) \le 4\pi^2 \operatorname{cap} A_1 \operatorname{cap} A_2 \le MR^2 \quad for \quad R < R_0$$

where the constants M and R_0 depend only on f_1 and f_2.

Proof. We choose R_0 such that $0 < R_0 \le \operatorname{dist}(J, f_j(z))$ for $|z| \le 1/3$ and $j = 1, 2$. For $0 < R < R_0$, let Γ_j denote the family of all curves in $\{w \in f_j(\mathbb{D}) : R < |w - a| < R_0\}$ that connect the two circles. Since $f_1(\mathbb{D}) \cap f_2(\mathbb{D}) = \emptyset$ it follows from Proposition 9.2, Proposition 9.3 and (9.2.7) that

(14) $$\operatorname{mod}\Gamma_1 + \operatorname{mod}\Gamma_2 = \operatorname{mod}(\Gamma_1 \cup \Gamma_2) \le 2\pi / \log \frac{R_0}{R} ,$$

and Theorem 9.17 shows that

$$\frac{\sqrt{3}}{4} \operatorname{cap} A_j \le \exp\left(-\frac{\pi}{\operatorname{mod}\Gamma_{A_j}(1/3)}\right) \le \exp\left(-\frac{\pi}{\operatorname{mod}\Gamma_j}\right)$$

because every curve in $f(\Gamma_{A_j}(1/3))$ contains some curve in Γ_j. Now we multiply the inequalities for $j = 1$ and $j = 2$. Since $1/a_1 + 1/a_2 \ge 4/(a_1 + a_2)$ for $a_1, a_2 > 0$ we conclude from (14) that

$$\frac{3}{16} \operatorname{cap} A_1 \operatorname{cap} A_2 \le \exp[-2\log(R_0/R)] = R^2/R_0^2$$

and (13) follows by (7).   □

The factor R^2 can often be replaced by $R^{2+\delta}$ with $\delta > 0$ if a little more is known about the geometry of J; see Roh91.

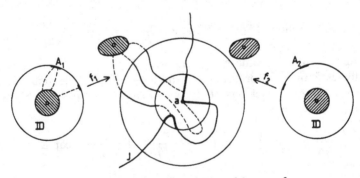

Fig. 9.6. Corollary 9.18 and its proof

Exercises 9.4

1. Let $E \subset \mathbb{T}$ be compact and cap $E > 0$. If μ is an equilibrium measure of E show that

$$f(z) = \exp\left(\int_E \log(z - \zeta)\, d\mu(\zeta) \right) \quad (|z| > 1)$$

maps $\mathbb{D}^*$ conformally onto a starlike domain in $\{|w| > \text{cap } E\}$.

2. Let E be an F_σ-set in $\mathbb{C}$ with cap $E = 0$. Construct a function u harmonic in $\mathbb{C} \setminus E$ with $u(z) = \log|z| + \dots$ ($z \to \infty$) such that $u(z) \to -\infty$ as $z \to \zeta$, $\zeta \in E$. (This is an "Evans function" of E.)

3. Use the function of Exercise 2 to generalize (Pri19) the Privalov uniqueness theorem: Let $E \subset \mathbb{C}$ be an F_σ-set of zero capacity and let g be meromorphic in $\mathbb{D}$. If the angular limit $g(\zeta)$ exists for each $\zeta \in A \subset \mathbb{T}$ and $g(\zeta) \in E$ then $\Lambda(A) = 0$.

4. Under the assumptions of Corollary 9.18, suppose that J lies in $\{|\arg(w - a)| < \pi\alpha\}$ near a where $\alpha < 1$. Show that cap A_1 cap $A_2 \leq M' R^{2/[\alpha(2-\alpha)]}$.

9.5 Applications to Conformal Mapping

We first generalize Theorem 1.7.

Theorem 9.19. *Let f map $\mathbb{D}$ conformally into $\mathbb{C}$. Then the angular limit $f(\zeta)$ exists and is finite for nearly all $\zeta \in \mathbb{T}$. If $R \geq 1$ then*

(1) $$|f(\zeta) - f(0)| \leq |f'(0)|R \quad for \quad \zeta \in \mathbb{T} \setminus E,$$

(2) $$\int_0^1 |f'(r\zeta)|\, dr \leq K_1 |f'(0)| R \log(6R) \quad for \quad \zeta \in \mathbb{T} \setminus E$$

with an absolute constant K_1 where the exceptional set E satisfies

(3) $$\Lambda(E)/(2\pi) \leq \text{cap } E \leq 1/\sqrt{R}.$$

By applying a suitable Möbius transformation we deduce (Beu40) at once that every conformal map of $\mathbb{D}$ into $\widehat{\mathbb{C}}$ has angular limits nearly everywhere (which is much stronger than almost everywhere).

Proof. We may assume that $f(0) = 0$ and $f'(0) = 1$. Let $0 < r < 1/3$. The Koebe distortion theorem shows that

(4) $$r(1+r)^{-2} \equiv r' \leq |f(\zeta)| \leq r'' \equiv r(1-r)^{-2} < 1 \quad for \quad |z| = r.$$

Let E denote the set of all $\zeta \in \mathbb{T}$ such that

(5) $$f(C_1) \not\subset D(0, R) \quad or \quad \int_{C_1} |f'(z)|\, |dz| \geq R \log(6R)$$

for all curves C_1 from $\{|z| = 1/3\}$ to ζ and let $\Gamma_E(r)$ be the family of all curves from $\{|z| = r\}$ to E. We define $a = \log(R/r'')$ and

(6) $\qquad \rho(z) = \left| \dfrac{f'(z)}{af(z)} \right| \quad$ if $\quad |f(z)| \leq R, \quad \rho(z) = 0 \quad$ if $\quad |f(z)| > R.$

Let $C \in \Gamma_E(r)$ begin at z_1 with $|z_1| = r$ and end at $\zeta \in E \subset \mathbb{T}$. If $f(C) \not\subset D(0, R)$ then there exists a first $z_2 \in C$ with $|f(z_2)| = R$ and we see from (6) and (4) that

$$\int_C \rho(z)|dz| \geq \left| \int_{z_1}^{z_2} \frac{f'(z)}{af(z)} \, dz \right| = \frac{1}{a} \left| \log \frac{f(z_2)}{f(z_1)} \right| \geq \frac{1}{a} \log \frac{R}{r''} = 1.$$

If however $f(C) \subset D(0, R)$ then we write $C = C_0 \cup C_1$ where C_1 runs from $\{|z| = 1/3\}$ to E. Then the second relation in (5) holds and thus, by (6) and Exercise 1.3.3,

$$a \int_C \rho|dz| \geq \int_{C_0} \frac{1 - |z|}{1 + |z|} \left| \frac{dz}{z} \right| + \int \frac{|f'(z)|}{R} |dz|$$

$$\geq \log \frac{3(1 + r)^2}{16r} + \log(6R) > \log \frac{R}{r'} > a$$

by (4). Hence ρ is admissible for $\Gamma_E(r)$ and it follows from (6) and (4) that, with $B = \{z \in \mathbb{D} : |z| \geq r, |f(z)| \leq R\}$,

$$\operatorname{mod} \Gamma_E(r) \leq \iint_B \left| \frac{f'(z)}{af(z)} \right|^2 dx\,dy \leq \frac{2\pi}{a^2} \int_{r'}^R \frac{ds}{s}$$

$$= \frac{2\pi \log(R/r')}{[\log(R/r'')]^2} = \frac{2\pi}{\log(R/r) + O(r)}$$

as $r \to 0$ by (4). Hence (3) follows from Theorem 9.17.

Now let $\zeta \in \mathbb{T} \setminus E$. Then (5) is false for some curve C_1 ending at ζ and it follows from Theorem 4.20 that (2) holds. Hence the angular limit $f(\zeta)$ exists. By Corollary 2.17 it is equal to the limit along C_1 so that $|f(\zeta)| \leq R$. Furthermore it follows that

(7) $\qquad \operatorname{cap} \left\{ \zeta \in \mathbb{T} : \int_0^1 |f'(r\zeta)|\,dr = \infty \right\} = 0$

and $f(\zeta) \neq \infty$ exists outside this Borel set. $\square$

The next result is more precise than Corollary 4.18. Let again $d_f(z) = \operatorname{dist}(f(z), \partial G)$.

Corollary 9.20. *Let f map $\mathbb{D}$ conformally onto $G \subset \mathbb{C}$ and let $\zeta \in \mathbb{T}$, $0 \leq r < 1$, $a > 0$ and $R \geq 1$. Then*

(8) $\qquad |f(z) - f(r\zeta)| \leq 4R d_f(r\zeta) \quad$ for $\quad z \in I \setminus E,$

where $I = \{e^{it} : |t - \arg \zeta| \leq a(1 - r)\}$ *and*

(9) $$\operatorname{cap} E \leq (1 + a^2)(1 - r)/\sqrt{R},$$

and if S is the non-euclidean segment from $r\zeta$ to z then

(10) $$\int_S |f'(s)|\,|ds| \leq K_2 d_f(r\zeta) R \log(6R) \quad \text{for} \quad z \in I \setminus E.$$

Proof. We write

(11) $$\varphi(s) = \frac{s + r\zeta}{1 + r\bar{\zeta}s}, \quad g(s) = f(\varphi(s)) \quad \text{for} \quad s \in \mathbb{D}.$$

Since $|g'(0)| = (1 - r^2)|f'(r\zeta)| \leq 4d_f(r\zeta)$ we see from Theorem 9.19 that (8) and (10) hold for $z = \varphi(s)$ and $s \in \mathbb{T} \setminus E^*$ where $\operatorname{cap} E^* \leq 1/\sqrt{R}$. Since

$$|\varphi'(s)| = \frac{|1 - r\bar{\zeta}\varphi(s)|^2}{1 - r^2} \leq (1 + a^2)(1 - r) \quad \text{for} \quad \varphi(s) \in I$$

it follows from (9.3.10) and (9.3.11) that $E = \varphi(E^*)$ satisfies (9). □

The next result (Pom75, p. 348; compare Kom42, Jen56) is an analogue of Löwner's lemma. See Ham88 and Theorem 10.11 for generalizations.

Theorem 9.21. *Let f map $\mathbb{D}$ conformally into $\mathbb{D}$ with $f(0) = 0$. If E is a Borel set on $\mathbb{T}$ such that the angular limit exists and $f(\zeta) \in \mathbb{T}$ for $\zeta \in E$ then*

(12) $$\operatorname{cap} f(E) \geq \frac{\operatorname{cap} E}{\sqrt{|f'(0)|}} \geq \operatorname{cap} E.$$

Proof. Let A_n be closed subsets of E such that $\operatorname{cap} A_n \to \operatorname{cap} E$ as $n \to \infty$. Then $f(A_n)$ is a Borel set on $\mathbb{T}$. Since $r'' \equiv \max\{|f(z)| : |z| = r\} \sim |f'(0)|r$ as $r \to 0$ and since

$$\operatorname{mod} \Gamma_{f(A_n)}(r'') \geq \operatorname{mod} f(\Gamma_{A_n}(r)) = \operatorname{mod} \Gamma_{A_n}(r)$$

by Corollary 2.17 and by Propositions 9.2 and 9.1, we see from Theorem 9.17 that

$$\operatorname{cap} f(E) \geq \operatorname{cap} f(A_n) \geq |f'(0)|^{-1/2} \operatorname{cap} A_n$$

and (12) follows for $n \to \infty$. □

We formulate a special case in terms of the harmonic measure $\omega(z, A)$ at $z \in \mathbb{D}$ of $A \subset \mathbb{T}$ with respect to $\mathbb{D}$ defined by (4.4.1) and the non-euclidean distance $\lambda(z, z')$ given by (1.2.8); see Fig. 9.7.

Corollary 9.22. *Let f map $\mathbb{D}$ conformally into $\mathbb{D} \setminus \{w\}$. If A is a Borel set in $\mathbb{T}$ and $f(A) \subset \mathbb{T}$ then, for $z \in \mathbb{D}$,*

(13) $$\sin[\pi\omega(z, A)/2] \le (1 - \exp[-4\lambda(f(z), w)])^{1/2} \ ;$$

if $f(A)$ is an arc of $\mathbb{T}$ there there is an additional factor $\sin[\pi\omega(f(z), f(A))/2]$
< 1 on the right-hand side.

If $G = f(\mathbb{D})$ and $E = f^{-1}(\mathbb{D} \cap \partial G)$ we obtain from (13) applied to $A = \mathbb{T} \setminus E$
and from $\omega(z, E) = 1 - \omega(z, A)$ that

(14) $$\sin\left[\frac{\pi}{2}\omega(z, E)\right] \ge \exp\left[-2\lambda(f(z), w)\right].$$

This is related to the *Carleman-Milloux problem*; see Nev70, p. 102.

Proof. Since the harmonic measure and the non-euclidean metric are invariant
under Möbius transformations of $\mathbb{D}$ onto $\mathbb{D}$ we may assume that $z = 0$ and
$f(0) = 0$, furthermore that $w > 0$. Since

$$f(z)/(1 + f(z))^2 = f'(0)z + \dots \quad (z \in \mathbb{D})$$

is analytic, univalent and $\ne w/(1 + w)^2$, we see that $|f'(0)| \le 4w/(1 + w)^2$.
Hence we conclude from (9.4.7) and from (12) that

$$\sin\left[\frac{\pi}{2}\omega(0, A)\right] = \sin\frac{\Lambda(A)}{4} \le \text{cap}\, A \le \frac{2\sqrt{w}}{1 + w} \text{cap}\, f(A).$$

We have $\text{cap}\, f(A) \le \text{cap}\, \mathbb{T} = 1$ and $\text{cap}\, f(A) = \sin[\pi\omega(0, f(A))/2]$ if $f(A)$ is
an arc, by (9.3.31). This implies (13) because $\exp[-2\lambda(0, w)] = (1-w)/(1+w)$.
□

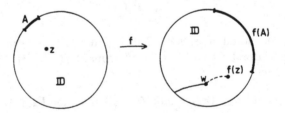

Fig. 9.7. The situation of Corollary 9.22

We now show that the estimates (12) and (13) are sharp. Let $0 < a \le 1$.
Then

(15) $$f(z) = 4az\left(1 - z + \sqrt{(1 - z)^2 + 4za}\right)^{-2} = az + \dots$$

maps $\mathbb{D}$ conformally onto $\mathbb{D} \setminus [-1, -b]$ where $b = a\left(1 + \sqrt{1 - a}\right)^{-2}$. If $0 < \alpha \le$
$\arcsin\sqrt{a}$ and $A = \{e^{it} : |t| \le 2\alpha\}$ then

$$f(A) = \{e^{i\vartheta} : |\vartheta| \le 2\beta\}, \quad \sin\beta = a^{-1/2}\sin\alpha.$$

Since $\operatorname{cap} A = \sin \alpha$ and $\operatorname{cap} B = \sin \beta$ we see that equality holds in (12) and furthermore

$$1 - \exp[-4\lambda(0, -b)] = 1 - [(1 - b)/(1 + b)]^2 = a$$

so that equality holds in (13) for $z = 0$.

The next result (Schi46, Pom68) deals with general conformal maps. It assumes its simplest form for the class Σ of univalent functions of the form $g(\zeta) = \zeta + b_0 + b_1\zeta^{-1} + \ldots$ for $|\zeta| > 1$.

Theorem 9.23. *If $g \in \Sigma$ and $A \subset \mathbb{T}$ is a Borel set such that $g(\zeta)$ exists for $\zeta \in A$ then*

$$(16) \qquad \operatorname{cap} g(A) \geq (\operatorname{cap} A)^2.$$

In particular it easily follows (Duf45) that, for all conformal maps f of $\mathbb{D}$ into $\widehat{\mathbb{C}}$,

$$(17) \qquad \operatorname{cap} A > 0 \quad \Rightarrow \quad \operatorname{cap} f(A) > 0.$$

Proof. We restrict ourselves to the special case that $g(A)$ is a continuum; see Pom75, p. 344 or Mak87, p. 50 for the general case. Let $h(\zeta) = c\zeta + \ldots$ with $c = \operatorname{cap} g(A)$ map $\mathbb{D}^*$ conformally onto the outer domain of $g(A)$; see Corollary 9.9. Then

$$f(z) = 1/h^{-1}(g(1/z)) = cz + \ldots$$

maps $\mathbb{D}$ conformally into $\mathbb{D}$ and Theorem 9.21 shows that, with $A^* = \{1/z : z \in A\}$,

$$\operatorname{cap} A = \operatorname{cap} A^* \leq c^{1/2} \operatorname{cap} f(A^*) \leq c^{1/2}. \qquad \square$$

If $g \in \Sigma$ has a κ-quasiconformal extension to $\mathbb{C}$ then (Küh71; see Pom75, p. 346)

$$(18) \qquad (\operatorname{cap} A)^{1+\kappa} \leq \operatorname{cap} g(A) \leq (\operatorname{cap} A)^{1-\kappa}$$

for Borel sets $A \subset \mathbb{T}$.

Theorem 9.24. *Let f map $\mathbb{D}$ conformally onto $G \subset \mathbb{C}$ and let $E \subset \mathbb{T}$. If $0 < \alpha \leq 1$ and if H is an open subset of G such that*

$$(19) \qquad \operatorname{dist}(f(0), H) \geq \alpha|f'(0)|, \quad \Lambda(f(C) \cap H) \geq b \geq 0$$

for every curve $C \subset \mathbb{D}$ from 0 to E then

$$(20) \qquad \Lambda(E) \leq 2\pi \operatorname{cap} E < \frac{15}{\sqrt{\alpha}} \exp\left(-\frac{\pi b^2}{\operatorname{area} H}\right).$$

This is a quantitative version of the principle that boundary sets that are difficult to reach have small harmonic measure; compare e.g. Corollary

6.26. This fact has severe consequences ("crowding") for numerical conformal mapping; see e.g. Gai72b, PaKoHo87, GaHa91. The situation is depicted in Fig. 9.8.

Proof. Let $\Gamma = \Gamma_E(\alpha/3)$ be defined as in Theorem 9.17. The Koebe distortion theorem shows that $f(z) \notin H$ for $|z| \leq \alpha/3$. Hence $\Lambda(f(C) \cap H) \geq b$ for $C \in \Gamma$ by (19) so that

$$\rho(w) = 1/b \quad \text{for} \quad w \in H, \quad \rho(w) = 0 \quad \text{for} \quad w \in G \setminus H,$$

is admissible for $f(\Gamma)$. Hence

$$\operatorname{mod} \Gamma = \operatorname{mod} f(\Gamma) \leq \iint_H \rho(w)^2 \, du \, dv = b^{-2} \operatorname{area} H$$

and therefore, by (9.4.8),

$$(\sqrt{3\alpha}/4) \operatorname{cap} E \leq \exp(-\pi b^2 / \operatorname{area} H)$$

which implies (20) by (9.4.7). □

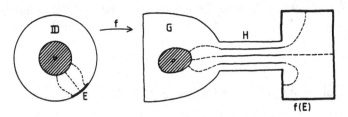

Fig. 9.8. Theorem 9.24 and its proof

Exercises 9.5

1. Let f map $\mathbb{D}$ conformally into $\mathbb{D}$ and let $b \geq 1$. Show that

$$\int_0^1 |f'(r\zeta)| \, dr \leq b \quad \text{for} \quad \zeta \in \mathbb{T} \setminus E$$

where $\operatorname{cap} E < \exp(-cb^2)$ for some constant $c > 0$.

2. Let f map $\mathbb{D}$ conformally into $\mathbb{C}$ and suppose that the angular limit $f(\zeta)$ exists. Show that, for $\varepsilon > 0$,

$$\delta^{-1} \operatorname{cap}\{z \in \mathbb{T} : |z - \zeta| \leq \delta, |f(z) - f(\zeta)| \geq \varepsilon\} \to 0 \quad \text{as} \quad \delta \to 0.$$

3. Let f_t map $\mathbb{D}$ conformally onto $G_t \subset \mathbb{C}$ such that $f_t(0) = 0$ for $t \in I$ and suppose that $G_t \subset G_\tau$ for $t < \tau$. Show that

$$|f_t'(0)|^{-1/2} \operatorname{cap} f_t^{-1}(B) \quad (t \in I)$$

is increasing if B is a fixed compact set with $B \subset \partial G_t$ for $t \in I$.

4. Let A and B be Borel sets. If $A \cup B$ is a continuum show that

$$\operatorname{cap}(A \cup B) \leq \operatorname{cap} A + \operatorname{cap} B$$

and show that this need not be true if $A \cup B$ is not connected.

5. If A is a Borel set on $\mathbb{T}$ show that $\operatorname{cap} f(A) \geq (\operatorname{cap} A)^2/16$ for $f \in S$.

6. For the "comb" domain defined by (2.2.2), show that the harmonic measure of $[1, 1 + i/n]$ with respect to $3i/4$ is less than $\exp(-\pi n^2)$ for large n.

Chapter 10. Hausdorff Measure

10.1 An Overview

Let $\alpha > 0$. The α-dimensional Hausdorff measure of a Borel set $E \subset \mathbb{C}$ is defined by

$$(1) \qquad \Lambda_\alpha(E) = \lim_{\varepsilon \to 0} \inf_{(B_k)} \sum_k (\operatorname{diam} B_k)^\alpha$$

where the infimum is taken over the covers (B_k) of E with $\operatorname{diam} B_k \leq \varepsilon$ for all k. In particular Λ_1 is the linear measure Λ discussed in Section 6.2.

We also introduce

$$(2) \qquad \Lambda'_\alpha(E) = \inf_{(B_k)} \sum_k (\operatorname{diam} B_k)^\alpha$$

where the infimum is taken of all covers of E. This quantity behaves more like a capacity than a measure and satisfies (Theorem 10.3)

$$(3) \qquad \Lambda'_\alpha(E) \leq K(\operatorname{cap} E)^\alpha .$$

The Hausdorff dimension is defined by

$$(4) \qquad \dim E = \inf\{\alpha : \Lambda_\alpha(E) = 0\} = \inf\{\alpha : \Lambda'_\alpha(E) = 0\} .$$

Sets of non-integer dimension are often called "fractals". The dimension of a set on $\mathbb{R}$ can often be determined by Theorem 10.5 (due to Hungerford in its final form).

In fact we consider the Hausdorff measure Λ_φ with respect to any continuous increasing function φ with $\varphi(0) = 0$. The α-dimensional Hausdorff measure is the special case $\varphi(t) = t^\alpha$. See e.g. Fal85 and Rog70 for a more thorough discussion.

N.G. Makarov has proved several important results about Hausdorff measures and conformal mapping. Two of these can be stated as:

Makarov Dimension Theorem. *Let f map $\mathbb{D}$ conformally into $\mathbb{C}$ and let $A \subset \mathbb{T}$. If $\dim A > 0$ then $\dim f(A) > (\dim A)/2$ and*

$$(5) \qquad \dim A = 1 \quad \Rightarrow \quad \dim f(A) \geq 1 .$$

The last implication solves an interesting problem; the best previous result, due to Carleson, was that $\Lambda(A) > 0$ implies $\dim f(A) > 1/2$. In Theorem 10.6 Makarov gives a result that is much more precise than (5); it is a consequence of his law of the iterated logarithm for Bloch functions (Theorem 8.10). His Theorem 10.8 gives lower estimates for $\dim f(A)$ in terms of upper estimates for the integral means of $1/|f'|$ studied in Section 8.3. These theorems should be contrasted with the results of McMillan and Makarov in Section 6.5. See also Mak89d for a survey and the connection with stochastic processes.

Hamilton has proved an analogue of Löwner's lemma (for α-capacity) which implies that

(6) $$f(\mathbb{D}) \subset \mathbb{D}, \quad f(A) \subset \mathbb{T} \quad \Rightarrow \quad \dim f(A) \geq \dim A\,;$$

see Makarov's Theorem 10.11 for a more precise statement.

Anderson-Pitt Theorem. *If f maps $\mathbb{D}$ conformally into $\mathbb{C}$ then*

(7) $$\dim\{\zeta \in \mathbb{T} : f'(\zeta) \text{ exists finitely}\} = 1\,.$$

This solves another problem that had been open for a long time. Makarov's Theorem 10.14 gives the best possible bound for the size of the set where the angular derivative $f'(\zeta)$ exists. These theorems are closely connected to results about Zygmund measures and Bloch functions; see Section 10.4.

In the definition of Hausdorff measure we allowed covers by sets of widely different size. We can also consider covers by sets of fixed size. Let $N(\varepsilon, E)$ denote the minimal number of disks of diameter ε that are needed to cover E. We define the (upper) Minkowski dimension by

(8) $$\text{mdim}\, E = \inf\{\alpha : N(\varepsilon, E) = O(\varepsilon^{-\alpha}) \text{ as } \varepsilon \to 0\}$$

so that $\dim E \leq \text{mdim}\, E$. The name "Minkowski dimension" is not standard; other names are "box dimension" or "upper entropy dimension". It was first systematically studied by Kolmogorov; see e.g. KoTi61, Haw74 and Fal90.

If f is a conformal map of $\mathbb{D}$ onto a quasidisk G then (Corollary 10.18)

$$\text{mdim}\, \partial G = \inf \alpha \quad \text{where} \quad \int_0^{2\pi} |f'(re^{it})|^\alpha \, dt = O((1-r)^{-\alpha+1}) \quad (r \to 1)\,.$$

10.2 Hausdorff Measures

1. We assume throughout this section that φ is a positive continuous increasing function in $(0, +\infty)$ and that $\varphi(t) \to 0$ as $t \to 0$. If E is any set in $\mathbb{C}$ we define

(1) $$\Lambda_\varphi(E) = \lim_{\varepsilon \to 0+} \inf_{(B_k)} \sum_k \varphi(\text{diam}\, B_k) \leq +\infty$$

where (B_k) runs through all ε-covers of E, that is,

$$E \subset \bigcup_k B_k, \quad \operatorname{diam} B_k \le \varepsilon \quad \text{for} \quad k = 1, 2, \dots .$$

It is clear that only the behaviour of $\varphi(t)$ near $t = 0$ matters. If $\alpha > 0$ and $\varphi(t) = t^\alpha$ we write Λ_α instead of Λ_φ. This is the most important case. In particular we see from (6.2.7) that $\Lambda_1 = \Lambda$.

We now use the concepts introduced in Section 6.2. The following basic result is proved in the same way as Proposition 6.1.

Proposition 10.1. *The quantity Λ_φ is a metric outer measure, and the Λ_φ-measurable sets form a σ-algebra that contains all Borel sets. In particular*

$$\tag{2} \Lambda_\varphi \left(\bigcup_n E_n \right) = \sum_n \Lambda_\varphi(E_n)$$

if E_n $(n = 1, 2, \dots)$ are disjoint Borel sets.

We call Λ_φ the *Hausdorff measure* with respect to φ and $\Lambda_\alpha \equiv \Lambda_{t^\alpha}$ the *α-dimensional* Hausdorff measure. The 1-dimensional measure is thus the linear measure studied in Section 6.2.

Proposition 10.2. *If $\varphi \le \psi$ then $\Lambda_\varphi(E) \le \Lambda_\psi(E)$, and if $\varphi(x)/\psi(x) \to 0$ as $x \to 0$ then $\Lambda_\varphi(E) = 0$ whenever E has σ-finite outer Λ_ψ-measure.*

Proof. The first assertion follows at once from (1). In order to prove the second assertion it is sufficient to consider the case that $\Lambda_\psi(E) < M < \infty$. Given $\varepsilon > 0$ we choose $\delta > 0$ such that $\varphi(x) < \varepsilon\psi(x)/M$ for $0 < x < \delta$ and that there is a cover (B_k) of E such that

$$\sum_k \psi(\operatorname{diam} B_k) < M, \quad \operatorname{diam} B_k < \delta \quad (k \in \mathbb{N}).$$

Then

$$\sum_k \varphi(\operatorname{diam} B_k) < (\varepsilon/M)M = \varepsilon \quad \text{so that} \quad \Lambda_\varphi(E) = 0. \qquad \square$$

Thus different Hausdorff measures make a very fine distinction between the sizes of sets. Furthermore we see that Λ_φ is not σ-finite (except for cases like $\varphi(t) = t^2$) which reduces its value for integration theory.

The *Hausdorff dimension* of $E \subset \mathbb{C}$ is defined by

$$\tag{3} \dim E = \inf\{\alpha : \Lambda_\alpha(E) = 0\}.$$

It follows from Proposition 10.2 that

$$\tag{4} \begin{aligned} \Lambda_\alpha(E) &= \infty \quad \text{for} \quad 0 < \alpha < \dim E, \\ \Lambda_\alpha(E) &= 0 \quad \text{for} \quad \alpha > \dim E \end{aligned}$$

while $0 \le \Lambda_{\dim E}(E) \le \infty$.

It is clear that $0 \leq \dim E \leq 2$ for $E \subset \mathbb{C}$. If $E \subset \mathbb{T}$ then $0 \leq \dim E \leq 1$, and $\Lambda_1(E) > 0$ implies $\dim E = 1$ but not vice versa.

As an example the snowflake curve has dimension $\log 4 / \log 3$; see e.g. Fal85 where also other examples of selfsimilar sets are discussed.

2. For $E \subset \mathbb{C}$ we also define

$$(5) \qquad \Lambda'_\varphi(E) = \inf_{(B_k)} \sum_k \varphi(\operatorname{diam} B_k)$$

where (B_k) runs through all covers of E, not only through the ε-covers as in the definition (1) of Λ_φ. Hence $0 \leq \Lambda'_\varphi(E) \leq \Lambda_\varphi(E) \leq \infty$ and

$$(6) \qquad \Lambda'_\varphi(E) = 0 \quad \Leftrightarrow \quad \Lambda_\varphi(E) = 0.$$

We write Λ'_α instead of Λ'_φ if $\varphi(t) = t^\alpha$ and $\alpha > 0$.

The quantity $\Lambda'_\varphi(E)$ has the advantage that it is finite for bounded sets and it is therefore often used in complex analysis. By (6) it has the same nullsets as Λ_φ so that, by (3),

$$(7) \qquad \dim E = \inf\{\alpha : \Lambda'_\alpha(E) = 0\}.$$

Though Λ'_φ is an outer measure, it is not a metric outer measure and does not lead to a useful concept of measure; see Exercise 6.2.4. In many respects Λ'_φ behaves more like a capacity than a measure. Let $M_1, M_2, \ldots$ denote suitable positive constants.

Theorem 10.3. *Suppose that*

$$(8) \qquad \varphi(xt) \leq M_1 x^\alpha \varphi(t) \quad (0 < x \leq 2, t > 0)$$

for some $\alpha > 0$. If E is a Borel set in $\mathbb{C}$ then

$$(9) \qquad \Lambda'_\varphi(E) \leq M_2 \varphi(\operatorname{cap} E).$$

In particular we see that

$$\Lambda'_\alpha(E) \leq M_2 (\operatorname{cap} E)^\alpha$$

so that $\operatorname{cap} E = 0$ implies $\dim E = 0$. Note that we cannot replace Λ'_α by Λ_α because $\Lambda_\alpha(E)$ will in general be infinite. More precise results can be found in Lan72, Chapt III; see also Nev70 and HayKe76.

Proof. We restrict ourselves to the case that E is compact; see Lan72, p. 203 for the general case. If $c > \operatorname{cap} E$ then, by Proposition 9.6, there exists n such that

$$(10) \qquad \max_{z \in E} \prod_{k=1}^{n} |z - z_k| < c^n$$

where $z_1, \ldots, z_n$ are nth Fekete points of E. We write

(11) $\qquad \eta = ce^{72/\alpha}, \quad \eta_\nu = \eta e^{-2\nu/\alpha} \quad$ for $\quad \nu = 0, 1, \ldots$

and partition $\mathbb{C}$ by the halfopen squares of a grid of mesh size η_ν. A square will be called "black" if it contains more than $n2^{-\nu}$ Fekete points.

Every point $z \in E$ lies in a black square or in one of its eight neighbours for some ν. This is trivial if z itself is a Fekete point. Otherwise, for every ν, there would be at most $9n2^{-\nu}$ Fekete points z_k with $|z - z_k| < \eta_\nu$ and therefore

$$\prod_{k=1}^{n} \frac{|z - z_k|}{\eta} \geq \prod_{\nu=0}^{\infty} \left(\frac{\eta_{\nu+1}}{\eta} \right)^{9n2^{-\nu}}$$
$$= \prod_{\nu=0}^{\infty} \exp \left(-\frac{2(\nu+1)}{\alpha} 9n2^{-\nu} \right) = \exp \left(-\frac{72n}{\alpha} \right) = \left(\frac{c}{\eta} \right)^n$$

by (11), which would contradict (10).

Thus E is covered by all black squares and its eight neighbours. Let Q be one of these squares. We see from (11) that

$$\varphi(\operatorname{diam} Q) \leq \varphi(2\eta_\nu) = \varphi(2ce^{72/\alpha}e^{-2\nu/\alpha}) \leq M_3 e^{-2\nu}\varphi(c)$$

where we have used (8) repeatedly. Since there are n Fekete points, there are at most 2^ν black squares of size η_ν. Hence we obtain that

$$\sum_{Q} \varphi(\operatorname{diam} Q) \leq M_4 \sum_{\nu=0}^{\infty} 2^\nu 2^{-2\nu} \varphi(c) = M_2 \varphi(c)$$

and it follows from (5) that $\Lambda'_\varphi(E) \leq M_2 \varphi(c)$ for every $c > \operatorname{cap} E$ which implies (8). $\qquad\qquad\qquad\qquad\qquad\qquad\qquad\qquad\qquad\qquad\qquad\qquad \square$

The following result (Fro35) is often useful: see e.g. HayKe76, p. 223 for the proof.

Proposition 10.4. *If $E \subset \mathbb{C}$ is compact and $\Lambda_\varphi(E) > 0$ then there exists a measure μ on E with $0 < \mu(E) < \infty$ such that*

(12) $\qquad \mu(D(a, r)) \leq \varphi(r) \quad$ for $\quad a \in \mathbb{C}, \quad 0 < r \leq 1.$

3. The final result (Hun88) can be used to estimate the Hausdorff measure Λ_α of sets on $\mathbb{T}$ from below. Let $\Lambda = \Lambda_1$ be the linear measure.

Theorem 10.5. *For $k = 0, 1, \ldots$ let $A_k \subset [0, 1]$ be a union of finitely many disjoint closed sets $B_{k,n}$ such that $\Lambda(A_0) > 0$ and $A_k \supset A_{k+1}$. We assume that $0 < a < b < 1$ and*

(13) $B_{k+1,n} \cap B_{k,m} \neq \emptyset \quad \Rightarrow \quad B_{k+1,n} \subset B_{k,m}, \operatorname{diam} B_{k+1,n} \leq a \operatorname{diam} B_{k,m},$

(14) $$\Lambda(A_{k+1} \cap B_{km}) \geq b\Lambda(B_{km}) \quad \text{for all} \quad m.$$

Then

(15) $$\Lambda_\alpha \left(\bigcap_{k=0}^\infty A_k \right) > 0, \quad \alpha = \frac{\log(b/a)}{\log(1/a)}.$$

Proof (Roh89). We may assume that $\Lambda(A_k) \to 0$ as $k \to \infty$. We first show that we can replace the sets B_{kn} and thus A_k by subsets with the additional property that equality always holds in (14). Suppose that this replacement has been made for the sets B_{jn} with $j \leq k$ and consider some set $B \equiv B_{km}$; we may assume that $\Lambda(B) > 0$.

Let $\mu \geq 2$ be first number such that $\Lambda(A_{k+\mu} \cap B) < b\Lambda(B)$; compare (14). Replacing $A_{k+\nu} \cap B$ by $A_{k+\nu+\mu-2} \cap B$ for $\nu \geq 2$ we may assume that $\mu = 2$. Then

$$\Lambda(A_{k+2} \cap B) < b\Lambda(B) \leq \Lambda(A_{k+1} \cap B).$$

Since the complement of a closed linear set is the union of countably many disjoint open intervals it is not difficult to see that there is a closed set E such that

$$A_{k+2} \cap B \subset E \subset A_{k+1} \cap B, \quad \Lambda(E) = b\Lambda(B).$$

We now replace the sets $B_{k+1,n} \subset B$ by $B_{k+1,n} \cap E$ together with the extreme points of $B_{k+1,n}$ so that the diameter is not changed. The union of the new sets has linear measure $b\Lambda(B)$. Making these replacements for all B_{mk} we obtain a new set A_{k+1} such that equality holds in (14). The sets A_j for $j > k+1$ are not affected because $A_{k+2} \cap B \subset E$.

We next show by induction on $j = 0, \ldots, k$ that

(16) $$\Lambda(A_k \cap B_{k-j,m}) = b^j \Lambda(B_{k-j,m}) \quad \text{for all} \quad m.$$

This is trivial if $j = 0$. Suppose that (16) holds for some $j \geq 0$. Then

$$\Lambda(A_k \cap B_{k-j-1,m}) = \sum_n \Lambda(A_k \cap B_{k-j,n}),$$

where we sum over all n such that $B_{k-j,n} \subset B_{k-j-1,m}$. By (16) and (14) (with equality), this is

$$= b^j \sum_n \Lambda(B_{k-j,n}) = b^j \Lambda(A_k \cap B_{k-j-1,m}) = b^{j+1} \Lambda(B_{k-j-1,m}),$$

which is (16) for $j+1$.

Let $A = \bigcap_k A_k$ be covered by the open intervals $I_1, \ldots, I_N$; by compactness we have to consider only finitely many. We choose k_ν such that $a^{k_\nu+1} < \Lambda(I_\nu) \leq a^{k_\nu}$. There exists $m > \max k_\nu$ such that $A_m \subset I_1 \cup \cdots \cup I_N$. We have

$$A_m \cap I_\nu \subset \bigcup_n B_{k_\nu n} \quad \text{where} \quad B_{k_\nu n} \cap I_\nu \neq \emptyset.$$

Since diam $B_{k_\nu n} \leq a^{k_\nu}$ by repeated application of (13), we see that the last union is contained in an interval of length $3a^{k_\nu}$. Hence it follows from (16) that

$$\Lambda(A_m \cap I_\nu) \leq \sum_n \Lambda(A_m \cap B_{k_\nu n}) = b^{m-k_\nu} \sum_n \Lambda(B_{k_\nu n})$$

$$\leq 3(a/b)^{k_\nu} b^m = 3a^{\alpha k_\nu} b^m \leq 3(\Lambda(I_\nu)/a)^\alpha b^m$$

because $a/b = a^\alpha$. We conclude by (14) that

$$\Lambda(A_0) \leq b^{-m}\Lambda(A_m) \leq 3a^{-\alpha} \sum_{\nu=1}^{N} \Lambda(I_\nu)^\alpha$$

which implies $\Lambda_\alpha(A) \geq 3^{-1}a^\alpha \Lambda(A_0) > 0$. $\square$

It is often easy to give upper estimates by constructing specific covers that, together with Theorem 10.5, lead to the precise determination of the Hausdorff dimension.

Example. (See Fig. 10.1) Let $p = 2, 3, \ldots$ and $0 < a < 1/p$. Let $A_0 = [0, 1]$. The set A_k consists of p^k disjoint closed intervals of length a^k, and A_{k+1} is obtained by replacing each interval by p disjoint subintervals of length a^{k+1}. Let $A = \bigcap A_k$. The p^k intervals of A_k cover A. Hence we see from (1) that

$$\Lambda_\alpha(A) \leq p^k a^{\alpha k} = 1 \quad \text{where} \quad \alpha = \frac{\log p}{\log(1/a)}.$$

On the other hand, it follows from Theorem 10.5 with $b = ap$ that $\Lambda_\alpha(A) > 0$. Hence we see that $\dim A = \alpha$. For the classical Cantor set we have $p = 2$ and $a = 1/3$ and therefore $\dim A = \log 2/\log 3$. See e.g. Fal90 for further examples.

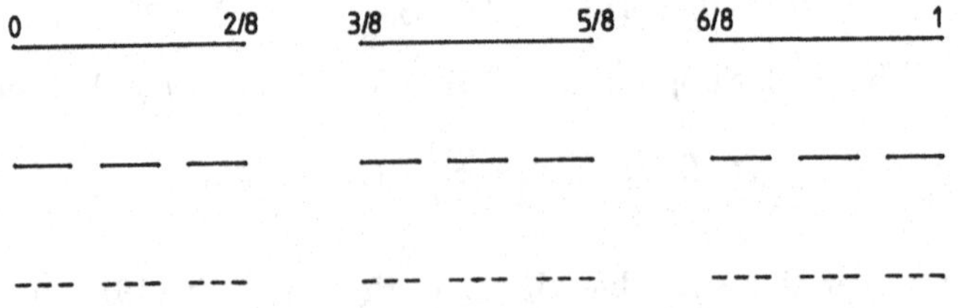

Fig. 10.1. Three stages of the construction for $p = 3$ and $a = 1/4$. The dimension is $\log 3/\log 4 \approx 0.792$

Exercises 10.2

1. Let h be a bilipschitz map of E, that is,
   $$M^{-1}|z_1 - z_2| \leq |h(z_1) - h(z_2)| \leq M|z_1 - z_2| \quad \text{for} \quad z_1, z_2 \in E.$$
   Show that $M^{-\alpha}\Lambda_\alpha(E) \leq \Lambda_\alpha(h(E)) \leq M^\alpha \Lambda_\alpha(E)$.

2. Let h satisfy a Hölder condition with exponent $\alpha \leq 1$ on E. Show that $\dim h(E) \leq \alpha^{-1} \dim E$.

3. Let $\varphi(t) = 1/\log(1/t)$ for $0 < t < 1/e$. Use Corollary 9.13 to show that $\Lambda_\varphi(E) = 0$ implies cap $E = 0$ for every compact set E. (This *logarithmic measure* φ is close to being best possible, see e.g. HayKe76, p. 229.)

4. Show that $\dim \bigcup_k E_k = \sup_k \dim E_k$ and deduce that, for compact $E \subset \mathbb{C}$, the function

$$\lim_{r \to 0} \dim \left[E \cap D(z,r) \right] \quad (z \in \mathbb{C})$$

is upper semicontinuous and has $\dim E$ as its maximum.

5. Suppose that there is a measure μ on the compact set E such that $\mu(E) > 0$ and (12) is satisfied. Show that $\Lambda_\varphi(E) > 0$.

6. Construct uncountably many disjoint compact sets on $\mathbb{T}$ of Hausdorff dimension one.

10.3 Lower Bounds for Compression

1. We have seen (Corollary 6.26) that a set $A \subset \mathbb{T}$ of positive measure is often mapped onto a set of zero linear measure in $\mathbb{C}$. We now show that A cannot be compressed much further. We tacitly assume that the angular limit $f(\zeta)$ exists and is finite for $\zeta \in A$; note that this is true except possibly for a set of zero capacity (Theorem 9.19).

Theorem 10.6. *Let*

$$(1) \quad \varphi(t) = t \exp\left(30 \sqrt{\log \frac{1}{t} \log \log \log \frac{1}{t}} \right) \quad \text{for} \quad 0 < t \le \exp(-e^e).$$

If f maps $\mathbb{D}$ conformally into $\mathbb{C}$ then, for all Borel sets $A \subset \mathbb{T}$,

$$(2) \quad \Lambda(A) > 0 \quad \Rightarrow \quad \Lambda_\varphi(f(A)) > 0.$$

This result (Mak85) is remarkably precise. It implies in particular that

$$(3) \quad \alpha < 1, \quad \Lambda(A) > 0 \quad \Rightarrow \quad \Lambda_\alpha(f(A)) > 0.$$

This solved a problem that had been open for many years (Beurling, Mat64); the best previous result (Car73) was that (3) holds for $\alpha < \alpha_0$ with some absolute constant $\alpha_0 > 1/2$. See Mak89d for further results.

The theorem is best possible except for the constant (Mak85) as the function f defined by (8.6.6) and the law of the iterated logarithm for lacunary series show: There is a set $A \subset \mathbb{T}$ with $\Lambda(A) = 2\pi$ such that $\Lambda_{\varphi_1}(f(A)) = 0$ where φ_1 is defined as in (1) but with 30 replaced by a smaller constant. See PrUrZd89 and Mak89d for examples from the theory of Julia sets.

We present an elementary proof (Roh88) of Theorem 10.6. It is based on the Makarov law of the iterated logarithm and on the following result. Let $K_1, K_2, \ldots$ denote suitable absolute constants.

Lemma 10.7. *Let $0 < \delta < \varepsilon$, $1/2 \le r < 1$ and $A \subset \mathbb{T}$. If diam $f(A) \le \varepsilon$ and*

(4) $$|f(r\zeta) - f(\zeta)| \le \varepsilon, \quad (1-r)|f'(r\zeta)| \ge \delta \quad \text{for} \quad \zeta \in A,$$

then A can be covered by at most $K_1(\varepsilon/\delta)^2$ sets of diameter at most $1 - r$.

Proof. We consider a square grid of mesh size $c\delta$ where the absolute constant c with $0 < c < 1$ will be determined below. Let Q_k $(k = 1, \ldots, m)$ be the halfopen squares that contain points of $f(rA)$. Since these squares are disjoint and since diam $f(rA) \le 3\varepsilon$ by (4), we see that

$$(c\delta)^2 m = \text{area}(Q_1 \cup \cdots \cup Q_m) \le K_2 \varepsilon^2.$$

Hence it suffices to show that $A_k = \{\zeta \in A : f(r\zeta) \in Q_k\}$ satisfies diam $A_k \le 1 - r$ for $k = 1, \ldots, m$.

Suppose that diam $A_k > 1 - r$ for some k. Then there are $\zeta, \zeta' \in A_k$ with $|\zeta - \zeta'| > 1 - r$ so that the non-euclidean distance satisfies $\lambda(r\zeta, r\zeta') \ge 1/K_3$. Hence, by (1.3.18) and (4),

$$\delta < (1 - r^2)|f'(r\zeta)| \le K_4|f(r\zeta) - f(r\zeta')| < 2K_4 c\delta$$

because $f(r\zeta), f(r\zeta') \in Q_k$, which is a contradiction if we choose $c = 1/(2K_4)$. $\qquad\square$

Proof of Theorem 10.6. Let $\Lambda(A) > 0$. By Corollary 8.11 there exists a Borel set $A' \subset A$ with $\Lambda(A') > 0$ such that

(5) $$-\psi(r) \le \log|f'(r\zeta)| \le \psi(r) \equiv 7\sqrt{\log \frac{1}{1-r} \log \log \log \frac{1}{1-r}}$$

for $\zeta \in A'$ and $r_0 \le r < 1$ with suitable $r_0 < 1$. Hence

(6) $$|f(\zeta) - f(r\zeta)| \le \int_r^1 |f'(s\zeta)| \, ds \le \int_r^1 e^{\psi(s)} \, ds \, ;$$

the angular limit $f(\zeta)$ exists because the integral is finite. Now

$$\frac{d}{dr}\left[\int_r^1 e^{\psi(s)} \, ds - 2(1 - r)e^{\psi(r)}\right] = (1 - 2(1-r)\psi'(r))e^{\psi(r)} > 0$$

for r close to 1 because $(1 - r)\psi'(r) \to 0$ as $r \to 1$. Since the quantity in the square bracket converges to 0 as $r \to 1$ we conclude from (6) that

(7) $$|f(\zeta) - f(r\zeta)| < 2(1 - r)e^{\psi(r)} \quad \text{for} \quad r_0 \le r < 1, \quad \zeta \in A',$$

increasing r_0 if necessary.

Let (B_k) be any open cover of $f(A')$ and let $A_k = \{\zeta \in A' : f(\zeta) \in B_k\}$ for $k = 1, 2, \ldots$. We define ε_k, r_k and δ_k by

(8) $$\varepsilon_k = \text{diam } B_k = K_1(1 - r_k)\exp[\psi(r_k)], \quad \delta_k = (1 - r_k)\exp[-\psi(r_k)]$$

where $K_1 > 2$ is as in Lemma 10.7. Since (4) holds by (7) and (5), we obtain from Lemma 10.7 that A_k can be covered by at most $K_2(\varepsilon_k/\delta_k)^2$ sets of diameter $1 - r_k$ so that, by (8), (1) and (5),

$$\Lambda(A') \leq \sum_k \Lambda(A_k) \leq K_6 \sum_k (1 - r_k)\exp[4\psi(r_k)] < K_7 \sum_k \varphi(\varepsilon_k);$$

we have used here that $\log 1/\varepsilon_k \sim \log 1/(1 - r_k)$ by (8). It follows that $\Lambda_\varphi(f(A)) \geq \Lambda_\varphi(f(A')) \geq \Lambda(A')/K_7 > 0.$ □

Remark. The measure Λ_φ defined by (1) (but with a different constant) plays also a role (Mak89c) for non-Smirnov domains: If μ is the singular measure in (7.3.1) and if E is a Borel set on $\mathbb{T}$ with $\mu(E) > 0$ then $\Lambda_\varphi(f(E)) > 0$, and this is best possible.

2. We now turn to small sets on $\mathbb{T}$; see Mak87. As in (8.3.1) we define $\beta_f(-q)$ as the infimum of all β such that

$$(9) \qquad \int_\mathbb{T} |f'(r\zeta)|^{-q}|d\zeta| = O\big((1-r)^{-\beta}\big) \qquad \text{as} \quad r \to 1-.$$

Theorem 10.8. *Let f map $\mathbb{D}$ conformally into $\mathbb{C}$. If A is a Borel set on $\mathbb{T}$ then*

$$(10) \qquad \dim f(A) \geq \frac{q \dim A}{\beta_f(-q) + q + 1 - \dim A} \qquad \text{for} \quad q > 0.$$

The following result (Mak87) is, in part, an easy consequence.

Theorem 10.9. *Let f map $\mathbb{D}$ conformally into $\mathbb{C}$. Let $A \subset \mathbb{T}$ be a Borel set and $\alpha = \dim A$. Then*

$$\dim f(A) > \frac{\alpha}{2} \qquad \text{for} \quad 0 < \alpha \leq \frac{11}{12},$$
$$(11)$$
$$\dim f(A) > \frac{\alpha}{1 + \sqrt{12(1-\alpha)}} \qquad \text{for} \quad \frac{11}{12} < \alpha < 1.$$

If f is close-to-convex then

$$(12) \qquad \dim f(A) \geq \alpha/(2 - \alpha) \qquad \text{for} \quad 0 < \alpha \leq 1$$

and this estimate is best possible.

Taking the limit $\alpha \to 1$ in (11) we obtain that, for all conformal maps,

$$(13) \qquad \dim A = 1 \quad \Rightarrow \quad \dim f(A) \geq 1$$

which is stronger than (2) because we may have $\Lambda(A) = 0$ but is also weaker because we do not have a function φ as precise as (1). The estimate (11) is not sharp for $0 < \alpha < 1$. It is however not far from being sharp near $\alpha = 0$ and near $\alpha = 1$; see Mak87 where a wealth of further information can be found.

To prove (11) for $11/12 < \alpha < 1$ we use the estimate

$$\beta_f(-q) \le \sqrt{4q^2 + q + 1/4} - q - 1/2 < 3q^2 \quad (q > 0)$$

of Theorem 8.5. Our assertion follows from (10) if we choose $q = \sqrt{(1-\alpha)/3}$. The estimate (12) follows at once from (10) because $\beta_f(-1) = 0$ if f is close-to-convex, by Theorem 8.4. See Mak87, p. 63 for the construction of a starlike function for which (12) is sharp.

While the estimate $\dim f(A) \ge \alpha/2$ is not difficult to establish (Mat64, see Exercise 4) the proof of $\dim f(A) > \alpha/2$ is much harder and uses a method of Carleson; see Mak87, p. 75. If the Brennan conjecture $\beta_f(-2) \le 1$ is true then we would obtain from (10) that

$$\dim f(A) \ge 2\alpha/(4-\alpha) > \alpha/2 \quad \text{for} \quad \alpha > 0.$$

The proof of Theorem 10.8 is based on the following technical result (Mak87). Let $K_1, K_2, \ldots$ be absolute constants and $d_f(z) = \text{dist}(f(z), \partial G)$. We consider the dyadic points

$$(14) \qquad z = (1 - 2^{-n})e^{i\pi(2\nu - 1)/2^n} \quad (\nu = 1, \ldots, 2^n; \, n = 1, 2, \ldots)$$

and the corresponding dyadic arcs $I(z)$; see (6.5.15).

Proposition 10.10. *Let f map $\mathbb{D}$ conformally onto G and let $0 < \alpha \le 1$. If $A \subset \mathbb{T}$ is a Borel set with*

$$(15) \qquad \text{diam } f(A) \le \varepsilon < \min(\alpha, d_f(0))/4,$$

then there are dyadic points $z_1, \ldots, z_N$ with

$$(16) \qquad N \le \frac{K_1}{\alpha} \log \frac{1}{\varepsilon}, \quad d_f(z_j) \le K_2\varepsilon \quad (j = 1, \ldots, N)$$

and a set $E \subset \mathbb{T}$ with $\Lambda'_\alpha(E) < \varepsilon^2$ such that

$$(17) \qquad A \subset E \cup I(z_1) \cup \cdots \cup I(z_N).$$

Proof (see Fig. 10.2). Let D_1 and D_2 be concentric disks of radii ε and 2ε with $D_1 \cap f(A) \ne \emptyset$. It follows from Proposition 2.13 that there are countably many crosscuts $B_j \subset f^{-1}(G \cap \partial D_2)$ of $\mathbb{D}$ such that $G \cap D_2$ is covered by the sets $f(V_j)$ where V_j is a component of $\mathbb{D} \setminus B_j$. Let

$$E_j = A \cap \overline{V}_j, \quad H_j = f(V_j) \setminus \overline{D}_1.$$

We see from (15) that $\text{dist}(f(0), H_j) \ge d_f(0) - 2\varepsilon \ge d_f(0)/2$, and if $C \subset \mathbb{D}$ is a curve from 0 to E_j then $\Lambda(f(C) \cap H_j) \ge \varepsilon$. Hence it follows from Theorem 9.24 that

$$\text{cap } E_j < K_3 \exp(-\pi\varepsilon^2/\text{ area } H_j).$$

We may assume that area $H_j \geq$ area H_{j+1}. Then j area $H_j \leq$ area $D_2 = 4\pi\varepsilon^2$ and thus, by Theorem 10.3,

$$\sum_{j>N} \Lambda'_\alpha(E_j) \leq K_4 \sum_{j>N} (\text{cap } E_j)^\alpha \leq K_5 \sum_{j>N} e^{-\alpha j/4}$$

$$< \frac{K_6}{\alpha} e^{-\alpha N/4} < \frac{K_6}{\varepsilon} e^{-\alpha N/4} < \varepsilon^2$$

if we choose N suitably subject to (16). Hence the set $E = \bigcup_{j>N} E_j$ satisfies $\Lambda'_\alpha(E) < \varepsilon^2$.

Now let $j = 1, \ldots, N$. Let I_j^* be the arc of $\mathbb{T}$ between the endpoints of B_j and let z_j^* be the point on the non-euclidean line S_j connecting these endpoints that is nearest to 0. Then, by the Gehring-Hayman theorem (4.5.6),

$$d_f(z_j^*) \leq \text{diam } f(S_j) \leq K_7 \text{diam } f(B_j) \leq K_7 \text{diam } \partial D_2 = 4K_7\varepsilon,$$

and it follows from Corollary 1.5 that there is a dyadic point z_j such that $d_f(z_j) < K_8 d_f(z_j^*) \leq K_2\varepsilon$ and $E_j \subset I_j^* \subset I(z_j)$. Hence (17) holds because $E = \bigcup_{j>N} E_j$. $\qquad\square$

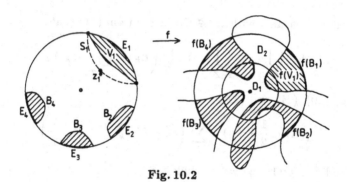

Fig. 10.2

Proof of Theorem 10.8. Let

(18) $\qquad \alpha < \dim A, \quad \beta > \beta_f(-q), \quad \gamma > (1 + \beta - \alpha)/q$

and let $M_1, M_2, \ldots$ be suitable constants. Let $m \in \mathbb{N}$ and (see (14))

(19) $\qquad Z = \{z \text{ dyadic point}: |z| \geq 1 - 2^{-m}, |f'(z)| < (1 - |z|)^\gamma\}.$

Since $|f'(|z|\zeta)| < M_1|f'(z)|$ for $\zeta \in I(z)$ by Corollary 1.6, we see from (9) and (18) that, with $r_n = 1 - 2^{-n}$, $n \geq m$,

$$(1 - r_n)^{1-\gamma q} \text{card } \{z \in Z : |z| = r_n\} < M_2 \int_{\mathbb{T}} \frac{|d\zeta|}{|f'(r_n\zeta)|^q} < \frac{M_3}{(1 - r_n)^\beta}$$

and therefore

$$\sum_{z \in Z} \Lambda(I(z))^\alpha \leq (2\pi)^\alpha \sum_{n=m}^\infty (1 - r_n)^\alpha \, \text{card} \{ z \in Z : |z| = r_n \}$$

(20)

$$< M_4 \sum_{n=m}^\infty 2^{-(\gamma q - 1 - \beta + \alpha)n} < \frac{1}{2} \Lambda'_\alpha(A)$$

by (18) if m is chosen sufficiently large; note that $\Lambda'_\alpha(A) > 0$ because $\alpha < \dim A$.

Let (B_k) be any cover of $f(A)$ with $\varepsilon_k \equiv \operatorname{diam} B_k$ sufficiently small. By Proposition 10.10 there exist dyadic points z_{kj} $(j = 1, \ldots, N_k)$ with $N_k < M_5 \log(1/\varepsilon_k)$ and $|z_{kj}| \geq 1 - 2^{-m}$ such that

(21) $$A_k \equiv \mathbb{T} \cap f^{-1}(B_k) \subset E_k \cup \sum_{j=1}^{N_k} I(z_{kj}), \quad \Lambda'_\alpha(E_k) < \varepsilon_k^2$$

and $d_f(z_{kj}) \leq K_2 \varepsilon_k$. If $|f'(z_{kj})| \geq (1 - |z_{kj}|)^\gamma$ then $(1 - |z_{kj}|)^{1+\gamma} \leq 4K_2 \varepsilon_k$ and therefore

(22) $$\sum_j \Lambda(I(z_{kj}))^\alpha < M_6 N_k \varepsilon_k^{\alpha/(1+\gamma)} < M_7 \varepsilon_k^{\alpha/(1+\gamma)} \log \frac{1}{\varepsilon_k}$$

where we sum over all j with $z_{kj} \notin Z$. We see from (21) that

$$A \subset \bigcup_k A_k \subset \bigcup_{z \in Z} I(z) \cup \bigcup_k \left(E_k \cup \bigcup_{z_{kj} \notin Z} I(z_{kj}) \right)$$

so that, by (20), (21) and (22),

$$\Lambda'_\alpha(A) < \frac{1}{2} \Lambda'_\alpha(A) + \sum_k \left(\varepsilon_k^2 + M_7 \varepsilon_k^{\alpha/(1+\gamma)} \log \frac{1}{\varepsilon_k} \right).$$

If $\delta \leq 2$ and $\delta < \alpha/(1+\gamma)$ we conclude that

$$M_8 \sum_k \varepsilon_k^\delta > \Lambda'_\alpha(A)/2 > 0, \quad \text{thus} \quad \Lambda_\delta(f(A)) > 0.$$

Hence $\dim f(A) \geq \delta$ and our assertion (10) follows if we let $\gamma \to (1+\beta-\alpha)/q$, $\beta \to \beta_f(-q)$ and $\alpha \to \dim A$; see (18). $\qquad\Box$

The next result (Mak89b) generalizes Löwner's lemma (except for the constant).

Theorem 10.11. *Suppose that* $\varphi(xt) \leq M x^\alpha \varphi(t)$ $(0 < x \leq 2, \, t > 0)$ *where* $\alpha > 0$. *If* f *maps* $\mathbb{D}$ *conformally into* $\mathbb{D}$ *with* $f(0) = 0$ *and if* $A \subset \mathbb{T}$ *is a Borel set such that* $f(A) \subset \mathbb{T}$ *then*

(23) $$\Lambda'_\varphi(f(A)) \geq c \Lambda'_\varphi(A)$$

where c *is a positive constant.*

Choosing $\varphi(t) = t^\alpha$ we see that (Ham88)

(24) $\qquad\qquad \dim f(A) \geq \dim A \quad$ if $\quad A \subset \mathbb{T}, \quad f(A) \subset \mathbb{T}$

for all conformal maps of $\mathbb{D}$ into itself.

Proof. Let (I_k) be a covering of $f(A)$ by arcs of $\mathbb{T}$ and define $A_k = A \cap f^{-1}(I_k)$. It follows from Theorem 10.3 and Theorem 9.21 that

$$c\Lambda'_\varphi(A_k) \leq \varphi(\operatorname{cap} A_k) \leq \varphi(\operatorname{cap} I_k) \leq \varphi(\operatorname{diam} I_k)$$

and thus

$$c\Lambda'_\varphi(A) \leq c \sum_k \Lambda'_\varphi(A_k) \leq \sum_k \varphi(\operatorname{diam} I_k)$$

which implies (23). $\qquad\qquad\qquad\qquad\qquad\qquad\qquad\qquad\qquad\qquad\qquad$ $\square$

Exercises 10.3

Let f map $\mathbb{D}$ conformally into $\mathbb{C}$ and let A be a Borel set on $\mathbb{T}$.

1. If $|f'(z)| \geq a(1 - |z|)^b$ with $a > 0, b < 1$ show that $\dim f(A) \geq (b+1)^{-1} \dim A$.

2. Show that $\dim f(A) \geq \dim A/(2.601 - \dim A)$.

3. Let g map $\mathbb{D}$ conformally onto a Jordan domain and let $|g'(z)| \geq a(\varepsilon)(1 - |z|)^\varepsilon$ $(r(\varepsilon) < |z| < 1)$ for every $\varepsilon > 0$. If $f(\mathbb{D}) \subset g(\mathbb{D})$ and $f(A) \subset \partial g(\mathbb{D})$ show that $\dim f(A) \geq \dim A$.

4. Use Theorem 9.23 and Proposition 10.10 to prove that $\Lambda_\alpha(A) > 0$ implies $\Lambda_{\alpha/2}(f(A)) > 0$.

The following two exercises are more difficult.

5. If f is conformal for all $\zeta \in A$ show that $\dim f(A) = \dim A$.

6. Let $0 < \alpha \leq 1$. Then

$$\frac{|f(\zeta) - f(r\zeta)|}{(1-r)|f'(r\zeta)|} < (1-r)^{-\frac{2(1-\alpha)}{\alpha}} \left(\log \frac{1}{1-r}\right)^{\frac{5}{\alpha}} \quad \text{for} \quad \zeta \in \mathbb{T} \backslash E, \quad r_0(\zeta) < r < 1$$

where the exceptional set satisfies $\Lambda_\alpha(E) = 0$ (Pom87).

10.4 Zygmund Measures and the Angular Derivative

In Section 7.2 we defined a Zygmund measure μ as a non-negative finite measure on $\mathbb{T}$ such that $|\mu(I) - \mu(I')| \leq M_1 \Lambda(I)$ for all adjacent arcs I and I' of equal length. Let $M_1, M_2, \ldots$ denote suitable positive constants.

Theorem 10.12. *Let μ be a Zygmund measure and let*

(1) $\qquad\qquad \psi(t) = t\sqrt{\log \frac{1}{t} \log \log \log \frac{1}{t}} \quad \text{for} \quad 0 < t \leq e^{-e^e}.$

If E is a Borel set on $\mathbb{T}$ *then*

$$(2) \qquad\qquad \Lambda_\psi(E) = 0 \quad \Rightarrow \quad \mu(E) = 0.$$

The function ψ in this result (Mak89b) is related to that of Theorem 10.6 but is not the same. Makarov has given an example of a nonzero Zygmund measure μ such that $\Lambda_\psi(E) < \infty$ where E is the support of μ. Hence it follows from Proposition 10.2 that ψ in (2) cannot be replaced by any function φ with $\varphi(t)/\psi(t) \to 0$ as $t \to 0$.

Proof. By Proposition 7.3, the function

$$(3) \qquad\qquad g(z) = \frac{1}{2\pi} \int_\mathbb{T} \frac{\zeta+z}{\zeta-z}\, d\mu(\zeta) \quad (z \in \mathbb{D})$$

has positive real part and belongs to $\mathcal{B}$. Since $d\mu(\zeta) \geq 0$ we see that

$$\operatorname{Re} g(z) \geq \frac{1}{2\pi} \int_{I(z)} \frac{1-|z|^2}{|\zeta-z|^2}\, d\mu(\zeta),$$

where $I(re^{i\vartheta}) = \{e^{it} : |t - \vartheta| \leq \pi(1-r)\}$. Since $|\zeta - z| \leq (1+\pi)(1-|z|)$ for $\zeta \in I$ we conclude that

$$(4) \qquad\qquad \mu(I(z)) = \int_{I(z)} d\mu(\zeta) \leq M_2 |g(z)| \Lambda(I(z)).$$

For $n = 1, 2, \ldots$ and $\nu = 1, \ldots, 2^n$, let $z_{n\nu}$ be the dyadic points defined by (6.5.14) and $I_{n\nu}$ the corresponding dyadic arcs. We have $|z_{n\nu}| = r_n = 1 - 2^{-n}$. Since

$$|g(r_n\zeta) - g(z_{n\nu})| \leq M_3 \quad \text{for} \quad \zeta \in I_{n\nu}$$

by (4.2.5), we obtain from (4) that, for $k = 1, 2, \ldots$,

$$\int_\mathbb{T} |g(r_n\zeta)|^{2k-1}\, d\mu(\zeta) \leq M_4^k \sum_{\nu=1}^{2^n} (1 + |g(z_{n\nu})|)^{2k-1} \mu(I_{n\nu})$$

$$\leq M_5^k \sum_{\nu=1}^{2^n} (1 + |g(z_{n\nu})|)^{2k} \Lambda(I_{n\nu})$$

$$\leq M_6^k \int_\mathbb{T} (1 + |g(r_n\zeta)|)^{2k} |d\zeta|.$$

Using again (4.2.5) and then Theorem 8.9 we conclude that

$$(5) \qquad\qquad \int_\mathbb{T} |g(r\zeta)|^{2k-1}\, d\mu(\zeta) \leq M_7^k \left(k! \left(\log \frac{1}{1-r} \right)^k + 1 \right)$$

for $0 < r < 1$ and $k = 1, 2, \ldots$. As in the proof of Theorem 8.10 we deduce that

(6) $$\limsup_{r \to 1} \frac{|g(r\zeta)|}{\sqrt{\log \frac{1}{1-r} \log \log \log \frac{1}{1-r}}} \le M_8 \quad \text{for } \mu\text{-almost all } \zeta \in \mathbb{T}.$$

Now let E be a Borel set on $\mathbb{T}$ with $\mu(E) > 0$. By (6) there exists $E' \subset E$ with $\mu(E') > 0$ such that

(7) $$|g(r\zeta)| < M_9 \frac{\psi(1-r)}{1-r} \quad \text{for } \zeta \in E', \quad r_0 < r < 1$$

with some $r_0 < 1$. Let (I_j) be a covering of E' by arcs with diam $I_j \le 1 - r_0$ each of which intersects E' at some point ζ_j. If r_j is defined by diam $I_j = 1 - r_j$ then $I_j \subset I(r_j \zeta_j)$ and thus, by (4) and (7),

$$\mu(I_j) \le \mu(I(r_j\zeta_j)) \le M_2 |g(r_j\zeta_j)| \Lambda(I(r_j\zeta_j)) \le M_{10}\psi(\text{diam } I_j)$$

so that $\Lambda_\psi(E') \ge \mu(E')/M_{10} > 0$. $\qquad\square$

The angular derivative was studied in Section 4.3 and 6.5.

Theorem 10.13. *Let f map $\mathbb{D}$ conformally into $\mathbb{C}$. Then f has a finite angular derivative on a set $E \subset \mathbb{T}$ with $\Lambda_\psi(E) > 0$ where ψ is given by (1).*

This result (Mak89b, compare AnPi89) solves a problem on which much work has been done starting with Hal68; see in particular AnPi88 and also Ber92 for a related result in a more general context. The answer is best possible (Mak89b): There is f such that $\{\zeta \in \mathbb{T} : f'(\zeta) \ne \infty \text{ exists}\}$ has σ-finite Λ_ψ-measure. The *Anderson conjecture* that

$$\int_0^1 |f''(r\zeta)| \, dr < \infty \quad \text{for some } \zeta \in \mathbb{T}$$

is however still open; see And71.

Proof. The case that f has a nonzero angular derivative on a set E with $\Lambda(E) > 0$ is trivial because then $\Lambda_\psi(E) = \infty$ by Proposition 10.2. Note that there are conformal maps without any finite nonzero angular derivative (Proposition 4.12).

The other case is a consequence of Proposition 4.7 and the next result applied to the Bloch function $g = -\log f'$. $\qquad\square$

Theorem 10.14. *If $g \in \mathcal{B}$ has a finite angular limit almost nowhere, then*

$$\operatorname{Re} g(r\zeta) \to +\infty \quad \text{as } r \to 1-, \quad \zeta \in E$$

for some Borel set $E \subset \mathbb{T}$ with $\Lambda_\psi(E) > 0$ where ψ is given by (1).

Proof. Let f map $\mathbb{D}$ conformally onto some component of $\{w \in \mathbb{D} : \operatorname{Re} g(w) > 0\}$; this set is non-empty because otherwise g would have finite radial limits almost everywhere by Fatou's theorem and the Riesz uniqueness theorem. Then $h = g \circ f$ has positive real part and is also a Bloch function as we see from (4.2.1) and $(1 - |z|^2)|f'(z)| \leq 1 - |f(z)|^2$. Hence, by Proposition 7.3,

$$(8) \qquad h(z) = ib + \int_{\mathbb{T}} \frac{s+z}{s-z} \, d\mu(s) \qquad \text{for} \quad z \in \mathbb{D}$$

where μ is a Zygmund measure on $\mathbb{T}$.

Suppose now that μ is absolutely continuous. Then we can find a set $A \subset \mathbb{T}$ with $\Lambda(A) > 0$ such that the radial limit $h(\zeta)$ exists for $\zeta \in A$ and $\operatorname{Re} h(\zeta) > 0$ because μ is a finite measure; see Dur70, p. 4. Thus $g(w) \to h(\zeta)$ for $w = f(r\zeta)$, $r \to 1$ so that the Bloch function g has the angular limit $h(\zeta)$ at $f(\zeta)$ by Theorem 4.3. Since $f(\mathbb{D})$ is a component of $\{\operatorname{Re} g(w) > 0\}$ we see that $f(\zeta) \in \mathbb{T}$, and since $\Lambda(A) > 0$ we conclude from Löwner's lemma that $\Lambda(f(A)) > 0$. This contradicts our assumption that g has a finite angular limit almost nowhere on $\mathbb{T}$.

Hence μ is not absolutely continuous. It follows (HeSt69, p. 296) that $\mu(B) > 0$ where B is the set on $\mathbb{T}$ such that the increasing function corresponding to μ has an infinite derivative. If $\zeta \in B$ then, by (8),

$$(9) \qquad \operatorname{Re} h(r\zeta) = \int_{\mathbb{T}} p(r\zeta, s) \, d\mu(s) \to +\infty \qquad \text{as} \quad r \to 1$$

which is proved similarly as in Dur70, p. 4. Hence $\operatorname{Re} g(w) \to +\infty$ for $w = f(r\zeta)$, $r \to 1$ so that, by (4.2.19),

$$(10) \qquad \operatorname{Re} g(rw) \to +\infty \qquad \text{as} \quad r \to 1, \quad \omega \in E \equiv f(B).$$

Finally it follows from $\mu(B) > 0$ by Theorem 10.12 that $\Lambda_\psi(B) > 0$. Since $f(B) \subset \mathbb{T}$ by (10) and the definition of f and since ψ satisfies the assumptions of Theorem 10.11 with $\alpha = 1/2$, we obtain that $\Lambda_\psi(f(B)) > 0$. $\qquad \square$

We now show that there are always many points where a conformal map is not twisting; compare Section 6.4.

Theorem 10.15. *If f maps $\mathbb{D}$ conformally into $\mathbb{C}$ then there is a set $E \subset \mathbb{T}$ with $\dim E = 1$ such that*

$$(11) \qquad \sup_r |\arg f'(r\zeta)| < \infty, \quad \sup_r |\arg[f(r\zeta) - f(\zeta)]| < \infty \qquad \text{for} \quad \zeta \in E.$$

This result (Mak89a) is best possible: There exists f such that the set where (11) holds has Λ_φ-measure 0 for every φ satisfying $\varphi(t) = O(t^\alpha)$ $(t \to 0)$ for each $\alpha < 1$.

Proof. For $k = 7, 8, \ldots$ we define f_k by $f_k(0) = 0$ and $\log f_k' = k^{-1} \log f'$ so that, by Proposition 1.2,

$$(1 - |z|^2)|f_k''(z)/f_k'(z)| \leq 6/k < 1 \qquad \text{for} \quad z \in \mathbb{D}.$$

Hence f_k maps $\mathbb{D}$ conformally onto the inner domain of a Jordan curve J_k by the Becker univalence criterion. It follows from (4.2.4) that $|\log f_k'(z)| \leq (3/k)\log[1/(1-|z|)] + \text{const}$ so that f_k satisfies a Hölder condition of exponent $1 - 3/k$ (Dur70, p. 74).

Let B_k be the set of all $\omega \in J_k$ such that $\{t\omega : t > 1\} \cap J_k = \emptyset$. A projection argument shows that $\Lambda_1(B_k) > 0$. It follows (Exercise 10.2.2) that

$$\dim f_k^{-1}(B_k) \geq (1 - 3/k)\dim B_k \geq 1 - 3/k$$

so that $E = \bigcup_k f_k^{-1}(B_k)$ satisfies $\dim E \geq 1$. The dimension is not decreased if we delete from E those points where the angular limit $f(\zeta)$ does not exist or is infinite because these points form a set of zero capacity by Theorem 9.19.

Now let $\zeta \in E$ and thus $\zeta \in f_k^{-1}(B_k)$ for some k. Then $|\arg[f_k(r\zeta) - f_k(\zeta)]| < 2\pi$ for $0 < r < 1$ by the definition of B_k. As in the proof (ii) $\Rightarrow$ (iii) of Theorem 11.1 below, it can be deduced that $|\arg f_k'(r\zeta)|$ is bounded. Hence $|\arg f'(r\zeta)|$ is bounded and thus also $|\arg[f(r\zeta) - f(\zeta)]|$; see Ost35. $\square$

In fact more is true (Mak89d, Mak89e): If $g \in \mathcal{B}$ then

(12)
$$\dim\{\zeta \in \mathbb{T} : \sup_r |g(r\zeta)| < \infty\} = 1.$$

For conformal maps this means that

(13)
$$\dim\{\zeta \in \mathbb{T} : \sup_r |\log f'(r\zeta)| < \infty\} = 1.$$

This can be further generalized (Roh89): Let $g \in \mathcal{B}$ have finite angular limits almost nowhere and consider a square grid of mesh size $a > 0$. We prescribe a sequence (Q_n) of closed squares of this grid with the only stipulation that Q_n and Q_{n+1} are adjacent; see Fig. 10.3. Then there exists a set $E \subset \mathbb{T}$ with

(14)
$$\dim E \geq 1 - a^{-1}K\|g\|_{\mathcal{B}}$$

(where K is an absolute constant) such that, for each $\zeta \in E$, we can find $r_n \to 1$ with

(15)
$$g(r\zeta) \subset Q_n \cup Q_{n+1} \quad \text{for} \quad r_n \leq r \leq r_{n+1}.$$

In other words practically all types of behaviour occur in sets of dimension almost one.

We will consider the subspace $\mathcal{B}_0$ in Section 11.2.

Exercises 10.4

Let f map $\mathbb{D}$ conformally into $\mathbb{C}$.

1. Suppose that f is conformal almost nowhere on $\mathbb{T}$. Show that the set where $\arg f'(r\zeta) \to +\infty$ and the set where $f'(\zeta) = 0$ have positive Λ_ψ-measure where ψ is given by (1).
2. Use Rohde's theorem stated above to show that there is a set $E \subset \mathbb{T}$ with $\dim E = 1$ such that $f'(\zeta) \neq \infty$ exists and $\sup_r |\arg f'(r\zeta)| < \infty$ for $\zeta \in E$.

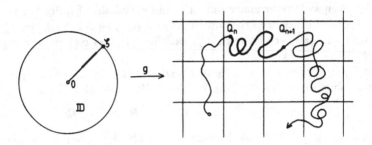

Fig. 10.3

10.5 The Size of the Boundary

In the definition of Hausdorff measure we allowed coverings by sets the diameter of which may vary a great deal. For many purposes coverings by sets (e.g. disks) of the same size is more natural and simpler to handle.

Let E be a bounded set in $\mathbb{C}$ and let $N(\varepsilon, E)$ denote the minimal number of disks of diameter ε that are needed to cover E. Up to bounded multiplies it is the same as the number of squares of a grid of mesh size ε that intersect E. We define the (upper) *Minkowski dimension* of E by

$$(1) \qquad \operatorname{mdim} E = \limsup_{\varepsilon \to 0} \frac{\log N(\varepsilon, E)}{\log(1/\varepsilon)}$$

whereas $\dim E$ was the Hausdorff dimension.

Proposition 10.16. *If E is any bounded set in $\mathbb{C}$ then*

$$(2) \qquad \dim E \le \liminf_{\varepsilon \to 0} \frac{\log N(\varepsilon, E)}{\log(1/\varepsilon)} \le \operatorname{mdim} E.$$

Proof. Let β be any number greater than the limes inferior in (2). Then there are $\varepsilon_n \to 0$ such that $N_n = N(\varepsilon_n, E) < \varepsilon_n^{-\beta}$. If E is covered by the disks $D_1, \ldots, D_{N_n}$ of diameter ε_n then, for $\alpha > \beta$,

$$\sum_{k=1}^{N_n} (\operatorname{diam} D_k)^\alpha = N_n \varepsilon_n^\alpha < \varepsilon_n^{\alpha - \beta} \to 0 \qquad \text{as} \quad n \to \infty$$

and thus $\dim E \le \alpha$ which implies (2). $\qquad \square$

Example. The (locally connected) continuum

$$E = [0,1] \cup \bigcup_{n=1}^{\infty} \bigcup_{\nu=1}^{[n^{3/4}]} \left[\frac{\nu}{n}, (1+i)\frac{\nu}{n} \right]$$

has σ-finite linear measure so that $\dim E = 1$. On the other hand, given $\varepsilon > 0$ choose n such that $1/(2n) \le \varepsilon < 1/n$. No disk of diameter ε can intersect two different segments $[\nu/n, (1+i)\nu/n]$. Hence we need at least $c_1 n^{3/2}$ such disks to cover E so that $N(\varepsilon, E) \ge c_2 \varepsilon^{-3/2}$. It follows that $\mathrm{mdim}\, E \ge 3/2 > \dim E = 1$.

There are however many cases for which the Hausdorff and Minkowski dimensions are identical; see e.g. Hut81 and MaVu87.

The Minkowski dimension of the boundary is related (Pom89) to the growth of the integral means of the derivative.

Theorem 10.17. *Let f map $\mathbb{D}$ conformally onto a bounded domain G and let $1 \le p \le 2$. If*

$$(3) \qquad N(\varepsilon, \partial G) = O(\varepsilon^{-p}) \qquad as \quad \varepsilon \to 0$$

then

$$(4) \qquad \int_{\mathbb{T}} |f'(r\zeta)|^p |d\zeta| = O\left(\frac{1}{(1-r)^{p-1}} \log \frac{1}{1-r} \right) \qquad as \quad r \to 1.$$

If (4) holds and if G is a John domain then

$$(5) \qquad N(\varepsilon, \partial G) = O\left(\frac{1}{\varepsilon^p} \left(\log \frac{1}{\varepsilon} \right)^2 \right) \qquad as \quad \varepsilon \to 0.$$

Note that (5) is slightly weaker than (3). Furthermore every (bounded) quasidisk is a John domain. Let $\beta_f(p)$ be defined by (8.3.1).

Corollary 10.18. *If f maps $\mathbb{D}$ conformally onto a John domain G then*

$$\mathrm{mdim}\, \partial G = p$$

where p is the unique solution of $\beta_f(p) = p - 1$.

Proof. Since G is a John domain it follows from (8.3.1) and Theorem 5.2 (iii) that $\beta_f(p + \delta) \le \beta_f(p) + q\delta$ ($\delta > 0$) for some $q < 1$. Hence $\beta_f(p) = p - 1$ has a unique solution p and

$$\int_{\mathbb{T}} |f'(r\zeta)|^{p+\delta} |d\zeta| = O((1-r)^{1-p-q\delta-\eta}) = O\left((1-r)^{1-p-\delta} \log \frac{1}{1-r} \right)$$

for suitable $\eta > 0$ so that (4) holds with p replaced by $p + \delta$. Therefore $\mathrm{mdim}\, \partial G \le p + \delta$ by (5) and thus $\mathrm{mdim}\, \partial G \le p$. The converse follows similarly from the fact that (3) implies (4). $\square$

The assumption that G is a John domain cannot be deleted as the following example shows: Let G be a bounded starlike domain with area $\partial G > 0$. Then mdim $E = 2$ but $\beta_f(p) \le p - 1$ for every $p \ge 1$; see the proof of Theorem 8.4 and use that f is bounded.

Corollary 10.19. *Let $f(z) = \Sigma a_n z^n$ map $\mathbb{D}$ conformally onto a bounded domain G with* mdim $\partial G = p$. *If $p = 1$ then*

(6) $$a_n = O(n^{\varepsilon-1}) \quad (n \to \infty) \quad \text{for every } \varepsilon > 0,$$

and if $p > 1$ then there is $\eta = \eta(p) > 0$ such that

(7) $$a_n = O(n^{-\eta-1/p}) \quad \text{as} \quad n \to \infty.$$

This statement (Pom89) is more precise than (8.4.5) which corresponds to $p = 2$; see also CaJo91.

Proof. The first part of Theorem 10.17 shows that $\beta_f(p) \le p - 1$ so that $\beta_f(1) = 0$ if $p = 1$. If $p > 1$ we see from Proposition 8.3 and Theorem 8.5 that, for $0 < \delta < 1$,

$$\beta_f(1) \le \frac{p-1}{p-\delta}\beta_f(\delta) + \frac{1-\delta}{p-\delta}\beta_f(p) \le \frac{p-1}{p-\delta}\left(3\delta^2 + 7\delta^3 + 1 - \delta\right) < 1 - \frac{1}{p} - \eta(p)$$

for some $\eta(p) > 0$ if δ is chosen small enough. Now (6) and (7) follow from (8.4.2). $\qquad\square$

Proof of Theorem 10.17. Let $M, M_1 \ldots$ be suitable constants. For $n = 1, 2, \ldots$ let $z_{n\nu}$ $(\nu = 1, \ldots, 2^n)$ denote the dyadic points (6.5.14) and $I_{n\nu}$ the dyadic arcs (6.5.15).

We first assume that (3) holds. By Corollary 1.6 it suffices to prove (4) for the case that $r = r_n = 1 - 2^{-n}$. For fixed $n = 1, 2, \ldots$ we write $\varepsilon_{-1} = 0$ and

(8) $$\varepsilon_k = 2^k(1 - r_n)^{1/p} = 2^{k-n/p} \quad \text{for} \quad k = 0, 1, \ldots.$$

Let m_k denote the number of points $z_{n\nu}$ $(\nu = 1, \ldots, 2^n)$ such that

(9) $$\varepsilon_{k-1} \le d_f(z_{n\nu}) \equiv \text{dist}(f(z_{n\nu}), \partial G) < \varepsilon_k.$$

Since $d_f(z)$ is bounded we have $m_k = 0$ for $k \ge Mn$. It follows from Corollary 1.6 and $(1 - |z|)|f'(z)| \le 4d_f(z)$ that

(10)
$$\int_0^{2\pi} |f'(r_n e^{it})|^p \, dt \le \frac{M_1}{2^n} \sum_{\nu=1}^{2^n} |f'(z_{n\nu})|^p \le \sum_{k<Mn} \frac{M_2 \varepsilon_k^p m_k}{2^n(1-r_n)^p}$$
$$= M_2 2^{(p-2)n} \sum_{k<Mn} 2^{kp} m_k$$

by (8). Now let $k \ge 1$. We see from (3) that ∂G can be covered by

(11) $$N_k = N(\varepsilon_k, \partial G) < M_3 \varepsilon_k^{-p}$$

disks of diameter $\varepsilon_k/3$; we may assume that each disk contains a point of ∂G. Let V_k be the union of these disks; see Fig. 10.4.

Consider an index ν such that (9) holds. Then $f(z_{n\nu}) \notin \overline{V}_k$. Hence we can find a connected set $A_{k\nu} \subset \mathbb{D}$ connecting $[0, e^{2i\pi(\nu-1)/2^n}]$ with $[0, e^{2i\pi\nu/2^n}]$ such that $f(A_{k\nu}) \subset \partial V_k$. It follows from Corollary 4.19 that $d_f(z_{n\nu}) < M_4 \Lambda(f(A_{k\nu})) + \varepsilon_k/3$. Summing over all ν such that (9) holds we see that

$$m_k \varepsilon_k/6 = m_k(\varepsilon_{k-1} - \varepsilon_k/3) < M_4 \sum_\nu \Lambda(f(A_{k\nu})) \leq M_4 \Lambda(\partial V_k) \leq \pi M_4 N_k \varepsilon_k/3$$

because the sets $A_{k\nu}$ are essentially disjoint. Hence $m_k < M_5 \varepsilon_k^{-p} = M_5 2^{-kp+n}$ by (11) and (8) so that, by (10),

$$\int_0^{2\pi} |f'(r_n e^{it})|^p \, dt < M_7 2^{(p-2)n} \left(2^n + \sum_{1 \leq k < Mn} 2^{kp} 2^{-kp+n} \right) < M_8 n 2^{(p-1)n}$$

which proves (4) because $1 - r_n = 2^{-n}$.

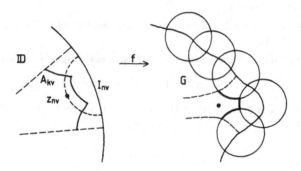

Fig. 10.4

Conversely we assume that (4) holds and that G is a John domain. Let $0 < \varepsilon < d_f(0)$ be given and let q_n denote the number of $\nu \in \{1, \ldots, 2^n\}$ such that

$$(12) \qquad \varepsilon/c < d_f(z_{n\nu}) < \varepsilon c$$

where c will be chosen below. Since G is a John domain we see from Theorem 5.2 (iii) with $z = 0$ that $q_n = 0$ for $n > M_9 \log(c/\varepsilon)$. Furthermore

$$\frac{2\pi}{2^n} \sum_{\nu=1}^{2^n} d_f(z_{n\nu})^p \leq M_{10} \int_0^{2\pi} (1 - r_n)^p |f'(r_n e^{it})|^p \, dt < \frac{M_{11}}{2^n} \log \frac{1}{1 - r_n}$$

by Corollary 1.6 and by (4) so that $q_n < M_{12} n (c/\varepsilon)^p$. It follows that

$$(13) \qquad \sum_{n=1}^\infty q_n < M_{13} (c/\varepsilon)^p [\log(c/\varepsilon)]^2 .$$

Let $\zeta \in \mathbb{T}$. Since $\varepsilon < d_f(0)$ there exists $r = r(\zeta)$ such that $d_f(r\zeta) = \varepsilon$ and thus $n = n(\zeta)$ and $\nu = \nu(\zeta)$ such that

$$\zeta \in I_{n\nu}, \quad M_{14}^{-1}\varepsilon < d_f(z_{n\nu}) < M_{14}\varepsilon,$$

where we have again used Corollary 1.6. Hence (12) holds if we choose $c = M_{14}$. Since G is a John domain it follows from Corollary 5.3 that $\operatorname{diam} f(I_{n\nu}) < M_{15}d_f(z_{n\nu})$ so that $f(I_{n\nu})$ lies in at most M_{16} disks of diameter ε. If we consider all $\zeta \in \mathbb{T}$ we deduce that $N(\varepsilon, \partial G) \leq M_{16}\sum_n q_n$ and (5) follows from (13). $\square$

Exercises 10.5

1. Show that the set indicated in Fig. 10.1 has the Minkowski dimension $\log 3/\log 4$.

2. Let the curve C be the limit of polygonal curves C_n where C_{n+1} is obtained from C_n by replacing each side S by p sides of length $a\Lambda(S)$ where $ap > 1$. Show that $\operatorname{mdim} C \leq \log p/\log(1/a)$.

3. Let $f(z) = \Sigma a_n z^n$ map $\mathbb{D}$ conformally onto G and let $\operatorname{mdim} \partial G \leq 1.5$. Show that $a_n = O(n^{0.331-1})$.

4. Let J be the image of $\mathbb{T}$ under an ε-quasiconformal map of $\mathbb{C}$ onto $\mathbb{C}$. Show that

$$\operatorname{mdim} J < 1 + K\varepsilon^2, \quad 0 < \varepsilon < 1$$

where K is an absolute constant (BePo87; use Exercise 8.3.4 and the Lehto majorant principle).

Chapter 11. Local Boundary Behaviour

11.1 An Overview

We first consider quasicircles J for which

$$|a - w| + |w - b| = (1 + o(1))|a - b| \quad \text{as} \quad a, b \in J, \quad |a - b| \to 0$$

where w lies on J between a and b. The conformal maps f of $\mathbb{D}$ onto the inner domain of J can be characterized by the property that

$$(1) \qquad\qquad (1 - |z|)f''(z)/f'(z) \to 0 \quad \text{as} \quad z \to 1,$$

i.e. $\log f'$ lies in the little Bloch space $\mathcal{B}_0$.

Then we study the behaviour of a general conformal map f of $\mathbb{D}$ near a given point $\zeta \in \mathbb{T}$ where f has a finite angular limit w. We study the case that w is well-accessible, i.e. accessible through a curvilinear sector in G. This is a local version of the concept of a John domain. In particular we consider a local version of (1), i.e.

$$(2) \qquad (z - \zeta)f''(z)/f'(z) \to 0 \quad \text{as} \quad z \to \zeta \text{ in every Stolz angle}.$$

This leads to a geometric characterization of isogonality. A main tool in the first two sections is the Carathéodory kernel theorem about convergence of domain sequences.

The main tool in the last two sections is the module of curve families introduced in Section 9.2. For the moment we only consider domains of the form

$$(3) \qquad\qquad G = \{w + \rho e^{i\vartheta} : 0 < \rho < c, \, \vartheta_-(\rho) < \vartheta < \vartheta_+(\rho)\}$$

and use the notation indicated in Fig. 11.1. A key fact (RoWa76, JenOi77) is that $\operatorname{mod} \Gamma(\rho_1, \rho_2)$ is approximately additive if and only if $f^{-1}(T(r))$ is approximately a circular arc in $\mathbb{D}$. This is used to establish conditions for f to be conformal at a point. The simplest is the Warschawski condition (Corollary 11.11): If $\vartheta_\pm(\rho) = \pm(\pi/2 - \delta(\rho))$ where $\delta(\rho)$ is increasing then f is conformal at ζ if and only if

$$\int_0^c \delta(\rho)\rho^{-1} \, d\rho < \infty.$$

If $\vartheta_\pm$ satisfies a certain regularity condition then, with $f(z) = \omega + \rho e^{i\vartheta}$,

$$(4) \quad \log(z - \zeta) = -\int_\rho^c \frac{\pi}{\vartheta_+(t) - \vartheta_-(t)} \frac{dt}{t} + i\pi \frac{\vartheta - \vartheta_-(\rho)}{\vartheta_+(\rho) - \vartheta_-(\rho)} + b + o(1)$$

as $z \to \zeta$ where b is a finite constant; see Theorem 11.16. The Ahlfors integral in (4) plays an important role in a wider context.

In the literature these results are often stated for conformal maps between strip domains and for the boundary point ∞ because the statement becomes simpler in this setting. We however consider $\mathbb{D}$ and a finite boundary point in accordance with the other chapters.

We can present only a selection of the great number of results that are known. Some of the early workers were Ahlfors, Ostrowski, Wolff, Warschawski and Ferrand, see e.g. GaOs49a, GaOs49b and LeF55. There are many interesting later publications, in particular a series of papers by Rodin and Warschawski.

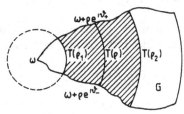

Fig. 11.1. The family $\Gamma(\rho_1, \rho_2)$ consists of all curves in the shaded domain that separate the circular arcs $T(\rho_1)$ and $T(\rho_2)$

11.2 Asymptotically Conformal Curves and $\mathcal{B}_0$

1. The Jordan curve $J \subset \mathbb{C}$ is called *asymptotically conformal* if

$$(1) \quad \max_{w \in J(a,b)} \frac{|a - w| + |w - b|}{|a - b|} \to 1 \quad \text{as} \quad a, b \in J, \quad |a, b| \to 0$$

where $J(a, b)$ is the smaller arc of J between a and b; note that the quotient is constant on ellipses with foci a and b. It follows from (5.4.1) that every asymptotically conformal curve is a quasicircle. Every smooth curve is asymptotically conformal but corners are not allowed.

We now give (Pom78) analytic characterizations; see PoWa82, RoWa84 and Bec87 for quantitative estimates.

Theorem 11.1. *Let f map $\mathbb{D}$ conformally onto the inner domain of the Jordan curve J. Then the following conditions are equivalent:*

(i) *J is asymptotically conformal;*

(ii) *$(1 - |z|^2)f''(z)/f'(z) \to 0$ as $|z| \to 1-$;*

(iii) $\dfrac{f(z) - f(\zeta)}{(z - \zeta)f'(z)} \to 1$ *as* $|z| \to 1-,\ \zeta \in \overline{\mathbb{D}},\ \dfrac{|z - \zeta|}{1 - |z|} \le a\ (a > 0).$

If $z \in \mathbb{D}$ and $a > 1$ then $\{\zeta \in \overline{\mathbb{D}} : |z - \zeta| \le a(1 - |z|)\}$ contains a proper arc of $\mathbb{T}$. Further equivalent conditions (BePo78, compare AgGe65) are

(iv) $\left(1 - |z|^2\right)^2 S_f(z) \to 0$ as $|z| \to 1-$;

(v) f has a quasiconformal extension to $\mathbb{C}$ such that

$$\frac{\partial f}{\partial \overline{z}} \Big/ \frac{\partial f}{\partial z} \to 0 \quad \text{as} \quad |z| \to 1+ .$$

Proof (i) $\Rightarrow$ (ii). Let (i) hold and $|z_n| \to 1-$. We write $z_n = \zeta_n r_n$ and $\delta_n = d_f(z_n) = \text{dist}(f(z_n), J)$. Taking a subsequence we may assume that $n\delta_n \to 0$ as $n \to \infty$. Let a_n and b_n be the first and last points where the (positively oriented) Jordan curve J intersects the circle $\{w : |w - f(z_n)| = n\delta_n\}$ and let J_n be the arc of J between a_n and b_n; see Fig. 11.2. Since $|a_n - b_n| \le 2n\delta_n$ we may assume, by (1), that

(2) $(|a_n - w| + |w - b_n|)/|a_n - b_n| < 1 + 1/(4n^4)$ for $w \in J_n$.

With $\alpha_n = \arg(b_n - a_n)$ it follows that $|\,\text{Im}[e^{-i\alpha_n}(w - a_n)]| < \delta_n/n$ for $w \in J_n$. Since $|w - f(z_n)| \ge n\delta_n$ for $w \in J \setminus J_n$ and $\delta_n = \text{dist}(f(z_n), J)$ we conclude that $|\,\text{Im}[e^{-i\alpha_n}(f(z_n) - a_n)] - \delta_n| < \delta_n/n$ so that

$$\left(-1 - \frac{2}{n}\right)\delta_n < \text{Im}\left[e^{-i\alpha_n}(w - f(z_n))\right] < \left(-1 + \frac{2}{n}\right)\delta_n \quad \text{for} \quad w \in J_n.$$

The univalent functions

(3) $$h_n(z) = \delta_n^{-1} e^{i\beta_n}\left[f\left(\zeta_n \frac{z + r_n}{1 + r_n z}\right) - f(z_n)\right] \quad (z \in \mathbb{D})$$

satisfy $h_n(0) = 0$ and we can determine β_n such that $h_n'(0) > 0$. Taking again a subsequence we deduce from the above results that $h_n(\mathbb{D})$ converges to a halfplane $\{w : \text{Im}[e^{i\gamma}w] > -1\}$ $(\gamma \in \mathbb{R})$ with respect to 0 in the sense of kernel convergence; see Section 1.4. Hence it follows from Theorem 1.8 that

(4) $h_n(z) \to 2z/(1 - e^{i\gamma}z)$ as $n \to \infty$ locally uniformly in $\mathbb{D}$.

Now $f(\mathbb{D})$ is a quasidisk and thus a John domain. Applying Theorem 5.2 (iii) to f we obtain from (3) after a short computation that

$$(1 + r_n x)^{a+1}|h_n'(x)| \ge c|h_n'(0)| \quad \text{for} \quad -r_n < x < 0$$

where a and c are positive constants. Hence we conclude from (4) that $(1 + x)^{a+1}|1 - e^{i\gamma}x|^{-2} \ge 2c$ for $-1 < x < 0$ so that $e^{i\gamma} = -1$ and thus

$$\zeta_n(1 - r_n^2)\frac{f''(z_n)}{f'(z_n)} = 2r_n + \frac{h_n''(0)}{h_n'(0)} \to 2 + 2e^{i\gamma} = 0$$

as $n \to \infty$ which implies (ii).

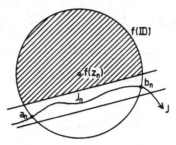

Fig. 11.2

(ii) $\Rightarrow$ (iii). Let $|z| = r < 1$ and $|\zeta| < 1$, $|z - \zeta| \le a(1 - r)$. By (ii) there exists $r_0 < 1$ such that

$$\left| \log \frac{f'(\zeta)}{f'(z)} \right| = \left| \int_z^\zeta \frac{f''(t)}{f'(t)} \, dt \right| \le \varepsilon \log \frac{1}{1 - |s|} \qquad \text{for} \quad r_0 < r < 1$$

where we have integrated over the non-euclidean segment from z to ζ and where $s = (z - \zeta)/(1 - \bar{z}\zeta)$. Hence $|f'(\zeta)/f'(z) - 1| \le (1 - |s|)^{-\varepsilon} - 1$. Since $|d\zeta/ds| = |1 - \bar{z}\zeta|^2/(1 - r^2) \le b(1 - r)$ with $b = (a + 2)^2$, we obtain by another integration that, for $r_0 < r < 1$,

$$\left| \frac{f(z) - f(\zeta)}{(z - \zeta)f'(z)} - 1 \right| \le \frac{b(1 - r)}{|z - \zeta|} \int_0^{|s|} \left[(1 - \sigma)^{-\varepsilon} - 1 \right] \, d\sigma$$

$$= \frac{b(1 - r)}{|1 - \bar{z}\zeta|} \left(\frac{1 - (1 - |s|)^{1-\varepsilon}}{(1 - \varepsilon)|s|} - 1 \right) \le \frac{b\varepsilon}{1 - \varepsilon}$$

because the last bracket increases with $|s|$. By continuity this estimate also holds for $|\zeta| \le 1$. Hence (iii) is true.

(iii) $\Rightarrow$ (i). Let $w_j = f(\zeta_j) \in J$ for $j = 1, 2$. If $|w_1 - w_2|$ is small then $J(w_1, w_2)$ corresponds to the smaller arc of $\mathbb{T}$ between ζ_1 and ζ_2. We determine $z \in \mathbb{D}$ such that $|\zeta_j - z| = 2(1 - |z|)$. It follows from (iii) that

$$f(z) = f(\zeta) + (z - \zeta)f'(z)(1 + o(1)) \quad \text{as} \quad |z| \to 1 \text{ uniformly in } |\zeta - z| \le 2(1 - |z|)$$

and thus $|w - w_j| = |f'(z)|(|\zeta - \zeta_j| + o(|\zeta_1 - \zeta_2|))$ for $w \in J(w_1, w_2)$ and $\zeta = f^{-1}(w)$. If $|w_1 - w_2| \to 0$ then $|\zeta_1 - \zeta_2| \to 0$ hence

$$\frac{|w_1 - w| + |w - w_2|}{|w_1 - w_2|} = \frac{|\zeta_1 - \zeta| + |\zeta - \zeta_2| + o(|\zeta_1 - \zeta_2|)}{|\zeta_1 - \zeta_2| + o(|\zeta_1 - \zeta_2|)} \to 1$$

because ζ lies between ζ_1 and ζ_2 on the circle $\mathbb{T}$. Hence (i) holds. $\square$

2. Let again $\mathcal{B}_0$ be the space of all Bloch functions g with $(1 - |z|^2)g'(z) \to 0$ as $|z| \to 1$. Then

(5) J asymptotically conformal $\Leftrightarrow$ $\log f' \in \mathcal{B}_0$

by Theorem 11.1.

Example. Let $b_k \in \mathbb{C}$ with $|b_k| < 1/3$ and $b_k \to 0$ as $k \to \infty$. The function f defined by $f(0) = 0$ and

$$\text{(6)} \qquad \log f'(z) = \sum_{k=0}^{\infty} b_k z^{2^k} \qquad \text{for} \quad z \in \mathbb{D}$$

maps $\mathbb{D}$ onto a quasidisk by Proposition 8.14 and $\log f' \in \mathcal{B}_0$ by Proposition 8.12. Hence $f(\mathbb{D})$ is bounded by an asymptotically conformal Jordan curve J. If $\Sigma |b_k|^2 = \infty$ then $\log f'$ has almost nowhere a finite angular limit; compare (8.6.12). Hence f is conformal almost nowhere on $\mathbb{T}$ so that J has a tangent only on a set of zero linear measure by Theorem 5.5, Proposition 6.22 and Theorem 6.24. Furthermore f is twisting almost everywhere on $\mathbb{T}$ by Theorem 6.18. This shows that asymptotically conformal curves can be rather pathological.

Theorem 11.2. *Let f map $\mathbb{D}$ conformally onto the inner domain of an asymptotically conformal curve. Then f is conformal on a set of Hausdorff dimension one.*

This result of Makarov is, by (5), equivalent to the assertion that

$$\text{(7)} \qquad \dim\{\zeta \in \mathbb{T} : g \text{ has a finite angular limit at } \zeta\} = 1$$

for $g \in \mathcal{B}_0$; see Mak89d for the proof. It is furthermore best possible.

The assertion that (7) holds for $g \in \mathcal{B}_0$ is related to (10.4.12) for $g \in \mathcal{B}$ and in fact follows from it by the following transference principle (CaCuPo88): For every $g \in \mathcal{B}_0$, there exists $g^* \in \mathcal{B}$ such that

$$\zeta \in \mathbb{T}, \sup_r |g^*(r\zeta)| < \infty \quad \Rightarrow \quad \lim_{r \to 1} g(r\zeta) \neq \infty \text{ exists.}$$

Now let $g \in \mathcal{B}_0$ be an inner function, that is $g(\mathbb{D}) \subset \mathbb{D}$ and $|g(\zeta)| = 1$ for almost all $\zeta \in \mathbb{T}$; see Exercise 4. Then

$$\dim\{\zeta \in \mathbb{T} : g(\zeta) = w\} = 1 \quad \text{for each } w \in \mathbb{D}.$$

This surprising result (Roh89, compare Hun88) incidentally shows that there exist uncountably many disjoint sets of dimension one on $\mathbb{T}$.

In Section 7.2 we considered

$$\text{(8)} \qquad h(z) = \int_0^z (g(\zeta) - g(0))\zeta^{-1} \, d\zeta \quad (z \in \overline{\mathbb{D}})$$

for $g \in \mathcal{B}$. As in Theorem 7.2 it can be shown that h is *Zygmund-smooth*, i.e. that

$$\max_{|z|=1} |h(e^{it}z) + h(e^{-it}z) - 2h(z)| = o(t) \quad \text{as} \quad t \to 0,$$

if and only if $g \in \mathcal{B}_0$; see Zyg45. The problem of differentiability has now been completely solved (Mak89d, Mak89e):

Let h be a Zygmund-smooth function defined on $\mathbb{T}$. If either h can be continuously extended to an analytic function in $\mathbb{D}$ (i.e. (8) holds with $g \in \mathcal{B}_0$), or if h is real-valued, then

$$\dim\{t : h(e^{it}) \text{ has a finite derivative}\} = 1 .$$

There is however a complex-valued Zygmund-smooth function that is nowhere differentiable.

3. Finally we mention a property that is the opposite to asymptotic conformality. Let J be a quasicircle such that

$$(9) \qquad \max_{w \in J(a,b)} \frac{|a - w| + |w - b|}{|a - b|} > 1 + c_1 \quad (a, b \in J)$$

for some $c_1 > 0$. This holds e.g. for the snowflake curve and for many Julia sets (see e.g. Bla84, Bea91).

If f maps $\mathbb{D}$ conformally onto the inner domain of J then (Jon89)

$$(10) \quad (1 - |z^*|) \left| \frac{f''(z^*)}{f'(z^*)} \right| > c_2 \text{ for all } z \in \mathbb{D} \text{ and some } z^* \text{ with } \left| \frac{z^* - z}{1 - \bar{z}z^*} \right| < \rho$$

with constants $c_2 > 0$ and $\rho < 1$, furthermore

$$(11) \qquad \limsup_{r \to 1} \frac{\log|f'(r\zeta)|}{\sqrt{\log \frac{1}{1-r} \log\log\log \frac{1}{1-r}}} > 0 \text{ for almost all } \zeta \in \mathbb{T}.$$

This is a lower estimate in the law of the iterated logarithm; compare Corollary 8.11 and (8.6.22). See Roh91 for conformal welding.

Exercises 11.2

1. Let f map $\mathbb{D}$ conformally onto the inner domain of the asymptotically conformal curve J. Show that f is α-Hölder-continuous for each $\alpha < 1$ and deduce that J has Hausdorff dimension 1.

2. Show that (1) is equivalent to the condition

$$\max_{w \in J(a,b)} \text{Im}[(w - a)/(b - a)] \to 0 \quad \text{as} \quad a, b \in J, \quad |a - b| \to 0 .$$

3. Prove that (ii) implies (iv) for all analytic locally univalent functions f.

4. Let g be a function in $\mathcal{B}_0$ that has an angular limit almost nowhere and let φ map $\mathbb{D}$ conformally onto a component of $\{z \in \mathbb{D} : |g(z)| < 1\}$. Use Löwner's lemma to show that $\{\zeta \in \mathbb{T} : |\varphi(\zeta)| = 1\}$ has zero measure and deduce that $g \circ \varphi$ is an inner function in $\mathcal{B}_0$.

11.3 The Visser-Ostrowski Quotient

We assume throughout this section that f maps $\mathbb{D}$ conformally onto $G \subset \mathbb{C}$ and that f has the finite angular limit $\omega = f(\zeta)$ at a given point $\zeta \in \mathbb{T}$. Let again $d_f(z) = \text{dist}(f(z), \partial G)$ for $z \in \mathbb{D}$ and let $M_1, M_2, \ldots$ denote suitable constants.

The point $f(\zeta)$ is called *well-accessible* (for $\zeta \in \mathbb{T}$) if there is a Jordan arc $C \subset \mathbb{D}$ ending at ζ such that

$$(1) \qquad \text{diam} f(C(z)) \leq M_1 d_f(z) \qquad \text{for} \quad z \in C,$$

where $C(z)$ is the arc of C from z to ζ. Every boundary point of a John domain is well-accessible by Corollary 5.3, and the notion of a well-accessible point can be considered as a local version of the notion of a John domain. The following characterization (Pom64) is a local version of Theorem 5.2.

Theorem 11.3. *The following three conditions are equivalent:*

(i) *$f(\zeta)$ is well-accessible;*
(ii) *there exists $\alpha > 0$ with*

$$|f'(\rho\zeta)| \leq M_2 |f'(r\zeta)| \left(\frac{1-\rho}{1-r}\right)^{\alpha-1} \qquad \text{for} \quad 0 \leq r \leq \rho < 1;$$

(iii) *there exists $\alpha > 0$ with*

$$|f(\rho\zeta) - f(\zeta)| \leq M_3 d_f(r\zeta) \left(\frac{1-\rho}{1-r}\right)^{\alpha} \qquad \text{for} \quad 0 \leq r \leq \rho < 1.$$

Proof. First let (i) hold. We may assume that $\zeta = 1$. If $0 < r < 1$ and r is close to 1 then, by Corollary 4.18, we can find a curve S from r to $\mathbb{T}$ that satisfies $\text{diam} f(S) \leq M_4 d_f(r)$ and intersects C, say at z. It follows from (1) that $\text{diam} f(C(z)) \leq M_1 d_f(z) \leq M_1 \text{diam} f(S)$ so that the part of S from r to z together with $C(z)$ forms a curve Γ from z to 1 with $\text{diam} f(\Gamma) \leq (1 + M_1) \text{diam} f(S)$. Hence we see from the Gehring-Hayman theorem (4.5.6) that

$$(2) \qquad \text{diam} f([r,1]) \leq M_5 \text{diam} f(\Gamma) \leq M_6 \text{diam} f(S) \leq M_4 M_6 d_f(r).$$

Now we proceed as in the proof of (ii) $\Rightarrow$ (iii) in Theorem 5.2. If $r \leq x < 1$ and $0 \leq y \leq 1 - x$ then $|f(x + iy) - f(1)| \leq M_7 d_f(r)$ by (2) and Corollary 1.6. Hence (5.2.8) is again true if $\alpha > 0$ is suitably chosen. This implies (5.2.9), i.e. our assertion (ii).

Next if (ii) holds then (iii) follows by an integration. Finally if (iii) holds then (1) is true with $C = [0, \zeta]$ and $M_1 = M_3$. $\qquad \square$

We consider the *Visser-Ostrowski quotient*

$$(3) \qquad q(z) = \frac{(z - \zeta) f'(z)}{f(z) - \omega} \qquad \text{for} \quad z \in \mathbb{D}.$$

If $\operatorname{Re} q(r\zeta) \geq c > 0$ for $r_0 \leq r < 1$ it is easy to see that $|f(r\zeta) - f(\zeta)|$ strictly decreases. Hence it follows from (3) that

$$\operatorname{diam} f([r\zeta, \zeta]) \leq |f(r\zeta) - f(\zeta)| \leq c^{-1}(1 - r)|f'(r\zeta)|$$

so that $f(\zeta)$ is well-accessible by (1).

We saw in Proposition 4.11 that f satisfies the *Visser-Ostrowski condition*

$$(4) \qquad q(z) = \frac{(z - \zeta)f'(z)}{f(z) - \omega} \to 1 \quad \text{as} \quad z \to \zeta \text{ in every Stolz angle}$$

if f is isogonal at ζ. If G is bounded by an asymptotically conformal curve then (4) holds uniformly for all $\zeta \in \mathbb{T}$ by Theorem 11.1 (iii). The conformal map f defined by (11.2.6) is isogonal almost nowhere but $\partial f(\mathbb{D})$ is asymptotically conformal. Hence (4) is much weaker than isogonality.

The next result is a partial local analogue of Theorem 11.1.

Proposition 11.4. *The Visser-Ostrowski condition at ζ is equivalent to*

$$(5) \qquad (z - \zeta)f''(z)/f'(z) \to 0 \quad \text{as} \quad z \to \zeta \text{ in every Stolz angle}.$$

Proof. If (4) holds then it follows from Proposition 4.8 that $(z - \zeta)q'(z) \to 0$ in every Stolz angle and thus

$$(z - \zeta)\frac{f''(z)}{f'(z)} = q(z) - 1 + (z - \zeta)\frac{q'(z)}{q(z)} \to 0 \quad \text{as} \quad z \to \zeta.$$

The converse is shown as in the proof of (ii) $\Rightarrow$ (iii) in Theorem 11.1.   □

Quantitative results on the Visser-Ostrowski condition can be found e.g. in RoWa84. We now give a geometric characterization (RoWa80); see Fig. 11.3.

Theorem 11.5. *The Visser-Ostrowski condition (4) holds at ζ if and only if there is a curve $C \subset \mathbb{D}$ ending at ζ such that*

$$(6) \qquad f(C) : \omega + te^{i\alpha(t)}, \quad 0 < t \leq t_1$$

and the following conditions are both satisfied:

$$(A) \qquad \begin{aligned} &\{w : \operatorname{Re}[e^{-i\alpha(t)}(w - \omega)] > \varepsilon t, |w - \omega| < t/\varepsilon\} \subset G \\ &\quad \text{for} \quad 0 < \varepsilon < 1, \quad 0 < t < t_0(\varepsilon); \end{aligned}$$

$$(B) \qquad \text{there are points } \omega_t^{\pm} \in \partial G \text{ with } e^{-i\alpha(t)}(\omega_t^{\pm} - \omega) \sim \pm it \text{ as } t \to 0.$$

Proof. We may assume that $\zeta = 1$ and $\omega = 0$.

(a) First let (4) be satisfied. It follows that $t = |f(r)|$ is strictly decreasing. With $C = [0, 1)$ we can thus parametrize $f(C)$ as in (6) with $\alpha(t) = \arg f(r)$.

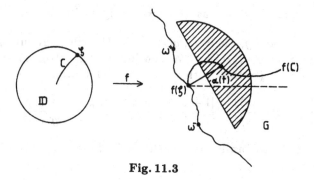

Fig. 11.3

We consider the univalent functions

(7) $$f_t(z) = f\left(\frac{z+r}{1+rz}\right) / f(r) \quad (z \in \mathbb{D}), \quad t = |f(r)|$$

for $0 < t \le t_1$ and write $G_t = f_t(\mathbb{D})$. With $s = (z+r)/(1+rz)$ we see from (4) that

(8) $$\frac{d}{dz} \log\left[\frac{1+z}{1-z} f_t(z)\right] = \frac{2}{1-z^2} + \frac{1+r}{(1+rz)(1-z)}(1-s)\frac{f'(s)}{f(s)} \to 0$$

as $r \to 1$ locally uniformly in $z \in \mathbb{D}$ and therefore that $f_t(z) \to (1-z)/(1+z)$.

Hence it follows from Theorem 1.8 that G_t converges to $\{\operatorname{Re} w > 0\}$ with respect to 1 in the sense of Carathéodory kernel convergence as $t \to 0$. Thus

$$\{w : \operatorname{Re} w > \varepsilon, \ |w| < 1/\varepsilon\} \subset G_t \quad \text{for} \quad 0 < t < t_0(\varepsilon),$$

and there exist $b_t^\pm \in \partial G_t$ with $b_t^\pm \to \pm i$ as $t \to 0$. Since $f(r) = te^{i\alpha(t)}$ and $G = \{f(r)w : w \in G_t\}$ we conclude that (A) and (B) hold where $\omega_t^\pm = te^{i\alpha(t)}b_t^\pm$.

(b) Conversely let there exist a curve C satisfying (A) and (B). For $r_n \to 1$ we consider the univalent functions

(9) $$g_n(z) = t_n^{-1}e^{-i\alpha(t_n)}f\left(\frac{z+r_n}{1+r_n z}\right), \quad t_n = |f(r_n)|.$$

Since $f(1) = 0$ is well-accessible by (A), we see from $d_f(r) \le |f(r)|$ and from Theorem 11.3 (iii) with $\rho = r$ that $c_1 < |g_n'(0)| < M_1$ where c_1 and M_1 are positive constants. Hence we obtain from (ii) and (iii) by a short calculation that

(10) $$|g_n'(x)| > \frac{c_2}{1+r_n x} \ (-r_n < x < 1), \quad |g_n(x)| < M_2(1-x)^{c_3} \ (0 < x < 1).$$

It follows from (A) and (B) first that

$$\alpha(\lambda t) - \alpha(t) \to 0 \quad (t \to 0) \text{ for each } \lambda > 0$$

and therefore second that the domains $g_n(\mathbb{D})$ converge to $H = \{\operatorname{Re} w > 0\}$ as $n \to \infty$. Hence, by the Carathéodory kernel theorem,

$$g_n(z) \to g(z) \quad \text{as} \quad n \to \infty \text{ locally uniformly in } \mathbb{D},$$

where $g(\mathbb{D}) = H$ and $g(-1) = \infty$, $g(1) = 0$ by (10), furthermore $|g(0)| = 1$ by (9), so that $g(z) = (1-z)/(1+z)$. We conclude that $g_n''(z)/g_n'(z) \to -2/(1+z)$. Hence we see from (9) that, with $s = (z + r_n)/(1 + r_n z)$,

$$\left| (1-s)\frac{f''(s)}{f'(s)} \right| = \frac{|1-z|\,|1+r_n z|}{1+r_n} \left| \frac{2r_n}{1+r_n z} + \frac{g_n''(z)}{g_n'(z)} \right| \to 0$$

as $n \to \infty$ locally uniformly in $\mathbb{D}$ which implies (5) and thus condition (4) by Proposition 11.4. $\qquad \square$

Next we give a geometric characterization of isogonality (Ost36; see e.g. Fer42, War67, JenOi77).

Theorem 11.6. *The function f is isogonal at ζ if and only if there is a curve $C \subset \mathbb{D}$ ending at ζ such that*

$$(11) \qquad\qquad f(C) : \omega + te^{i\alpha}, \quad 0 < t \le t_1$$

satisfies (A) *and* (B) *with* $\alpha(t) \equiv \alpha$.

Proof. (a) First we assume that f is isogonal at ζ so that (4) is satisfied. We proceed as in the proof of Theorem 11.5. Since f is isogonal we see from Proposition 4.11 that $\alpha(r) = \arg f'(r) \to \alpha$ as $r \to 1$. We finally replace $[0,1)$ by a curve $C \subset \mathbb{D}$ ending at 1 such that (11) holds, and (A) and (B) continue to hold if we replace $\alpha(t)$ by α.

(b) Conversely, let (A) and (B) be satisfied with $\alpha(t) \equiv \alpha$. We define g_n as in (9) with α instead of $\alpha(t_n)$ and deduce as in the proof of Theorem 11.5 that $g_n(z) \to (1 - z)/(1 + z)$. Hence

$$\arg f'\left(\frac{z + r_n}{1 + r_n z}\right) = \alpha + \arg\left[(1 + r_n z)^2 g_n'(z)\right] \to \alpha + \pi$$

as $n \to \infty$ locally uniformly in $\mathbb{D}$. Hence $\arg f'$ has the angular limit $\alpha + \pi$ at 1 so that f is isogonal by Proposition 4.11. $\qquad \square$

Finally we prove a regularity theorem (Ost36; see e.g. LeF55, Chapt. IV and EkWa69) that we will use in the next section. If f satisfies the Visser-Ostrowski condition (4) then $|f(r\zeta) - f(\zeta)|$ is decreasing for r close to 1. Hence we see from Proposition 2.14 that, for small $\rho > 0$, there is a unique crosscut $Q(\rho)$ of $\mathbb{D}$ that intersects $[0, \zeta)$ and is mapped onto an arc of $\partial D(f(\zeta), \rho)$ where $\rho = |f(r\zeta) - f(\zeta)|$; see Fig. 11.4.

Theorem 11.7. *If f satisfies the Visser-Ostrowski condition at ζ then*

$$(12) \qquad |\zeta - z| \sim 1 - r \quad \text{as} \quad z \in Q(\rho), \quad |f(r\zeta) - \omega| = \rho \to 0.$$

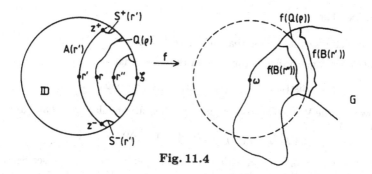

Fig. 11.4

Proof. We may assume that $\zeta = 1$ and $\omega = 0$. Let $\varepsilon > 0$, $0 < \delta < 1$ and

(13) $A(r) = \{z \in \mathbb{D} : |1 - z| = 1 - r,\ 1 - |z| \geq \delta(1 - r)\}$ for $0 < r < 1$.

It follows from (4) by integration over part of $A(r)$ that

(14) $\left| \log \dfrac{f(z)}{f(r)} - \log \dfrac{1-z}{1-r} \right| \leq \dfrac{\varepsilon}{\pi} \displaystyle\int_r^z \dfrac{|ds|}{|1-s|} < \varepsilon$ for $z \in A(r)$

if $r > r_1(\varepsilon, \delta)$ and thus $e^{-\varepsilon}|f(r)| < |f(z)| < e^{\varepsilon}|f(r)|$. If $z^{\pm}$ are the endpoints of $A(r)$ then

$$\frac{d_f(z^{\pm})}{|f(z^{\pm})|} < \frac{2(1 - |z^{\pm}|)}{|1 - z^{\pm}|} \left| \frac{(1 - z^{\pm})f'(z^{\pm})}{f(z^{\pm})} \right| < 3\delta$$

by (4) and (13) if $r > r_2(\varepsilon, \delta) > r_1$. Hence, by (13) and Corollary 4.18, there are non-euclidean segments $S^{\pm}(r)$ from $z^{\pm}$ to $\mathbb{T} \cap D(z^{\pm}, K_1\delta(1-r))$ such that

$$|f(z) - f(z^{\pm})| < K_2 d_f(z^{\pm}) < 3\delta K_2 |f(z^{\pm})| \text{ for } z \in S^{\pm}(r)$$

where K_1 and K_2 are absolute constants. If $\delta = \delta(\varepsilon) > 0$ is chosen sufficiently small we conclude that

(15) $B(r) \equiv S^-(r) \cup A(r) \cup S^+(r) \subset \left\{ z \in \mathbb{D} : e^{-\varepsilon} > \dfrac{|1-z|}{1-r} < e^{\varepsilon} \right\}$

for $r > r_3(\varepsilon)$ and that

(16) $e^{-2\varepsilon}|f(r)| < |f(z)| < e^{2\varepsilon}|f(r)|$ for $z \in B(r)$.

Let $r' = 1 - e^{4\varepsilon}(1 - r)$ and $r'' = 1 - e^{-4\varepsilon}(1 - r)$. As in (14) we deduce that $|f(r')| > e^{2\varepsilon}|f(r)|$ and $|f(r'')| < e^{-2\varepsilon}|f(r)|$ for $r > r_4(\varepsilon) > r_3$. Hence it follows from (16) that $Q(\rho) \cap B(r') = \emptyset$ and $Q(\rho) \cap B(r'') = \emptyset$. Since $r' < r < r''$ we therefore see from (15) that

$$e^{-5\varepsilon} < |1-z|/(1-r) < e^{5\varepsilon}   \text{ for } z \in Q(\rho), \quad r > r_4(\varepsilon). \qquad \square$$

Exercises 11.3

1. If ω is well-accessible show that $\inf_r |q(r\zeta)| > 0$.

2. Let G have a generalized cusp at ω in the sense that

$$\text{dist}(w, \partial G) = o(|w - \omega|) \quad \text{as} \quad w \to \omega, \quad w \in G.$$

   Show that the Visser-Ostrowski quotient q has the angular limit 0 at ζ and deduce that

$$\log[f(\zeta) - f(r\zeta)] = o(\log[1/(1-r)]) \quad \text{as} \quad r \to 1.$$

3. Let G be a Jordan domain with a corner of opening $\pi\alpha$ at $f(\zeta)$; see Section 3.4. Show that q has the angular limit α at ζ.

4. Show that f satisfies the Visser-Ostrowski condition at 1 if and only if

$$\left[f\left(\frac{z+r}{1+rz} \right) - f(r) \right] / \left[(1-r^2) f'(r) \right] \to \frac{z}{1+z} \quad \text{as} \quad r \to 1.$$

5. Construct a domain G such that f both is twisting and satisfies the Visser-Ostrowski condition at some ζ.

6. Let G be a quasidisk. Use Theorem 11.6 to give a geometrical proof of the fact (Theorem 5.5 (ii)) that f is isogonal at ζ if and only if ∂G has a tangent at $f(\zeta)$.

11.4 Module and Conformality

Throughout this section we assume again that f maps $\mathbb{D}$ conformally onto $G \subset \mathbb{C}$ and has the angular limit $\omega = f(\zeta) \neq \infty$ at a given point $\zeta \in \mathbb{T}$.

We consider a system of circular crosscuts $T(\rho)$ of G with

$$(1) \qquad\qquad T(\rho) \subset \partial D(\omega, \rho) \quad (0 < \rho \le \rho_0)$$

that forms a null-chain defining the prime end corresponding to ζ; more precisely if $\rho_n \searrow 0$ then $(T(\rho_n))$ is a null-chain (see Section 2.4). Then, by Theorem 2.15,

$$(2) \qquad\qquad Q(\rho) = f^{-1}(T(\rho)) \quad (0 < \rho \le \rho_0)$$

are crosscuts of $\mathbb{D}$ that separate 0 from ζ if ρ_0 is small enough. For $0 < \rho_1 < \rho_2 \le \rho_0$, let $\Gamma(\rho_1, \rho_2)$ denote the family of all crosscuts C of G that separate $T(\rho_1)$ and $T(\rho_2)$; see Fig. 11.5. In particular we have $T(\rho) \in \Gamma(\rho_1, \rho_2)$ for $\rho_1 < \rho < \rho_2$. First we prove (RoWa76, JenOi77):

Proposition 11.8. *The condition*

$$(3) \qquad\qquad \sup_{z \in Q(\rho)} |z - \zeta| / \inf_{z \in Q(\rho)} |z - \zeta| \to 1 \quad \text{as} \quad \rho \to 0$$

holds if and only if, for every $\varepsilon > 0$, there exists $\delta > 0$ such that

$(4) \quad \text{mod}\, \Gamma(\rho_1, \rho_3) < \text{mod}\, \Gamma(\rho_1, \rho_2) + \text{mod}\, \Gamma(\rho_2, \rho_3) + \varepsilon \text{ for } \rho_1 < \rho_2 < \rho_3 < \delta.$

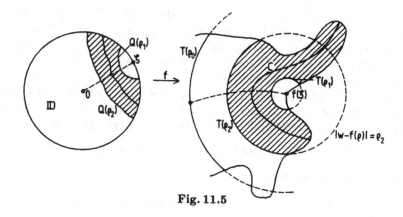

Fig. 11.5

We saw in Theorem 11.7 that (3) is a consequence of the Visser-Ostrowski condition. The module of a curve family was introduced in Section 9.2. It follows from Proposition 9.3 that

(5) $$\operatorname{mod} \Gamma(\rho_1, \rho_3) \geq \operatorname{mod} \Gamma(\rho_1, \rho_2) + \operatorname{mod} \Gamma(\rho_2, \rho_3)$$

holds for all $\rho_1 < \rho_2 < \rho_3 \leq \rho_0$. Hence (4) means that $\operatorname{mod} \Gamma(\rho_1, \rho_2)$ is "approximately additive".

Proof. (a) First let (3) hold. The function

(6) $$s = \frac{1}{\pi} \log \frac{\zeta + z}{\zeta - z} + \frac{i}{2}$$

maps $\mathbb{D}$ conformally onto the strip $S = \{0 < \operatorname{Im} s < 1\}$ and $Q(\rho)$ onto a crosscut $A(\rho)$ of S. Now (3) is equivalent to

(7) $$\sup \operatorname{Re} s - \inf \operatorname{Re} s \to 0 \quad \text{as} \quad \rho \to 0 \quad \text{where} \quad s \in A(\rho).$$

The family of curves in S separating $\{\operatorname{Re} s = \lambda_1\}$ and $\{\operatorname{Re} s = \lambda_2\}$ has module $\lambda_2 - \lambda_1$ by (9.2.5). Since the module is conformally invariant and monotone, we deduce that, for $0 < \rho_1 < \rho_2 \leq \rho_0$,

$$\inf \operatorname{Re} s_1 - \sup \operatorname{Re} s_2 \leq \operatorname{mod} \Gamma(\rho_1, \rho_2) \leq \sup \operatorname{Re} s_1 - \inf \operatorname{Re} s_2,$$

where $s_1 \in A(\rho_1)$ and $s_2 \in A(\rho_2)$. Hence (7) shows that

(8) $$\operatorname{mod} \Gamma(\rho_1, \rho_2) = \operatorname{Re}(s_1 - s_2) + o(1) \quad \text{as} \quad \rho_1 < \rho_2 \to 0$$

which implies (4).

(b) Conversely let (4) hold; see Fig. 11.6. We extend S by reflecting upon $\{\operatorname{Im} s = 1\}$ and then upon $\{\operatorname{Im} s = 0\}$ to obtain $S' = \{|\operatorname{Im} s| < 2\}$. The crosscut $A'(\rho)$ of S' obtained from $A(\rho)$ is now symmetric with respect to $\mathbb{R}$.

Applying Proposition 9.4 twice we see that the family $\Gamma'(\rho_1, \rho_2)$ of curves in S' separating $A'(\rho_1)$ and $A'(\rho_2)$ satisfies $\operatorname{mod}\Gamma(\rho_1, \rho_2) = 4 \operatorname{mod}\Gamma'(\rho_1, \rho_2)$.

Now let $\rho_1 < \rho_2 < \rho_3 < \rho_0$. By (9.2.7) there is an (exponential) conformal map of the part of $\{|\operatorname{Im} s| \leq 2\}$ between $A'(\rho_3)$ and $A'(\rho_1)$ (with the upper and lower sides identified) onto the annulus $\{1 < |w| < R\}$ where $R = \exp[(\pi/2) \operatorname{mod}\Gamma(\rho_1, \rho_3)]$. Then $A'(\rho_2)$ is mapped onto a Jordan curve $A''(\rho_2)$ separating $\{|w| = 1\}$ and $\{|w| = R\}$. If Γ_1'' and Γ_3'' denote the families of Jordan curves separating $A''(\rho_2)$ from these two circles then, by (4),

$$(2\pi)^{-1} \log R = 4^{-1} \operatorname{mod}\Gamma(\rho_1, \rho_3) < \operatorname{mod}\Gamma_1'' + \operatorname{mod}\Gamma_3'' + \varepsilon/4$$

so that, by Proposition 9.5,

$$(9) \qquad \sup|w|/\inf|w| < 1 + K\varepsilon^{1/3} \qquad \text{where} \quad w \in A''(\rho_2)$$

for small $\varepsilon > 0$.

Consider sequences $(\rho_{j,n})$ for $j = 1, 2, 3$ with $\rho_{1n} < \rho_{2n} < \rho_{3n} < \rho_0$ and $\rho_{3n} \to 0$, $\rho_{1n}/\rho_{2n} \to 0$, $\rho_{2n}/\rho_{3n} \to 0$ as $n \to \infty$. Let h_n map $\mathbb{D}$ conformally onto the part of S' between $A'(\rho_{1n})$ and $A'(\rho_{3n})$ translated by $-s_n$ where $s_n \in \mathbb{R} \cap A'(\rho_{2n})$, such that $h_n(0) = 0$ and $h_n'(0) > 0$. It follows that $h_n(\mathbb{D})$ converges to S' in the sense of Carathéodory kernel convergence. Hence Theorem 1.8 shows that h_n converges locally uniformly in $\mathbb{D}$ to the conformal map of $\mathbb{D}$ onto S'; see GaHa91 for good estimates. Since $h_n^{-1}(A(\rho_{2n}))$ lies in a fixed compact subset of $\mathbb{D}$ by our construction of S' and by (9), we conclude from (9) that (7) and therefore (3) is satisfied. □

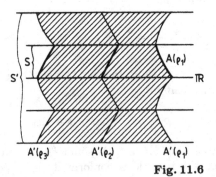 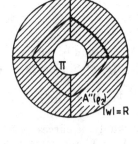

Fig. 11.6

In Section 4.3 we defined f to be conformal at $\zeta \in \mathbb{T}$ if f has a finite nonzero angular derivative $f'(\zeta)$. No strictly geometric characterization of conformality is known but there is a characterization in terms of modules (RoWa76, JenOi77):

Theorem 11.9. *The function f is conformal at $\zeta \in \mathbb{T}$ if and only if*

$$(10) \qquad \operatorname{mod}\Gamma(\rho_1, \rho_2) - \frac{1}{\pi}\log\frac{\rho_2}{\rho_1} \to 0 \qquad \text{as} \quad 0 < \rho_1 < \rho_2 \to 0$$

and if furthermore there is α such that

(11) $\{w : |\arg[w - \omega] - \alpha| < \pi/2 - \varepsilon, |w - \omega| < \delta\} \subset G$

for every $\varepsilon > 0$ and some $\delta = \delta(\varepsilon) > 0$.

Proof. First let f be conformal at ζ. Then f is isogonal at ζ so that (11) follows from Proposition 4.10. Since isogonality implies the Visser-Ostrowski condition by Proposition 4.11, it follows from Theorem 11.7 that, as $\rho = |f(r\zeta) - \omega| \to 0$,

$$\sup \left| \frac{\zeta + z}{\zeta - z} \right| \sim \inf \left| \frac{\zeta + z}{\zeta - z} \right| \sim \frac{2|f'(\zeta)|}{\rho} \qquad \text{where} \quad z \in Q(\rho).$$

Using the map (6) we therefore obtain (10) easily from (9.2.5), Proposition 9.1 and Proposition 9.2.

Conversely assume that the conditions of the theorem are satisfied. It follows from (10) that (4) holds and thus (3) by Proposition 11.8. As in part (a) of the proof of Proposition 11.8 we deduce that (8) holds which, by (6), can be rewritten

(12) $\mod \Gamma(\rho_1, \rho_2) = \dfrac{1}{\pi} \log \left| \dfrac{z_2 - \zeta}{z_1 - \zeta} \right| + o(1) \qquad \text{as} \quad \rho_1 < \rho_2 \to 0,$

where $z_j \in Q(\rho_j)$ and thus $\rho_j = |f(z_j) - \omega|$ for $j = 1, 2$. Hence we obtain from (10) that

(13) $\displaystyle\lim_{z \to \zeta} \log \left| \dfrac{f(z) - f(\zeta)}{z - \zeta} \right| \neq \infty \quad \text{exists where} \quad z \in \bigcup Q(\rho).$

Using (11) and (13) it can be shown (see RoWa76 or JenOi77) that f is isogonal at ζ. Together with (13) this implies that $\log[(f(r\zeta) - f(\zeta))/(r\zeta - \zeta)]$ has a finite limit as $r \to 1$ so that $f'(\zeta) \neq 0, \infty$ exists.   □

There are geometric characterizations of conformality for several important cases; see e.g. Fer44, LeF55, Hali65, Eke71 and RoWa77. We restrict ourselves to the case that f maps $\mathbb{D}$ conformally onto $G \subset \mathbb{D}$ and that the angular limit ω at $\zeta \in \mathbb{T}$ satisfies $\omega \in \mathbb{T}$.

We consider sequences (ρ_n) with $\rho_n \searrow 0$ as $n \to \infty$ and define $\delta_n > 0$ as the smallest number such that (see Fig. 11.7)

(14) $\{w \in \mathbb{D} : \rho_{n+1} < |w - \omega| \leq \rho_n, |\arg(1 - \overline{\omega}w)| < \pi/2 - \delta_n\} \subset G = f(\mathbb{D}).$

Theorem 11.10. *If $G \subset \mathbb{D}$ and $\omega \in \mathbb{T}$ then the following three conditions are equivalent.*

(i) *f is conformal at ζ;*

(ii) $\sum_n \dfrac{\delta_n(\rho_n - \rho_{n+1})}{\rho_n} < \infty$ *for all (ρ_n) with* $\sum_n \left(\dfrac{\rho_n - \rho_{n+1}}{\rho_n} \right)^2 < \infty;$

(iii) *there exists a sequence* (ρ_n) *such that*

$$\sum_n \delta_n(\rho_n - \rho_{n+1})/\rho_n < \infty, \qquad \sum_n \delta_n^2 < \infty.$$

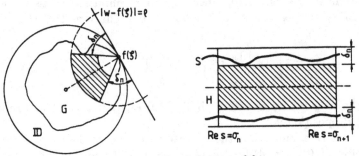

Fig. 11.7. The definition of δ_n

Proof. We change the definition of δ_n by replacing (14) by

$$(15) \qquad \left\{ w \in \mathbb{D} : \rho_{n+1} < \left| \frac{\omega - w}{\omega + w} \right| \le \rho_n, \left| \arg \frac{\omega - w}{\omega + w} \right| < \frac{\pi}{2} - \delta_n \right\} \subset G.$$

It is easy to see that conditions (ii) and (iii) remain unchanged. The function

$$s = \log[(\omega + w)/(\omega - w)]$$

maps $\mathbb{D}$ conformally onto the strip $S = \{|\operatorname{Im} s| < \pi/2\}$. If $H \subset S$ is the image of G then δ_n is the smallest number such that

$$(16) \qquad \{s \in S : \sigma_n \le \operatorname{Re} s < \sigma_{n+1}, |\operatorname{Im} s| < \pi/2 - \delta_n\} \subset H$$

where $\sigma_n = \log(1/\rho_n)$; see Fig. 11.7.

(i) $\Rightarrow$ (ii). Let Γ^* be the family of curves in H connecting $\{\operatorname{Re} s = \sigma_n\}$ and $\{\operatorname{Re} s = \sigma_{n+1}\}$. It follows from Proposition 9.1 and from (9.2.5) that $\operatorname{mod} \Gamma^* = 1/\operatorname{mod} \Gamma(\rho_n, \rho_{n+1})$. There exists $s_n \in \partial H$ with $\operatorname{Im} s_n = \pi/2 - \delta_n$, say. Let $a = \sigma_{n+1} - \sigma_n$ and define $\rho^*(s) = 1/a$ for those $s \in H$ with $\sigma_n < \operatorname{Re} s < \sigma_{n+1}$ for which either $\operatorname{Im} s < \pi/2 - \delta_n$ or $\operatorname{dist}(s_n, s) < a$. Otherwise we define $\rho^*(s) = 0$. Then $\int_C \rho^*(w)|dw| \ge 1$ for $C \in \Gamma^*$ so that ρ^* is admissible and therefore

$$\operatorname{mod} \Gamma^* \le \iint \rho^*(s)^2 \, d\sigma \, d\tau \le a^{-2} \left(a(\pi - \delta_n) + \pi a^2 \right).$$

Since $0 \le \delta_n < \pi/2$ for large n we obtain that

$$\pi \operatorname{mod} \Gamma(\rho_n, \rho_{n+1}) \ge \frac{a\pi}{\pi - \delta_n + \pi a} \ge a + \frac{\delta_n a}{\pi} - a^2$$

and thus, for large $j < k$,

$$\pi \bmod \Gamma(\rho_j, \rho_k) \geq \sum_{n=j}^{k-1} \pi \bmod \Gamma(\rho_n, \rho_{n+1})$$

$$\geq \sigma_k - \sigma_j + \frac{1}{\pi} \sum_{n=j}^{k-1} \delta_n(\sigma_{n+1} - \sigma_n) - \sum_{n=1}^{\infty} (\sigma_{n+1} - \sigma_n)^2 ;$$

the convergence of the last series is equivalent to the assumption that $\Sigma(\rho_n - \rho_{n+1})^2 \rho_n^{-2} < \infty$. If (i) holds then (10) is true by Theorem 11.9 and it follows that

(17) $$\qquad\qquad \Sigma \delta_n(\sigma_{n+1} - \sigma_n) < \infty$$

which is equivalent to $\Sigma \delta_n(\rho_n - \rho_{n+1})/\rho_n < \infty$.

(ii) $\Rightarrow$ (iii). We refer to RoWa77, p. 9 for the purely geometric construction of (ρ_n).

(iii) $\Rightarrow$ (i). Let Γ be the family of crosscuts of H separating $\{\operatorname{Re} s = \sigma'\}$ and $\{\operatorname{Re} s = \sigma''\}$ where $\sigma' < \sigma''$. Let

$$R_n^\pm = \{\sigma_n \leq \operatorname{Re} s \leq \sigma_{n+1}, \pi/2 - \delta_n \leq \pm \operatorname{Im} s < \pi/2\},$$

$$V_n^\pm = \{\sigma_n - \delta_n < \operatorname{Re} s < \sigma_{n+1} + \delta_n, \pi/2 - 2\delta_n < \pm \operatorname{Im} s < \pi/2\}.$$

We define $\rho(s) = 1/\pi$ for $s \in \{s \in S : \sigma' < \operatorname{Re} s < \sigma''\} \setminus \bigcup(V_n^+ \cup V_n^-)$ and $\rho(s) = 2/\pi$ for $s \in \bigcup(V_n^+ \cup V_n^-)$. If $C \in \Gamma$ then C intersects ∂R_n^+ and ∂R_m^- for suitable $m, n > k$ where $k \to \infty$ as $\sigma' \to \infty$. It follows that

$$\int_C \rho(w)|dw| \geq \frac{1}{\pi}(\pi - 2\delta_n - 2\delta_m) + \frac{2}{\pi}(\delta_n + \delta_m) = 1$$

so that ρ is admissible for Γ. Hence

$$\bmod \Gamma \leq \int_{\sigma'}^{\sigma''} \int_{-\pi/2}^{\pi/2} \rho(\sigma + i\tau)^2 \, d\sigma \, d\tau \leq \frac{\sigma'' - \sigma'}{\pi} + \frac{4}{\pi^2} \sum_{n=k}^{\infty} \operatorname{area}\left(V_n^+ \cup V_n^-\right)$$

$$= \frac{\sigma'' - \sigma'}{\pi} + \frac{16}{\pi^2} \sum_{n=k}^{\infty} \delta_n(\sigma_{n+1} - \sigma_n + 2\delta_n) = \frac{\sigma'' - \sigma'}{\pi} + o(1)$$

as $\sigma'' > \sigma' \to \infty$ because of (iii), compare (17).

Furthermore $\bmod \Gamma \geq (\sigma'' - \sigma')/\pi$ because $H \subset S$. Since $\bmod \Gamma(\rho', \rho'') = \bmod \Gamma$ if $\sigma' = \log(1/\rho')$ and $\sigma'' = \log(1/\rho'')$ we conclude that condition (10) of Theorem 11.9 holds, and (11) is true because $\delta_n \to 0$ as $n \to \infty$. It follows that f is conformal at ζ. $\qquad\square$

We deduce a simple sufficient condition (War32; see e.g. War67 and RoWa77 for generalizations) which is a local version of Theorem 3.5; compare also Theorem 4.14.

Corollary 11.11. *Let f map $\mathbb{D}$ conformally into $\mathbb{D}$ and let $\zeta \in \mathbb{T}$, $\omega = f(\zeta) \in \mathbb{T}$. We define*

(18) $$G_0 = \left\{ (1 - \rho e^{i\vartheta})\omega : 0 < \rho < \rho_0, |\vartheta| < \pi/2 - \delta(\rho) \right\},$$

where $\delta(\rho)$ $(0 < \rho < \rho_0)$ is a continuous increasing function. If $f(\mathbb{D}) \supset G_0$ and if

(19) $$\int_0^{\rho_0} \delta(\rho)\rho^{-1} \, d\rho < \infty$$

then f is conformal at ζ. Conversely, if $f(\mathbb{D}) = G_0$ and if f is conformal at ζ then (19) holds.

Proof. First suppose that $f(\mathbb{D}) \supset G_0$ and that (19) is satisfied. We choose ρ_n such that $\delta(\rho_n) = 1/n$ for $n \geq m$ and suitable m. Then $\delta_n \leq \delta(\rho_n) \leq 2\delta(\rho_{n+1})$ in (14) and thus, since δ is increasing,

$$\sum_{n=m}^{\infty} \frac{\delta_n(\rho_n - \rho_{n+1})}{\rho_n} \leq 2 \sum_{n=m}^{\infty} \int_{\rho_{n+1}}^{\rho_n} \frac{\delta(\rho)}{\rho} \, d\rho = 2 \int_0^{\rho_m} \frac{\delta(\rho)}{\rho} \, d\rho < \infty.$$

Hence condition (iii) of Theorem 11.10 holds so that f is conformal at ζ.

Conversely assume that f is conformal at ζ and that $f(\mathbb{D}) = G_0$. We choose $\rho_n = 1/n$ for $n \geq m$. Since δ increases it follows that $\delta_n = \delta(\rho_n)$ and thus

$$\int_0^{1/m} \frac{\delta(\rho)}{\rho} \, d\rho = \sum_{n=m}^{\infty} \int_{1/(n+1)}^{1/n} \frac{\delta(\rho)}{\rho} \, d\rho \leq \sum_{n=m}^{\infty} \frac{2\delta_n(\rho_n - \rho_{n+1})}{\rho_n}$$

and this series converges by Theorem 11.10 (ii). $\square$

Exercises 11.4

1. Let φ be continuous and increasing with $\varphi(0) = 0$ and let

$$f(\mathbb{D}) = \{u + iv : -\infty < u < \infty, v > \varphi(|u|)\}.$$

If $f(\zeta) = 0$ show that $f'(\zeta)$ exists and furthermore that

$$|f'(\zeta)| < \infty \quad \Leftrightarrow \quad \int_0^1 u^{-2}\varphi(u) \, du < \infty.$$

2. Let f map $\mathbb{D}$ conformally onto G with $f(1) = \infty$. Suppose that

$$\mathbb{C} \setminus G \subset \{u + iv : u > 0, |v| < u^\beta\}$$

where $0 \leq \beta < 1$. Show that (compare Hay59b, p. 100)

$$\lim_{x \to 1} (1 - x)^2 f(x) \neq 0, \infty \text{ exists}.$$

11.5 The Ahlfors Integral

Let f map $\mathbb{D}$ conformally onto $G \subset \mathbb{C}$ and let f have the finite angular limit ω at $\zeta \in \mathbb{T}$. As in the preceding section, we consider the circular crosscuts

$$T(\rho) \subset \partial D(\omega, \rho) \quad (0 < \rho \leq \rho_0)$$

of G and the crosscuts $Q(\rho) = f^{-1}(T(\rho))$ of $\mathbb{D}$. In addition, we write

$$(1) \qquad \lambda(\rho) = \Lambda(T(\rho)) \quad (0 < \rho \leq \rho_0)$$

so that $0 < \lambda(\rho) \leq 2\pi\rho$, furthermore

$$(2) \qquad B = \bigcup_{0 < \rho < \rho_0} T(\rho), \quad A = \bigcup_{0 < \rho < \rho_0} Q(\rho) = f^{-1}(B).$$

Again let $\Gamma(\rho_1, \rho_2)$ be the family of crosscuts of G that separate $T(\rho_1)$ and $T(\rho_2)$. See Fig. 11.8.

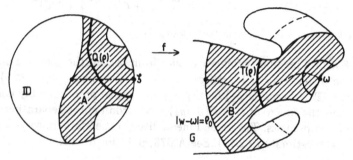

Fig. 11.8. The notation of this section

Proposition 11.12. *If* $0 < \rho_1 < \rho_2 \leq \rho_0$ *then*

$$(3) \qquad \operatorname{mod} \Gamma(\rho_1, \rho_2) - \int_{\rho_1}^{\rho_2} \frac{d\rho}{\lambda(\rho)} \geq 0$$

and the difference is decreasing in ρ_1.

Proof. Let φ be admissible for $\Gamma(\rho_1, \rho_2)$. Since $T(\rho) \in \Gamma(\rho_1, \rho_2)$ for $\rho_1 < r < \rho_2$ we conclude from (1) that

$$1 \leq \left(\int_{T(\rho)} \varphi(w) |dw| \right)^2 \leq \lambda(\rho) \int_{T(\rho)} \varphi(w)^2 |dw|$$

by the Schwarz inequality and thus, with (2),

$$\int_{\rho_1}^{\rho_2} \frac{d\rho}{\lambda(\rho)} \leq \iint_B \varphi(w)^2 \, du \, dv$$

so that (3) follows from the definition of the module. Furthermore, if $0 < \rho_1' < \rho_1$ then, by Proposition 9.3,

$$
\text{(4)} \quad
\begin{aligned}
\operatorname{mod}\Gamma(\rho_1',\rho_2) - \int_{\rho_1'}^{\rho_2} \lambda(\rho)^{-1}\,d\rho &\geq \operatorname{mod}\Gamma(\rho_1,\rho_2) - \int_{\rho_1}^{\rho_2} \lambda(\rho)^{-1}\,d\rho \\
&\quad + \operatorname{mod}\Gamma(\rho_1',\rho_1) - \int_{\rho_1'}^{\rho_1} \lambda(\rho)^{-1}\,d\rho
\end{aligned}
$$

and the expression on the last line is non-negative by (3). $\qquad\square$

The basic result (Ahl30, Tei38) is the *Ahlfors distortion theorem*.

Theorem 11.13. *If $0 < \rho_1 < \rho_2 \leq \rho_0$ and $z_1 \in Q(\rho_1)$, $z_2 \in Q(\rho_2)$ then*

$$
\text{(5)} \quad
\begin{aligned}
\exp\left(\int_{\rho_1}^{\rho_2} \frac{\pi}{\lambda(\rho)}\,d\rho\right) &\leq \exp[\pi \operatorname{mod}\Gamma(\rho_1,\rho_2)] \\
&< 32 \inf\left|\frac{z_2 - \zeta}{z_2 + \zeta}\right| \bigg/ \sup\left|\frac{z_1 - \zeta}{z_1 + \zeta}\right| ;
\end{aligned}
$$

if the last quotient is < 1 it is replaced by 1.

Proof. The first inequality follows at once from (3). Furthermore we map $\mathbb{D}$ by (11.4.6) onto $\{0 < \operatorname{Im} s < 1\}$, reflect upon $\mathbb{R}$ and then map $\{|\operatorname{Im} s| < 1\}$ by $e^{\pi s}$. The part of $\mathbb{D}$ between $Q(\rho_1)$ and $Q(\rho_2)$ is mapped onto a doubly connected domain and the image of $C \in \Gamma(\rho_1,\rho_2)$ is a Jordan curve separating the two boundary components. The second inequality (5) therefore follows easily from Teichmüller's estimate (9.2.19). See Ahl73, p. 77 for details. $\qquad\square$

Now we prove an asymptotic version (Eke67).

Theorem 11.14. *If A is given by (2) then*

$$
\text{(6)} \quad \lim_{z\to\zeta,z\in A} |z - \zeta| \exp\left(\int_{|f(z)-\omega|}^{\rho_0} \frac{\pi}{\lambda(\rho)}\,d\rho\right) \neq \infty \ \text{ exists.}
$$

Proof (RoWa76). It follows from Proposition 11.12 that

$$
\text{(7)} \quad \operatorname{mod}\Gamma(\rho_1,\rho_2) - \int_{\rho_1}^{\rho_2} \lambda(\rho)^{-1}\,d\rho \to a \quad \text{as} \quad \rho_1 \to 0
$$

where $0 \leq a \leq +\infty$. First we consider the case that $a < +\infty$. We obtain from (3) and (4) (with ρ_1, ρ_2, ρ_0 instead of ρ_1', ρ_1, ρ_2) that

$$
\text{(8)} \quad \operatorname{mod}\Gamma(\rho_1,\rho_2) - \int_{\rho_1}^{\rho_2} \lambda(\rho)^{-1}\,d\rho \to 0 \quad \text{as} \quad \rho_1 < \rho_2 \to 0.
$$

Hence (11.4.4) holds and therefore also (11.4.3) by Proposition 11.8. As in part (a) of its proof we deduce that

$$\pi \bmod \Gamma(\rho_1,\rho_2) - \log\left|\frac{z_2 - \zeta}{z_1 - \zeta}\right| \to 0 \quad \text{as} \quad \rho_1 < \rho_2 \to 0,$$
$$z_1 \in Q(\rho_1), \quad z_2 \in Q(\rho_2).$$

Together with (8) this shows that the limit (6) exists and is positive. Now we consider the case that $a = +\infty$. Then, with $\rho_1 = |f(z) - w|$,

$$|z - \zeta| \exp\left(\int_{\rho_1}^{\rho_0} \frac{\pi}{\lambda(\rho)}\, d\rho\right) = o\left(|z - \zeta| \exp[\pi \bmod \Gamma(\rho_1,\rho_0)]\right) \to 0$$

as $z \to \zeta$, $z \in A$ by (7) and the second inequality (5). This proves (6), and we have shown moreover that the limit is positive if and only if a is finite.   □

We now turn to domains of a special type in order to complement (3) by an upper estimate (War42, RoWa76).

Proposition 11.15. *Let (see Fig. 11.1)*

(9) $$G = \{w + \rho e^{i\vartheta} : 0 < \rho < c,\ \vartheta_-(\rho) < \vartheta < \vartheta_+(\rho)\}$$

where the functions $\vartheta_\pm$ are locally absolutely continuous and satisfy

(10) $$\int_0^c \frac{\vartheta_+'(\rho)^2 + \vartheta_-'(\rho)^2}{\vartheta_+(\rho) - \vartheta_-(\rho)}\, \rho\, d\rho < \infty.$$

Then, for $\rho_1 < \rho_2 < \rho_0$,

(11) $$\bmod \Gamma(\rho_1,\rho_2) \leq \int_{\rho_1}^{\rho_2} \frac{1}{\vartheta_+ - \vartheta_-}\, \frac{d\rho}{\rho} + \frac{1}{2}\int_{\rho_1}^{\rho_2} \frac{\vartheta_+'^2 + \vartheta_-'^2}{\vartheta_+ - \vartheta_-}\, \rho\, d\rho.$$

In particular it follows that the limit a in (7) is finite. See e.g. Ahl30, LeF55, JenOi71, RoWa79a and RoWa79b for further sufficient geometric conditions and RoWa83 for a necessary and sufficient condition.

Proof. We consider the non-analytic homeomorphism

(12) $$s = \sigma + i\tau \ \mapsto \ w = \omega + \exp[-\sigma + i\tau\vartheta_+(e^{-\sigma}) + i(1 - \tau)\vartheta_-(e^{-\sigma})]$$

of $S = \{-\log c < \sigma < +\infty,\ 0 < \tau < 1\}$ onto G. Writing

(13) $$\psi(s) = e^{-\sigma}[\tau\vartheta_+'(e^{-\sigma}) + (1 - \tau)\vartheta_-'(e^{-\sigma})], \qquad \theta(s) = \vartheta_+(e^{-\sigma}) - \vartheta_-(e^{-\sigma})$$

we see that

(14)
$$|dw|^2 = e^{-2\sigma}|-(1 + i\psi)\, d\sigma + i\theta\, d\tau|^2$$
$$= (1 + \psi^2)e^{-2\sigma}\left(d\sigma - \frac{\psi\theta}{1 + \psi^2}\, d\tau\right)^2 + \frac{e^{-2\sigma}\theta^2}{1 + \psi^2}\, d\tau^2 \geq \frac{e^{-2\sigma}\theta^2}{1 + \psi^2}\, d\tau^2.$$

With s and w related by (12) we define

(15) $$\varphi(w) = (1 + \psi^2)^{1/2} e^\sigma \theta^{-1};$$

note that $e^{-\sigma} = |w - \omega| = \rho$. If $C \in \Gamma(\rho_1, \rho_2)$ then the corresponding curve connects the horizontal sides of S. Hence it follows from (14) that φ is admissible so that, by (9.2.2), (13) and (15),

$$
\begin{aligned}
\operatorname{mod} \Gamma(\rho_1, \rho_2) &\le \int_{\rho_1}^{\rho_2} \int_{\vartheta_-}^{\vartheta_+} \varphi(\omega + \rho e^{i\vartheta})^2 \rho \, d\vartheta \, d\rho \\
&= \int_{\rho_1}^{\rho_2} \int_0^1 \frac{1 + \rho^2 (\tau \vartheta_+' + (1-\tau)\vartheta_-')^2}{\rho(\vartheta_+ - \vartheta_-)} \, d\tau \, d\rho \\
&\le \int_{\rho_1}^{\rho_2} \frac{1 + \rho^2(\vartheta_+'^2 + \vartheta_-'^2)/2}{\rho(\vartheta_+ - \vartheta_-)} \, d\rho \, . \qquad \square
\end{aligned}
$$

Finally we present an asymptotic formula (War42) for $f(z)$ as $z \to \zeta$ or rather for the inverse function $f^{-1}(w)$ as $w \to \omega$.

Theorem 11.16. *Let G be defined by (9) with locally absolutely continuous $\vartheta_\pm$ and let (10) be satisfied. Then, with $f(z) = \omega + \rho e^{i\vartheta}$,*

(16) $$\log(z - \zeta) = -\int_\rho^c \frac{\pi}{\vartheta_+(t) - \vartheta_-(t)} \frac{dt}{t} + i\pi \frac{\vartheta - \vartheta_-(\rho)}{\vartheta_+(\rho) - \vartheta_-(\rho)} + b + o(1)$$

as $z \to \zeta$, $z \in \mathbb{D}$, where $-\infty < \operatorname{Re} b < +\infty$ and $\operatorname{Im} b = \arg\zeta + \pi/2$.

Proof. The real part of (16) can be written

(17) $$|z - \zeta| \exp\left(\int_{|f(z)-\omega|}^c \frac{\pi}{\vartheta_+(t) - \vartheta_-(t)} \frac{dt}{t} \right) \to e^{\operatorname{Re} b} \quad \text{as} \quad z \to \zeta, z \in \mathbb{D}.$$

We know from Theorem 11.14 that this limit always exists; in the present situation we have $\lambda(\rho) = \rho(\vartheta_+(\rho) - \vartheta_-(\rho))$, and A is identical with $\mathbb{D}$ near ζ. Since (10) is satisfied we see from Proposition 11.15 that the limit a in (7) is finite so that the limit in (6) is positive as we remarked in the proof of Proposition 11.14. Hence (17) holds for some $b \in \mathbb{C}$.

We refer to War42 or RoWa76 for the assertion about the imaginary part. In fact it holds under a condition weaker than (10). $\qquad \square$

It is clear that Theorem 11.16 also holds if G has the form (9) only in a neighbourhood of ω.

We can use Theorem 11.16 to treat the case of an outward pointing cusp left out in Theorem 3.9. Let G be given by (9) where

(18) $$\vartheta_+(\rho) - \vartheta_-(\rho) \sim a\rho \quad (\rho \to 0), \qquad \int_0^c \vartheta_\pm'(\rho)^2 \, d\rho < \infty$$

with $a > 0$. Then (10) holds and we obtain from (16) that $\log|z - \zeta| = -(\pi a^{-1} + o(1))\rho^{-1}$ and thus

(19) $$|f(z) - f(\zeta)| = \left(\frac{\pi}{a} + o(1) \right) \Big/ \log\frac{1}{|z - \zeta|} \quad \text{as} \quad z \to \zeta, \quad z \in \mathbb{D}.$$

Exercises 11.5

1. Construct a function f that is conformal at ζ but satisfies $\lambda(\rho) = 2\pi\rho$ for almost all $\rho \in (0, 1)$ and deduce that the limit in (6) is zero.

2. Suppose that $f(\zeta) = f(\zeta^*) = \omega$. If $f'(\zeta) = 0$ use Theorem 11.14 to show that $f'(\zeta^*) = \infty$.

In the last two exercises we assume that G is given by (9).

3. Let $0 < \alpha \leq 2$ and $\beta < 1$. If $\vartheta_{\pm}$ is absolutely continuous and

$$\vartheta_+(\rho) - \vartheta_-(\rho) \to \pi\alpha \quad (\rho \to 0), \qquad \int_0^c \vartheta_{\pm}'(\rho)^2 \rho^\beta \, d\rho < \infty,$$

show that, for some $a \in \mathbb{C}$ with $a \neq 0$,

$$\frac{f(z) - f(\zeta)}{(z - \zeta)^\alpha} \to a \quad \text{as} \quad z \to \zeta, \quad z \in \mathbb{D}$$

(corner or inward-pointing cusp; compare Theorem 3.9).

4. If $\vartheta_+(\rho) - \vartheta_-(\rho) \to 0$ as $3^{-n} < \rho < 2 \cdot 3^{-n}$, $n \to \infty$, show that

$$\frac{\log |f(z) - \omega|}{\log |z - \zeta|} \to 0 \quad \text{as} \quad z \to \zeta, \quad z \in \mathbb{D};$$

compare Exercise 11.3.2.

References

AgGe65
    Agard, S.B., Gehring, F.W.: Angles and quasiconformal mappings. Proc. London Math. Soc. (3) **14A** (1965) 1–21

Ahl30
    Ahlfors, L.V.: Untersuchungen zur Theorie der konformen Abbildung und der ganzen Funktionen. Acta Soc. Sci. Fenn. **1**, no. 9 (1930) 1–40

Ahl38
    Ahlfors, L.V.: An extension of Schwarz's lemma. Trans. Amer. Math. Soc. **43** (1938) 359–364

Ahl63
    Ahlfors, L.V.: Quasiconformal reflections. Acta Math. **109** (1963) 291–301

Ahl66a
    Ahlfors, L.V.: Complex analysis, 2nd edn. McGraw-Hill Book Co, New York 1966

Ahl66b
    Ahlfors, L.V.: Lectures on quasiconformal mappings. Van Nostrand Co, Princeton 1966

Ahl73
    Ahlfors, L.V.: Conformal invariants: Topics in geometric function theory. McGraw-Hill Book Co, New York 1973

Ahl74
    Ahlfors, L.V.: Sufficient conditions for quasiconformal extension. Ann. Math. Studies 79. Princeton Univ. Press 1974, pp. 23–29

AhBer60
    Ahlfors, L.V., Bers, L.: Riemann's mapping theorem for variable metrics. Ann. Math. (2) **72** (1960) 385–404

AhBeu50
    Ahlfors, L.V., Beurling, A.: Conformal invariants and function-theoretic null-sets. Acta Math. **83** (1950) 101–129

AhWe62
    Ahlfors, L.V., Weill, G.: A uniqueness theorem for Beltrami equations. Proc. Amer. Math. Soc. **13** (1962) 975–978

Ale89
    Alexander, H.: Linear measure on plane continua of finite linear measure. Ark. Mat. **27** (1989) 169–177

And71
    Anderson, J.M.: Category theorems for certain Banach spaces of analytic functions. J. Reine Angew. Math. **249** (1971) 83–91

AnClPo74
    Anderson, J.M., Clunie, J., Pommerenke, Ch.: On Bloch functions and normal functions. J. Reine Angew. Math. **270** (1974) 12–37

AnPi88
    Anderson, J.M., Pitt, L.D.: The boundary behaviour of Bloch functions and univalent functions. Michigan Math. J. **35** (1988) 313–320

AnPi89
Anderson, J.M., Pitt, L.D.: Probabilistic behaviour of functions in the Zygmund spaces Λ^* and λ^*. Proc. London Math Soc. (3) **59** (1989) 558–592

ArFiPe85
Arazy, J., Fisher, S.D., Peetre, J.: Möbius invariant function spaces. J. Reine Angew. Math. **363** (1985) 110–145

Ast88a
Astala, K.: Selfsimilar zippers. Holomorphic functions and moduli I. Springer, New York 1988, pp. 61–73

Ast88b
Astala, K.: Calderon's problem for Lipschitz classes and the dimension of quasi-circles. Rev. Mat. Iber.-Amer. **4** (1988) 469–486

AsGe86
Astala, K., Gehring, F.W.: Injectivity, the BMO norm, and universal Teichmüller space. J. Analyse Math. **46** (1986) 16–57

AsZi91
Astala, K., Zinsmeister, M.: Teichmüller spaces and BMOA. Math. Ann. **289** (1991) 613–625

Bae74
Baernstein II, A.: Integral means, univalent functions and circular symmetrization. Acta Math. **133** (1974) 139–169

Bae80
Baernstein II, A.: Analytic functions of bounded mean oscillation. Aspects of Contemporary Complex Analysis, ed. by D.A. Brannan and J. Clunie. Academic Press, London 1980, pp. 3–36

Bae86
Baernstein II, A.: Coefficients of univalent functions with restricted maximum modulus. Complex Variables **5** (1986) 225–236

Bae89
Baernstein II, A.: A counterexample concerning integrability of derivatives of conformal mapping. J. Analyse Math. **53** (1989) 253–268

Bag55
Bagemihl, F.: Curvilinear cluster sets of arbitrary functions. Proc. Nat. Acad. Sci. USA **41** (1955) 379–382

Bañ86
Bañuelos, R.: Brownian motion and area functions. Indiana Univ. Math. J. **35** (1986) 643–668

Bea91
Beardon, A.F.: Iteration of rational functions. Springer, New York 1991

BeaPo78
Beardon, A.F., Pommerenke, Ch.: The Poincaré metric of plane domains. J. London Math. Soc. (2) **18** (1978) 475–483

BeBe65
Beckenbach, E.F., Bellmann, R.: Inequalities. Springer, Berlin 1965

Bec72
Becker, J.: Löwnersche Differentialgleichung und quasikonform fortsetzbare schlichte Funktionen. J. Reine Angew. Math. **255** (1972) 23–43

Bec87
Becker, J.: On asymptotically conformal extension of univalent functions. Complex Variables **9** (1987) 109–120

BePo78
Becker, J., Pommerenke, Ch.: Über die quasikonforme Fortsetzung schlichter Funktionen. Math. Z. **161** (1978) 69–80

BePo82
  Becker, J., Pommerenke, Ch.: Hölder continuity of conformal mappings and non-quasiconformal curves. Comment. Math. Helv. **57** (1982) 221–225
BePo84
  Becker, J., Pommerenke, Ch.: Schlichtheitskriterien und Jordangebiete. J. Reine Angew. Math. **354** (1984) 74–94
BePo87
  Becker, J., Pommerenke, Ch.: On the Hausdorff dimension of quasicircles. Ann. Acad. Sci. Fenn. Ser. AI Math. **12** (1987) 329–334
Ber92
  Berman, R.D.: Boundary limits and an asymptotic Pragmén-Lindelöf theorem for analytic functions of slow growth. Indiana Univ. Math. J. (1992) (to appear)
BerNi92
  Berman, R.D., Nishiura, T.: Interpolation by radial cluster set functions and a Bagemihl-Seidel conjecture. Preprint
BeRo86
  Bers, L., Royden, H.L.: Holomorphic families of injections. Acta Math. **157** (1986) 259–286
Bes38
  Besicovitch, A.S.: On the fundamental geometric properties of linearly measurable plane sets of points II. Math. Ann. **115** (1938) 296–329
Beu40
  Beurling, A.: Ensembles exceptionels. Acta Math. **72** (1940) 1–13
BeuAh56
  Beurling, A., Ahlfors, L.V.: The boundary correspondence under quasiconformal mappings. Acta Math. **96** (1956) 125–142
Bie16
  Bieberbach, L.: Über die Koeffizienten derjenigen Potenzreihen, welche eine schlichte Abbildung des Einheitskreises vermitteln. S.-B. Preuss. Akad. Wiss. 1916, S. 940–955
Bin86
  Bingham, N.H.: Variants of the law of the iterated logarithm. Bull. London Math. Soc. **18** (1986) 443–467
Binm69
  Binmore, K.G.: Analytic functions with Hadamard gaps. Bull. London Math. Soc. **1** (1969) 211–217
Bis88
  Bishop, C.J.: A counterexample in conformal welding concerning Hausdorff dimension. Michigan Math. J. **35** (1988) 151–159
Bis90
  Bishop, C.J.: Bounded functions in the little Bloch space. Pacific J. Math. **142** (1990) 209–225
BiCaGaJo89
  Bishop, C.J., Carleson, L., Garnett, J.B., Jones, P.W.: Harmonic measures supported on a curve. Pacific J. Math. **138** (1989) 233–236
BiJo90
  Bishop, C.J., Jones, P.W.: Harmonic measure and arc length. Ann. Math. (2) **132** (1990) 511–547
BiJo92
  Bishop, C.J., Jones, P.W.: Harmonic measure, L^2 estimates and the Schwarzian derivative. Preprint
Bla84
  Blanchard, P.: Complex analytic dynamics on the Riemann sphere. Bull. Amer. Math. Soc. **11** (1984) 85–141

Bon90
  Bonk, M.: On Bloch's constant. Proc. Amer. Math. Soc. **110** (1990) 889–894
BrCh51
  Brelot, M., Choquet, G.: Espaces et lignes de Green. Ann. Inst. Fourier (Grenoble)
  3 (1951) 199–263
Bre78
  Brennan, J.E.: The integrability of the derivative in conformal mapping. J. London
  Math. Soc. (2) **18** (1978) 261–272
BrWe63
  Browder, A., Wermer, J.: Some algebras of functions on an arc. J. Math. Mech.
  12 (1963) 119–130

Car12
  Carathéodory, C.: Untersuchungen über die konformen Abbildungen von festen
  und veränderlichen Gebieten. Math. Ann. **72** (1912) 107–144
Car13a
  Carathéodory, C.: Über die gegenseitige Beziehung der Ränder bei der Abbildung
  des Innern einer Jordanschen Kurve auf einen Kreis. Math. Ann. **73** (1913) 305–
  320
Car13b
  Carathéodory, C.: Über die Begrenzung einfach zusammenhängender Gebiete.
  Math. Ann. **73** (1913) 323–370
Car63
  Carathéodory, C.: Conformal representation. Cambridge Univ. Press, Cambridge
  1963
Car67
  Carleson, L.: Selected problems on exceptional sets. Van Nostrand, Princeton 1967
Car73
  Carleson, L.: On the distortion of sets on a Jordan curve under conformal mapping.
  Duke Math. J. **40** (1973) 547–559
CaJo91
  Carleson, L., Jones, P.W.: On coefficient problems for univalent functions and
  conformal dimension. Preprint
CaCuPo88
  Carmona, J.J., Cufí, J., Pommerenke, Ch.: On the angular limits of Bloch func-
  tions. Publ. Mat. (Barcelona) **32** (1988) 191–198
Cho45
  Choquet, G.: Sur un type de transformation analytique généralisant la représen-
  tation conforme et définie au moyen de fonctions harmoniques. Bull. Sci. Math.
  (2) **69** (1945) 156–165
Cho55
  Choquet, G.: Theory of capacities. Ann. Inst. Fourier (Grenoble) **5** (1955) 131–295
ClPo67
  Clunie, J., Pommerenke, Ch.: On the coefficients of univalent functions. Michigan
  Math. J. **14** (1967) 71–78
ClShS84
  Clunie, J., Sheil-Small, T.: Harmonic univalent functions. Ann. Acad. Sci. Fenn.
  Ser. AI Math. **9** (1984) 3–25
CoFe74
  Coifman, R.R., Fefferman, C.: Weighted norm inequalities for maximal functions
  and singular integrals. Studia Math. **51** (1974) 241–250
Col57
  Collingwood, E.F.: On sets of maximum indetermination of analytic functions.
  Math. Z. **67** (1957) 377–396

Col60
  Collingwood, E.F.: Cluster sets of arbitrary functions. Proc. Nat. Acad. Sci. USA
  46 (1960) 1236–1242
CoLo66
  Collingwood, E.F., Lohwater, A.J.: The theory of cluster sets. Cambridge Univ.
  Press, Cambridge 1966
CoPi59
  Collingwood, E.F., Piranian, G.: The structure and distribution of prime ends.
  Arch. Math. **10** (1959) 379–386
CoPi64
  Collingwood, E.F., Piranian, G.: The mapping theorems of Carathéodory and
  Lindelöf. J. Math. Pures Appl. (9) **43** (1964) 187–199

Dav84
  David, G.: Opérateurs intégraux singuliers sur certaines courbes du plan complexe.
  Ann. Sci. École Norm. Sup. (4) **17** (1984) 157–189
deB85
  de Branges, L.: A proof of the Bieberbach conjecture. Acta Math. **154** (1985)
  137–152
Die69
  Dieudonné, J.: Foundations of modern analysis. Academic Press, New York 1969
DoEa86
  Douady, A., Earle, C.J.: Conformally natural extension of homeomorphisms of the
  circle. Acta Math. **157** (1986) 23–48
Duf45
  Dufresnoy, J.: Sur les fonctions méromorphes et univalentes dans le cercle unité.
  Bull. Sci. Math. **69** (1945) 21–31, 117–121
duP70
  du Plessis, N.: An introduction to potential theory. Oliver & Boyd, Edinburgh
  1970
Dur70
  Duren, P.L.: Theory of H^p spaces. Academic Press, New York 1970
Dur72
  Duren, P.L.: Smirnov domains and conjugate functions. J. Approx. Theory **5**
  (1972) 393–400
Dur83
  Duren, P.L.: Univalent functions. Springer, New York 1983
DuShSh66
  Duren, P.L., Shapiro, H.S., Shields, A.L.: Singular measures and domains not of
  Smirnov type. Duke Math. J. **33** (1966) 247–254
Durr84
  Durrett, R.: Brownian motion and martingales in analysis. Wadworth, Belmont,
  Calif. 1984

Ear89
  Earle, C.J.: Angular derivatives of the barycentric extension. Complex Variables
  11 (1989) 189–195
Eke67
  Eke, B.G.: Remarks on Ahlfors' distortion theorem. J. Analyse Math. **19** (1967)
  97–134
Eke71
  Eke, B.G.: On the differentiability of conformal maps at the boundary. Nagoya
  Math. J. **41** (1971) 43–53

274   References

EkWa69
    Eke, B.G., Warschawski, S.E.: On the distortion of conformal maps at the boundary. J. London Math. Soc. **44** (1969) 625–630
Eps87
    Epstein, C.L.: Univalence criteria and surfaces in hyperbolic space. J. Reine Angew. Math. **380** (1987) 196–214

Fab07
    Faber, G.: Einfaches Beispiel einer stetigen nirgends differenzierbaren Funktion. Jahresber. Deutsch. Math.-Ver. **16** (1907) 538–540
Fal85
    Falconer, K.J.: The geometry of fractal sets. Cambridge Univ. Press, Cambridge 1985
Fal90
    Falconer, K.J.: Fractal geometry. Wiley & Sons, Chichester 1990
Fat19
    Fatou, P.: Sur les équations fonctionelles. Bull. Soc. Math. France **47** (1919) 161–271
Fat20
    Fatou, P.: Sur les équations fonctionelles. Bull. Soc. Math. France **48** (1920) 33–94, 208–314
Fef71
    Fefferman, C.: Characterizations of bounded mean oscillation. Bull. Amer. Math. Soc. **77** (1971) 587–588
FeSz33
    Fekete, M., Szegö, G.: Eine Bemerkung über ungerade schlichte Funktionen. J. London Math. Soc. **8** (1933) 85–89
FeMG76
    Feng, J., MacGregor, T.H.: Estimates of integral means of the derivatives of univalent functions. J. Analyse Math. **29** (1976) 203–231
Fern84
    Fernández, J.L.: On the coefficients of Bloch functions. J. London Math. Soc. (2) **29** (1984) 94–102
FeHa87
    Fernández, J.L., Hamilton, D.H.: Lengths of curves under conformal mappings. Comment. Math. Helv. **62** (1987) 122–134
FeHeMa89
    Fernández, J.L., Heinonen, J., Martio, O.: Quasilines and conformal mappings. J. Analyse Math. **52** (1989) 117–132
FeZi87
    Fernández, J.L., Zinsmeister, M.: Ensembles de niveau des représentations conformes. C.R. Acad. Sci. Paris Sér. I **305** (1987) 449–452
Fer42
    Ferrand, J.: Étude de la représentation conforme au voisinage de la frontière. Ann. Sci. École Norm. Sup. (3) **59** (1942) 43–106
Fer44
    Ferrand, J.: Sur l'inégalité d'Ahlfors et son application au probleme de la dérivée angulaire. Bull. Soc. Math. France **72** (1944) 178–192
FiLe86
    FitzGerald, C.H., Lesley, F.D.: Boundary regularity of domains satisfying a wedge condition. Complex Variables **5** (1986) 141–154
FiLe87
    FitzGerald, C.H., Lesley, D.F.: Integrability of the derivative of the Riemann mapping function for wedge domains. J. Analyse Math. **49** (1987) 271–292

Fro35
Frostman, O.: Potentiel d'equilibre et capacité des ensembles avec quelques applications à la théorie des fonctions. Meddel. Lunds Univ. Mat. Sem. **3** (1935) 1–118

Gai64
Gaier, D.: Konstruktive Methoden der konformen Abbildung. Springer, Berlin 1964
Gai72a
Gaier, D.: Estimates of conformal mappings near the boundary. Indiana Univ. Math. J. **21** (1972) 581–595
Gai72b
Gaier, D.: Ermittlung des konformen Moduls von Vierecken mit Differenzenmethoden. Numer. Math. **19** (1972) 179–194
Gai83
Gaier, D.: Numerical methods in conformal mapping, Computational aspects of complex analysis. Reidel-Dordtrecht, Boston 1983, pp. 51–78
GaHa91
Gaier, D., Hayman, W.K.: On the computation of modules of long quadrilaterals. Constr. Approx. **7** (1991) 453–467
GaPo67
Gaier, D., Pommerenke, Ch.: On the boundary behavior of conformal maps. Michigan Math. J. **14** (1967) 79–82
GaSch55a
Garabedian, P.R., Schiffer, M.: A proof of the Bieberbach conjecture for the fourth coefficient. J. Rational Mech. Anal. **4** (1955) 428–465
GaSch55b
Garabedian, P.R., Schiffer, M.: A coefficient inequality for schlicht functions. Ann. Math. (2) **61** (1955) 116–136
Gar81
Garnett, J.B.: Bounded analytic functions. Academic Press, New York 1981
GaOs49a
Gattegno, C., Ostrowski, A.: Représentation conforme à la frontière: domaines généraux. Mém. Sci. Math. vol. 109, Paris 1949
GaOs49b
Gattegno, C., Ostrowski, A.: Représentation conforme à la frontière: domaines particuliers. Mém. Sci. Math. vol. 110, Paris 1949
Geh62
Gehring, F.W.: Rings and quasiconformal mappings in space. Trans. Amer. Math. Soc. **103** (1962) 353–393
Geh73
Gehring, F.W.: The L^p-integrability of the partial derivatives of a quasiconformal mapping. Acta Math. **130** (1973) 265–277
Geh78
Gehring, F.W.: Spirals and the universal Teichmüller space. Acta Math. **141** (1978) 99–113
Geh87
Gehring, F.W.: Uniform domains and the ubiquitous quasidisk. Jahresber. Deutsch. Math.-Verein. **89** (1987) 88–103
GeHaMa89
Gehring, F.W., Hag, K., Martio, O.: Quasihyperbolic geodesics in John domains. Math. Scand. **65** (1989) 72–92
GeHa62
Gehring, F.W., Hayman, W.K.: An inequality in the theory of conformal mapping. J. Math. Pures Appl. (9) **41** (1962) 353–361

276   References

GeOs79
   Gehring, F.W., Osgood, B.G.: Uniform domains and the quasihyperbolic metric.
   J. Analyse Math. **36** (1979) 50–74
GePo84
   Gehring, F.W., Pommerenke, Ch.: On the Nehari univalence criterion and quasi-
   circles. Comment. Math. Helv. **59** (1984) 226–242
GnPo83
   Gnuschke, D., Pommerenke, Ch.: On the absolute convergence of power series
   with Hadamard gaps. Bull. London Math. Soc. **15** (1983) 507–512
GnHPo86
   Gnuschke-Hauschild, D., Pommerenke, Ch.: On Bloch functions and gap series. J.
   Reine Angew. Math. **367** (1986) 172–186
Gol57
   Golusin, G.M.: Geometrische Funktionentheorie. Deutscher Verlag Wiss., Berlin
   1957
Gru32
   Grunsky, H.: Neue Abschätzungen zur konformen Abbildung ein- und mehrfach
   zusammenhängender Bereiche. Schr. Math. Seminar Univ. Berlin **1** (1932) 93–140

Hali65
   Haliste, K.: Estimates of harmonic measures. Ark. Mat. **6** (1965) 1–31
Hal68
   Hall, R.L.: On the asymptotic behavior of functions holomorphic in the unit disc.
   Math. Z. **107** (1968) 357–362
HaSa90
   Hallenbeck, D.J., Samotij, K.: On radial variation of bounded analytic functions.
   Complex Variables **15** (1990) 43–52
Ham86
   Hamilton, D.H.: Fine structure of prime-ends. J. Reine Angew. Math. **372** (1986)
   87–95
Ham88
   Hamilton, D.H.: Conformal distortion of boundary sets. Trans. Amer. Math. Soc.
   308 (1988) 69–91
Ham90
   Hamilton, D.H.: The closure of Teichmüller space. J. Analyse Math. **55** (1990)
   40–50
HaLi26
   Hardy, G.H., Littlewood, J.E.: A further note on the converse of Abel's theorem.
   Proc. London Math. Soc. **25** (1926) 219–236
HaLiPo67
   Hardy, G.H., Littlewood, J.E., Pólya, G.: Inequalities, 2nd edn. Cambridge Uni-
   versity Press, Cambridge 1967
Haw74
   Hawkes, J.: Hausdorff measure, entropy, and the independence of small sets. Proc.
   London Math. Soc. (3) **28** (1974) 700–724
Hay58a
   Hayman, W.K.: Bounds for the large coefficients of univalent functions. Ann.
   Acad. Sci. Fenn. Ser. AI **250**, 1–13
Hay58b
   Hayman, W.K.: Multivalent functions. Cambridge Univ. Press, Cambridge 1958
Hay70
   Hayman, W.K.: Tauberian theorems for multivalent functions. Acta Math. **125**
   (1970) 269–298
Hay89
   Hayman, W.K.: Subharmonic functions, vol. 2. Academic Press, London 1989

HayHu86
Hayman, W.K., Hummel, J.A.: Coefficients of powers of univalent functions. Complex Variables **7** (1986) 51–70

HayKe76
Hayman, W.K., Kennedy, P.B.: Subharmonic functions, vol. 1. Academic Press, London 1976

HayWu81
Hayman, W.K., Wu, J.-M.: Level sets of univalent functions. Comment. Math. Helv. **56** (1981) 366–403

He91
He, Z.X.: An estimate for hexagonal circle packings. J. Diff. Geom. **33** (1991) 395–412

HeNä92
Heinonen, J., Näkki, R.: Quasiconformal distortion on arcs. J. Analyse Math. (to appear)

Hem79
Hempel, J.A.: The Poincaré metric on the twice punctured plane and the theorems of Landau and Schottky. J. London Math. Soc. (2) **20** (1979) 435–445

Hem80
Hempel, J.A.: Precise bounds in the theorems of Schottky and Picard. J. London Math. Soc. (2) **21** (1980) 279–286

Hem88
Hempel, J.A.: On the uniformization of the n-punctured sphere. Bull. London Math. Soc. **20** (1988) 97–115

Hen86
Henrici, P.: Applied and computational complex analysis, vol. 3. Wiley & Sons, New York 1986

Her72
Herold, H.: Ein Vergleichssatz für komplexe lineare Differentialgleichungen. Math. Z. **126** (1972) 91–94

HeSt69
Hewitt, E., Stromberg, K.: Real and abstract analysis. Springer, Berlin 1969

Hil62
Hille, E.: Analytic function theory, vol. II. Ginn & Comp., Boston 1962

Hin89
Hinkkanen, A.: The sharp form of certain majorization theorems for analytic functions. Complex Variables **12** (1989) 39–66

HoPa83
Hough, D.M., Papamichael, N.: An integral equation method for the numerical conformal mapping of interior, exterior and doubly-connected domains. Numer. Math. **41** (1983) 287–307

Hun88
Hungerford, G.J.: Boundaries of smooth sets and singular sets of Blaschke products in the little Bloch class. Thesis, California Institute of Technology, Pasadena 1988

Hut81
Hutchinson, J.E.: Fractals and self similarity. Indiana Univ. Math. J. **30** (1981) 713–747

Jae68
Jaenisch, S.: Length distortion of curves under conformal mappings. Michigan Math. J. **15** (1968) 121–128

Jen56
Jenkins, J.A.: Some theorems on boundary distortion. Trans. Amer. Math. Soc. **81** (1956) 477–500

Jen60
  Jenkins, J.A. : On certain coefficients of univalent functions. Analytic functions.
  Princeton Univ. Press, Princeton 1960, pp. 159–194
Jen65
  Jenkins, J.A.: Univalent functions and conformal mappings, 2nd edn. Springer,
  Berlin 1965
JenOi71
  Jenkins, J.A., Oikawa, K.: On results of Ahlfors and Hayman. Illinois J. Math. **15**
  (1971) 664–671
JenOi77
  Jenkins, J.A., Oikawa, K.: Conformality and semi-conformality at the boundary.
  J. Reine Angew. Math. **291** (1977) 92–117
JeKe82a
  Jerison, D.S., Kenig, C.E.: Boundary behavior of harmonic functions in non-
  tangentially accessible domains. Adv. Math. **46** (1982) 80–147
JeKe82b
  Jerison, D.S., Kenig, C.E.: Hardy spaces, A_∞ and singular integrals on chord-arc
  domains. Math. Scand. **50** (1982) 221–247
Jon89
  Jones, P.W.: Square functions, Cauchy integrals, analytic capacity and Hausdorff
  measure. Lecture Notes in Mathematics, vol. 1384. Springer, Berlin 1989, pp. 24–
  68
Jon90
  Jones, P.W.: Rectifiable sets and the traveling salesman problem. Invent. Math.
  102 (1990) 1–15
JoMa92
  Jones, P.W., Makarov, N.G.: Density properties of harmonic measure. Preprint
JoWo88
  Jones, P.W., Wolff, T.H.: Hausdorff dimension of harmonic measures in the plane.
  Acta Math. **161** (1988) 131–144
JoZi82
  Jones, P.W., Zinsmeister, M.: Sur la transformation conforme des domaines de
  Lavrentiev. C.R. Acad. Sci. Paris Sér. I **295** (1982) 563–566
Jør56
  Jørgensen, V.: On an inequality for the hyperbolic measure and its applications
  in the theory of functions. Math. Scand. **4** (1956) 113–124

Kah85
  Kahane, J.P.: Some random series of functions, 2nd edn. Cambridge Univ. Press,
  Cambridge 1985
Kap52
  Kaplan, W.: Close-to-convex schlicht functions. Michigan Math. J. **1** (1952) 169–
  185
KeLa37
  Keldyš, M.V., Lavrentiev, M.A.: Sur la représentation conforme des domaines
  limités par des courbes rectifiables. Ann. Sci. École Norm. Sup. **54** (1937) 1–38
Kel12
  Kellogg, O.D.: Harmonic functions and Green's integral. Trans. Amer. Math. Soc.
  13 (1912) 109–132
Keo59
  Keogh, F.R.: Some theorems on conformal mapping of bounded starshaped do-
  mains. Proc. London Math. Soc. (3) **9** (1959) 481–491
KoTi61
  Kolmogorov, A.N., Tihomirov, V.M.: ε-entropy and ε-capacity of sets in functional
  spaces. Amer. Math. Soc. Transl. **17** (1961) 277–364

Kom42
Komatu, Y.: Über eine Verschärfung des Löwnerschen Hilfssatzes. Proc. Imperial Acad. Japan 18 (1942) 354–359

KoSt59
Koppenfels, W.v., Stallmann, F.: Praxis der konformen Abbildung. Springer, Berlin 1959

Kru89
Krushkal', S.L.: On the question of the structure of the universal Teichmüller space. Soviet Math. Dokl. 38 (1989) 435–437

Küh67
Kühnau, R.: Randverzerrung bei konformer Abbildung in der euklidischen, elliptischen und hyperbolischen Ebene. Math. Nachr. 34 (1967) 317–325

Küh69a
Kühnau, R.: Herleitung einiger Verzerrungseigenschaften konformer und allgemeinerer Abbildungen mit Hilfe des Argumentprinzips. Math. Nachr. 39 (1969) 249–275

Küh69b
Kühnau, R.: Werteannahmeprobleme bei quasikonformen Abbildungen mit ortsabhängiger Dilatationsbeschränkung. Math. Nachr. 40 (1969) 1–11

Küh71
Kühnau, R.: Verzerrungssätze und Koeffizientenbedingungen vom Grunskyschen Typ für quasikonforme Abbildungen. Math. Nachr. 48 (1971) 77–105

Kur52
Kuratowski, C.: Topologie, vol. I, 3rd edn. Polskie Towarzystwo Matematyczne, Warszawa 1952

Kuz80
Kuz'mina, G.V.: Moduli of families of curves and quadratic differentials. Procedings Steklov Inst. Math. 139 (1980) 1–231 (translated 1982)

LaGa86
Landau, E., Gaier, D.: Darstellung und Begründung einiger neuerer Ergebnisse der Funktionentheorie, 3rd edn. Springer, Berlin 1986

Lan72
Landkof, N.S.: Foundations of modern potential theory. Springer, Berlin 1972

Lav36
Lavrentiev, M.: Boundary problems in the theory of univalent functions (Russian), Mat. Sb. (N.S.) 1 (1936) 815–844; Amer. Math. Soc. Transl. Ser. 2, 32 (1963) 1–35

LaSc67
Lawrentjew, M.A., Schabat, B.W.: Methoden der komplexen Funktionentheorie. Deutscher Verlag Wiss., Berlin 1967

Lehm57
Lehman, R.S.: Development of the mapping function at an analytic corner. Pacific J. Math. 7 (1957) 1437–1449

Leh76
Lehto, O.: On univalent functions with quasiconformal extensions over the boundary. J. Analyse Math. 30 (1976) 349–354

Leh87
Lehto, O.: Univalent functions and Teichmüller spaces. Springer, New York 1987

LeVi57
Lehto, O., Virtanen, K.I.: Boundary behaviour and normal meromorphic functions. Acta Math. 97 (1957) 47–65

LeVi73
Lehto, O., Virtanen, K.I.: Quasiconformal mappings in the plane. Springer, Berlin 1973

Lej34
  Leja, F.: Sur les suites des polynômes, les ensembles fermés et la fonction de Green.
  Ann. Soc. Polon. Math. **12** (1934) 57–71
Lej61
  Leja, F.: Sur les moyennes arithmétriques, géométriques et harmoniques des dis-
  tances mutuelles des points d'un ensemble. Ann. Polon. Math. **9** (1961) 211–218
LeF55
  Lelong-Ferrand, J.: Représentation conforme et transformations à intégrale de
  Dirichlet bornée. Gauthier-Villars, Paris 1955
Lew58
  Lewandowski, Z.: Sur l'identité de certaines classes de fonctions univalentes I.
  Ann. Univ. Mariae Curie-Skłodowska Sect. A **12** (1958) 131–146
Lew60
  Lewandowski, Z.: Sur l'identité de certaines classes de fonctions univalentes II.
  Ann. Univ. Mariae Curie-Skłodowska Sect. A **14** (1960) 19–46
Lin15
  Lindelöf, E.: Sur un principe général de l'analyse et ses applications à la représen-
  tation conforme. Acta Soc. Sci. Fenn. Nova Ser. A **46** (1920) no. 4 (1915)
Lit25
  Littlewood, J.E.: On inequalities in the theory of functions. Proc. London Math.
  Soc. (2) **23** (1925) 481–519
LiPa32
  Littlewood, J.E., Paley, R.E.A.C.: A proof that an odd function has bounded
  coefficients. J. London Math. Soc. **7** (1932) 167–169
Löw23
  Löwner, K.: Untersuchungen über schlichte konforme Abbildungen des Einheits-
  kreises. Math. Ann. **89** (1923) 103–121
Lyo90
  Lyons, T.: A synthetic proof of Makarov's law of the iterated logarithm. Bull.
  London Math. Soc. **22** (1990) 159–162
Lyu86
  Lyubich, M.Yu.: The dynamics of rational transforms: the topological picture.
  Russ. Math. Surv. **41**, no. 4 (1986) 43–117

MacL63
  MacLane, G.R.: Asymptotic values of holomorphic functions. Rice University
  Studies **49**, no. 1, Houston 1963
MaObSo66
  Magnus, W., Oberhettinger, F., Soni, R.P.: Formulas and theorems for the special
  functions of mathematical physics, 3rd edn. Springer, Berlin 1966
Mak84
  Makarov, N.G.: Defining subsets, the support of harmonic measure, and pertur-
  bations of the spectra of operators in Hilbert space. Soviet Math. Dokl. **29** (1984)
  103–106
Mak85
  Makarov, N.G.: On the distortion of boundary sets under conformal mappings.
  Proc. London Math. Soc. (3) **51** (1985) 369–384
Mak86
  Makarov, N.G.: A note on the integral means of the derivative in conformal map-
  ping. Proc. Amer. Math. Soc. **96** (1986) 233–236
Mak87
  Makarov, N.G.: Conformal mapping and Hausdorff measures. Ark. Mat. **25** (1987)
  41–89

Mak89a
Makarov, N.G.: On a class of exceptional sets for conformal mappings. Mat. Sbornik **180** (1989) 1171–1182 (Russian) [English transl.: Math. USSR Sbornik **68** (1991) 19–30]

Mak89b
Makarov, N.G.: Smooth measures and the law of the iterated logarithm. Izv. Akad. Nauk SSSR Ser. Mat. **53** (1989) 439–446 (Russian) [English transl.: Math. USSR Izv. **34** (1990) 455–463]

Mak89c
Makarov, N.G.: The size of the set of singular points on the boundary of a non-Smirnov domain. Zap. Naučn. Sem. LOMI **170** (1989) 176–183, 323 (Russian)

Mak89d
Makarov, N.G.: Probability methods in the theory of conformal mappings. Algebra i Analiz **1** (1989) 3–59 (Russian) [English transl.: Leningrad Math. J. **1** (1990) 1–56]

Mak89e
Makarov, N.G.: On the radial behavior of Bloch functions. Dokl. Akad. Nauk SSSR **309** (1989) 275–278 (Russian) [English transl.: Soviet Math. Dokl. **40** (1990) 505–508]

MaSaSu83
Mañé, R., Sad, P., Sullivan, D.: On the dynamics of rational maps. Ann. Sci. École Norm. Sup. **16** (1983) 193–217

MaSa79
Martio, O., Sarvas, J.: Injectivity theorems in plane and space. Ann. Acad. Sci. Fenn. Ser. AI Math. **4** (1978/79) 383–401

MaVu87
Martio, O., Vuorinen, M.: Witney cubes, p-capacity and Minkowski content. Expo. Math. **5** (1987) 17–40

Mat64
Matsumoto, K.: On some boundary problems in the theory of conformal mappings of Jordan domains. Nagoya Math. J. **24** (1964) 129–142

Maz36
Mazurkiewicz, S.: Über die Definition der Primenden. Fund. Math. **26** (1936) 272–279

McM66
McMillan, J.E.: Boundary properties of functions continuous in a disc. Michigan Math. J. **13** (1966) 299–312

McM67
McMillan, J.E.: Arbitrary functions defined on plane sets. Michigan Math. J. **14** (1967) 445–447

McM69a
McMillan, J.E.: Minimum convexity of a holomorphic function II. Michigan Math. J. **16** (1969) 53–58

McM69b
McMillan, J.E.: Boundary behavior of a conformal mapping. Acta Math. **123** (1969) 43–67

McM71
McMillan, J.E.: Distortion under conformal and quaisconformal mappings. Acta Math. **126** (1971) 121–141

McMPi73
McMillan, J.E., Piranian, G.: Compression and expansion of boundary sets. Duke Math. J. **40** (1973) 599–605

Mil65
Milnor, J.W.: Topology from the differentiable viewpoint. University Press of Virginia, Charlottesville 1965

Min87
  Minda, D.: A reflection principle for the hyperbolic metric and applications to
  geometric function theory. Complex Variables **8** (1987) 129–144
Moo28
  Moore, R.L.: Concering triods in the plane and the junction points of plane con-
  tinua. Proc. Nat. Acad. Sci. USA **14** (1928) 85–88
Muc72
  Muckenhoupt, B.: Weighted norm inequalities for Hardy maximal functions.
  Trans. Amer. Math. Soc. **165** (1972) 207–226
Mur81
  Murai, T.: The value-distribution of lacunary series and a conjecture of Paley.
  Ann. Inst. Fourier (Grenoble) **31** (1981) 136–156
Mur83
  Murai, T.: The boundary behaviour of Hadamard lacunary series. Nagoya Math.
  J. **89** (1983) 65–76
Myr33
  Myrberg, P.J.: Über die Existenz der Greenschen Funktionen auf einer gegebenen
  Riemannschen Fläche. Acta Math. **61** (1933) 39–79

NaRuSh82
  Nagel, A., Rudin, W., Shapiro, J.H.: Tangential boundary behavior of functions
  in Dirichlet-type spaces. Ann. Math. (2) **116** (1982) 331–360
Näk90
  Näkki, R.: Conformal cluster sets and boundary cluster sets coincide. Pacific J.
  Math. **141** (1990) 363–382
NäPa83
  Näkki, R., Palka, B.: Lipschitz conditions, b-arcwise connectedness and conformal
  mappings. J. Analyse Math. **42** (1983) 38–50
NäPa86
  Näkki, R., Palka, B.: Extremal length and Hölder continuity of conformal map-
  pings. Comment. Math. Helv. **61** (1986) 389–414
NäVä91
  Näkki, R., Väisälä, J.: John disks. Expo. Math. **9** (1991) 3–43
Neh49
  Nehari, Z.: The Schwarzian derivative and schlicht functions. Bull. Amer. Math.
  Soc. **55** (1949) 545–551
Neh52
  Nehari, Z.: Conformal mapping. McGraw-Hill Book Co, New York 1952
Nev60
  Nevanlinna, R.: On differentiable mappings. Analytic Functions, ed. by L. Ahlfors.
  Princeton Univ. Press 1960, pp. 3–9
Nev70
  Nevanlinna, R.: Analytic functions. Springer, Berlin 1970
New64
  Newman, M.H.A.: Elements of the topology of plane sets of points. Cambridge
  Univ. Press, Cambridge 1964
Nos60
  Noshiro, K.: Cluster sets. Springer, Berlin 1960

Oht70
  Ohtsuka, M.: Dirichlet problem, extremal length and prime ends. Van Nostrand
  Reinhold Co, New York 1970
Ost35
  Ostrowski, A.: Beiträge zur Topologie der orientierten Linienelemente I. Compos.
  Math. **2** (1935) 26–49

Ost36
Ostrowski, A.: Zur Randverzerrung bei konformer Abbildung. Prace Mat.-Fiz. **44** (1936) 371–471
Oxt80
Oxtoby, J.C.: Measure and category. Springer, New York 1980
Øxe81
Øxendal, B.: Brownian motion and sets of harmonic measure zero. Pacific J. Math. **95** (1981) 179–192

Paa31
Paatero, V.: Über die konforme Abbildung von Gebieten, deren Ränder von beschränkter Drehung sind. Ann. Acad. Sci. Fenn. Ser. A **33**, no. 9 (1931) 1–11
Pap89
Papamichael, N.: Numerical conformal mapping onto a rectangle with application to the solution of Laplacian problems. J. Comput. Appl. Math. **28** (1989) 63–83
PaKoHo87
Papamichael, N., Kokkinos, C.A., Hough, M.K.: Numerical techniques for conformal mapping onto a rectangle. J. Comput. Appl. Math. **20** (1987) 349–358
PaWaHo86
Papamichael, N., Warby, M.K., Hough, D.M.: The treatment of corner and pole-type singularities in numerical conformal mapping techniques. J. Comput. Appl. Math. **14** (1986) 163–191
Ped68
Pederson, R.N.: A proof of the Bieberbach conjecture for the sixth coefficient. Arch. Rat. Mech. Anal. **31** (1968) 331–351
PeSch72
Pederson, R.N., Schiffer, M.: A proof of the Bieberbach conjecture for the fifth coefficient. Arch. Rat. Mech. Anal. **45** (1972) 161–193
Pfl55
Pfluger, A.: Extremallängen und Kapazität. Comment. Math. Helv. **29** (1955) 120–131
PhSt75
Philipp, W., Stout, W.: Almost sure invariance principles for partial sums of weakly dependent random variables. Mem. Amer. Math. Soc. **2**, no. 161 (1975)
Pic17
Pick, G.: Über die konforme Abbildung eines Kreises auf ein schlichtes und zugleich beschränktes Gebiet. S.-B. Kaiserl. Akad. Wiss. Wien, Math.-Naturwiss. Kl. Abt. II a **126** (1917) 247–263
Pir58
Piranian, G.: The boundary of a simply connected domain. Bull. Amer. Math. Soc. **64** (1958) 45–55
Pir60
Piranian, G.: The distribution of prime ends. Michigan Math. J. **7** (1960) 83–95
Pir62
Piranian, G.: Jordan domains and absolute convergence of power series. Michigan Math. J. **9** (1962) 125–128
Pir66
Piranian, G.: Two monotonic, singular, uniformly almost smooth functions. Duke Math. J. **33** (1966) 255–262
Pir72
Piranian, G.: Totally disconnected sets, Jordan curves, and conformal maps. Period. Math. Hungar. **2** (1972) 15–19

Ple27
   Plessner, A.I.: Über das Verhalten analytischer Funktionen am Rande ihres Definitionsbereichs. J. Reine Angew. Math. **158** (1927) 219–227
Pól28
   Pólya, G.: Beitrag zur Verallgemeinerung des Verzerrungssatzes auf mehrfach zusammenhängende Gebiete I, II, III. S.-B. Preuss. Akad. Wiss. Berlin 1928, 228–232, 280–282; 1929, 55–62
Pom62
   Pommerenke, Ch.: On starlike and convex functions. J. London Math. Soc. **37** (1962) 209–224
Pom63
   Pommerenke, Ch.: On starlike and close-to-convex functions. Proc. London Math. Soc. **13** (1963) 290–304
Pom64
   Pommerenke, Ch.: Linear-invariante Familien analytischer Funktionen I, II. Math. Ann. **155** (1964) 108–154; **156** (1964) 226–262
Pom68
   Pommerenke, Ch.: On the logarithmic capacity and conformal mapping. Duke Math. J. **35** (1968) 321–326
Pom75
   Pommerenke, Ch.: Univalent functions, with a chapter on quadratic differentials by Gerd Jensen. Vandenhoeck & Ruprecht, Göttingen 1975
Pom78
   Pommerenke, Ch.: On univalent functions, Bloch functions and VMOA. Math. Ann. **236** (1978) 199–208
Pom82
   Pommerenke, Ch.: One-sided smoothness conditions and conformal mapping. J. London Math. Soc. (2) **26** (1982) 77–88
Pom84
   Pommerenke, Ch.: On uniformly perfect sets and Fuchsian groups. Analysis **4** (1984) 299–321
Pom85
   Pommerenke, Ch.: On the integral means of the derivative of a univalent function. J. London Math. Soc. (2) **32** (1985) 254–258
Pom86
   Pommerenke, Ch.: On the Epstein univalence criterion. Results Math. **10** (1986) 143–146
Pom87
   Pommerenke, Ch.: On graphical representation and conformal mapping. J. London Math. Soc (2) **35** (1987) 481–488
Pom89
   Pommerenke, Ch.: On boundary size and conformal mapping. Complex Variables **12** (1989) 231–236
PoWa82
   Pommerenke, Ch., Warschawski, S.E.: On the quantitative boundary behavior of conformal maps. Comment. Math. Helv. **57** (1982) 107–129
Pra27
   Prawitz, H.: Über die Mittelwerte analytischer Funktionen. Arkiv Mat. Astr. Fys. **20** (1927/28) 1–12
Pri19
   Privalov, I.I.: The Cauchy integral. Dissertation, Saratov 1919 (Russian)
Pri56
   Privalov, I.I.: Randeigenschaften analytischer Funktionen. Deutscher Verlag Wiss., Berlin 1956

PrUrZd89
Przytycki, F., Urbański, M., Zdunik, A.: Harmonic, Gibbs and Hausdorff measures on repellers for holomorphic maps I. Ann. Math. (2) **130** (1989) 1–40

Rad23
Radó, T.: Sur la représentation conforme de domaines variables. Acta Sci. Math. (Szeged) **1** (1923) 180–186

Rea55
Reade, M.O.: On close-to-convex univalent functions. Michigan Math. J. **3** (1955) 59–62

ReRy75
Reimann, H.M., Rychener, T.: Funktionen beschränkter mittlerer Oszillation. Lecture Notes in Mathematics, vol. 487. Springer, Berlin 1975

Rie16
Riesz, F., Riesz, M.: Über die Randwerte einer analytischen Funktion. 4. Cong. Scand. Math. Stockholm 1916, 27–44

Rob36
Robertson, M.S.: On the theory of univalent functions. Ann. Math. (2) **37** (1936) 374–408

RoSu87
Rodin, B., Sullivan, D.: The convergence of circle packings to the Riemann mapping. J. Diff. Geom. **26** (1987) 349–360

RoWa76
Rodin, B., Warschawski, S.E.: Extremal length and the boundary behavior of conformal mapping. Ann. Acad. Sci. Fenn. Ser. AI Math. **2** (1976) 467–500

RoWa77
Rodin, B., Warschawski, S.E.: Extremal length and univalent functions I. The angular derivative. Math. Z. **153** (1977) 1–17

RoWa79a
Rodin, B., Warschawski, S.E.: Extremal length and univalent functions II. Integral estimates of strip mappings. J. Math. Soc. Japan **31** (1979) 87–99

RoWa79b
Rodin, B., Warschawski, S.E.: Estimates for conformal maps of strip domains without boundary regularity. Proc. London Math. Soc. (3) **39** (1979) 356–384

RoWa80
Rodin, B., Warschawski, S.E.: On the derivative of the Riemann mapping function near a boundary point and the Visser-Ostrowski problem. Math. Ann. **248** (1980) 125–137

RoWa83
Rodin, B., Warschawski, S.E.: A necessary and sufficient condition for the asymptotic version of Ahlfors' distortion property. Trans. Amer. Math. Soc. **276** (1983) 281–288

RoWa84
Rodin, B., Warschawski, S.E.: Conformal mapping and locally asymptotically conformal curves. Proc. London Math. Soc. (3) **49** (1984) 255–273

Rog70
Rogers, C.A.: Hausdorff measures. Cambridge University Press, Cambridge 1970

Roh88
Rohde, S.: On an estimate of Makarov in conformal mapping. Complex Variables **10** (1988) 381–386

Roh89
Rohde, S.: Hausdorffmaß und Randverhalten analytischer Funktionen. Thesis, Technische Universität Berlin 1989

Roh91
  Rohde, S.: On conformal welding and quasicircles. Michigan Math. J. **38** (1991) 111–116
RuShTa75
  Rubel, L.A., Shields, A.L., Taylor, B.A.: Mergelyan sets and the modulus of continuity of analytic functions. J. Approx. Theory **15** (1975) 23–40
Rus75
  Ruscheweyh, St.: Duality for Hadamard products with applications to extremal problems for functions regular in the unit disc. Trans. Amer. Math. Soc. **210** (1975) 63–74
Rus78
  Ruscheweyh, St.: Some convexity and convolution theorems for analytic functions. Math. Ann. **238** (1978) 217–228
RuShS73
  Ruscheweyh, S., Sheil-Small, T.: Hadamard products of schlicht functions and the Pólya-Schoenberg conjecture. Comment. Math. Helv. **48** (1973) 119–135

Sak64
  Saks, S.: Theory of the integral, 2nd edn. Hafner Publ. Co, New York 1964
Sar75
  Sarason, D.: Functions of vanishing mean oscillation. Trans. Amer. Math. Soc. **207** (1975) 391–405
Schi38
  Schiffer, M.: Sur un problème d'extrémum de la représentation conforme. Bull. Soc. Math. France **66** (1938) 48–55
Schi46
  Schiffer, M.: Hadamard's formula and variation of domain-functions. Amer. J. Math. **68** (1946) 417–448
SchLa91
  Schinzinger, R., Laura, P.: Conformal mapping: methods and applications. Elsevier-North-Holland, Amsterdam 1991
Scho75
  Schober, G.: Univalent functions – Selected topics. Lecture Notes in Mathematics, vol. 478. Springer, Berlin 1975
Sem88
  Semmes, S.W.: Quasiconformal mappings and chord-arc curves. Trans. Amer. Math. Soc. **306** (1988) 233–263
ShS69
  Sheil-Small, T.: On convex univalent functions. J. London Math. Soc. (2) **1** (1969) 483–492
ShS70
  Sheil-Small, T.: Starlike univalent functions. Proc. London Math. Soc. (3) **21** (1970) 577–613
ShS78
  Sheil-Small, T.: The Hadamard product and linear transformations of classes of analytic functions. J. Analyse Math. **34** (1978) 204–239
Sid27
  Sidon, S.: Verallgemeinerung eines Satzes über die absolute Konvergenz von Fourierreihen mit Lücken. Math. Ann. **97** (1927) 675–676
SmSt87
  Smith, W., Stegenga, D.A.: A geometric characterization of Hölder domains. J. London Math. Soc. (2) **35** (1987) 471–480
SmSt90
  Smith, W., Stegenga, D.A.: Hölder domains and Poincaré domains. Trans. Amer. Math. Soc. **319** (1990) 67–100

SmSt91
Smith, W., Stegenga, D.A.: Exponential integrability of the quasi-hyperbolic metric on Hölder domains. Ann. Acad. Sci. Fenn. Ser. AI Math. **16** (1991) 345–360

Sta89
Staples, S.G.: L^p-averaging domains and the Poincaré inequality. Ann. Acad. Sci. Fenn. Ser AI Math. **14** (1989) 103–127

StSt81
Stegenga, D.A., Stephenson, K.: A geometric characterization of analytic functions with bounded mean oscillation. J. London Math. Soc. (2) **24** (1981) 243–254

Str84
Strebel, K.: Quadratic differentials. Springer, Berlin 1984

Suf70
Suffridge, T.J.: Some remarks on convex maps of the unit disk. Duke Math. J. **37** (1970) 775–777

SuThu86
Sullivan, D., Thurston, W.P.: Extending holomorphic motions. Acta Math. **157** (1986) 243–257

Tam78
Tammi, O.: Extremum problems for bounded univalent functions. Lecture Notes in Mathematics, vol. 646. Springer, Berlin 1978

Tamr73
Tamrazov, P.M.: Contour and solid structure properties of holomorphic functions of a complex variable. Russ. Math. Surv. **28** (1973) 141–173

Tei38
Teichmüller, O.: Untersuchungen über konforme und quasikonforme Abbildungen. Deutsche Math. **3** (1938) 621–678

Tho67
Thomas, D.K.: On starlike and close-to-convex functions. J. London Math. Soc. **42** (1967) 427–435

Thu86
Thurston, W.P.: Zippers and univalent functions. The Bieberbach conjecture, Math. Surv. no. 21. Amer. Math. Soc., Providence 1986, pp. 185–197

Tre86
Trefethen, L.N. (ed.): Numerical conformal mapping. North-Holland, Amsterdam 1986

Tuk80
Tukia, P.: The planar Schönflies theorem for Lipschitz maps. Ann. Acad. Sci. Fenn. Ser. AI Math. **5** (1980) 49–72

Tuk81
Tukia, P.: Extension of quasisymmetric and Lipschitz embeddings of the real line into the plane. Ann. Acad. Sci. Fenn. Ser. AI Math. **6** (1981) 89–94

TuVä80
Tukia, P., Väisälä, J.: Quasisymmetric embeddings of metric spaces. Ann. Acad. Sci. Fenn. AI Math. **5** (1980) 97–114

Two86
Twomey, J.B.: Tangential limits of starlike univalent functions. Proc. Amer. Math. Soc. **97** (1986) 49–54

Väi71
Väisälä, J.: Lectures on n-dimensional quasiconformal mappings. Lecture Notes in Mathematics, vol. 229. Springer, Berlin 1971

Väi81
Väisälä, J.: Quasi-symmetric embeddings in euclidean spaces. Trans. Amer. Math. Soc. **264** (1981) 191–204

Väi84
    Väisälä, J.: Quasimöbius maps. J. Analyse Math. **44** (1984/85) 218–234
Vis33
    Visser, C.: Über beschränkte analytische Funktionen und die Randverhältnisse bei
    konformen Abbildungen. Math. Ann. **107** (1933) 28–39
Vuo88
    Vuorinen, M.: Conformal geometry and quasiregular mappings. Lecture Notes in
    Mathematics, vol. 1319. Springer, Berlin 1988

WaGa55
    Walsh, J.L., Gaier, D.: Zur Methode der variablen Gebiete bei der Randverzer-
    rung. Arch. Math. **6** (1955) 77–86
War30
    Warschawski, S.E.: Über einige Konvergenzsätze aus der Theorie der konformen
    Abbildung. Nachr. Ges. Wiss. Göttingen 1930, 344–369
War32
    Warschawski, S.E.: Über das Randverhalten der Ableitung der Abbildungsfunk-
    tion bei konformer Abbildung. Math. Z. **35** (1932) 321–456
War35a
    Warschawski, S.E.: On the higher derivatives at the boundary in conformal map-
    ping. Trans. Amer. Math. Soc. **38** (1935) 310–340
War35b
    Warschawski, S.E.: Zur Randverzerrung bei konformer Abbildung. Compos. Math.
    1 (1935) 314–342
War42
    Warschawski, S.E.: On conformal mapping of infinite strips. Trans. Amer. Math.
    Soc. **51** (1942) 280–335
War55
    Warschawski, S.E.: On a theorem of Lichtenstein. Pacific J. Math. **5** (1955) 835–
    839
War61
    Warschawski, S.E.: On differentiability at the boundary in conformal mapping.
    Proc. Amer. Math. Soc. **12** (1961) 614–620
War67
    Warschawski, S.E.: On the boundary behavior of conformal maps. Nagoya Math.
    J. **30** (1967) 83–101
War71
    Warschawski, S.E.: Remarks on the angular derivative. Nagoya Math. J. **41** (1971)
    19–32
Wei59
    Weiss, M.: On the law of the iterated logarithm for lacunary trigonometric series.
    Trans. Amer. Math. Soc. **91** (1959) 444–469
Wei86
    Weitsman, A.: Symmetrization and the Poincaré metric. Ann. Math. (2) **124**
    (1986) 159–169
Why42
    Whyburn, G.T.: Analytic topology. Amer. Math. Soc. Colloq. Publ., vol. 28. Prov-
    idence 1942
Wig65
    Wigley, N.M.: Development of the mapping function at a corner. Pacific J. Math.
    15 (1965) 1435–1461

Zin84
    Zinsmeister, M.: Répresentation conforme et courbes presque lipschitziennes. Ann.
    Inst. Fourier (Grenoble) **34** (1984) 29–44

Zin85
   Zinsmeister, M.: Domaines de Lavrentiev. Publications Math. Orsay 1985 no. 3
Zin86
   Zinsmeister, M.: A distortion theorem for quasiconformal mappings. Bull. Soc.
   Math. France **114** (1986) 123–133
Zin89
   Zinsmeister, M.: Les domaines de Carleson. Michigan Math. J. **36** (1989) 213–220
Zyg45
   Zygmund, A.: Smooth functions. Duke Math. J. **12** (1945) 47–76
Zyg68
   Zygmund, A.: Trigonometric series, vols. I and II. Cambridge Univ. Press, Cam-
   bridge 1968

Author Index

Subject Index